Textbooks in Telecommunication Engineering

Era of the Internet and Network Science/Engineering

Series Editor

Tarek S. El-Bawab, Ph.D., Boston, MA, USA

Telecommunication and networks have evolved to embrace all aspects of our everyday life, including smart cities and infrastructures, healthcare, banking and businesses, manufacturing, space and aviation, meteorology and climate change, oceans and marine life, Internet of Things, defense, homeland security, education, research, social media, entertainment, and many others. Network applications and services continue to expand, virtually without limits. Therefore, specialized telecommunication and network engineering programs are recognized as a necessity to accelerate the pace of advancement in this field, and to prepare a new generation of engineers for imminent needs in our modern life. These programs need curricula, courses, labs, and textbooks of their own.

The IEEE Communications Society's Telecommunication Engineering Education (TEE) movement, led by Tarek S. El-Bawab-the editor of this Series, resulted in recognition of this field of engineering by the Accreditation Board for Engineering and Technology (ABET) on November 1, 2014. This Springer Series was launched to capitalizes on this milestone. The Series goal is to produce high-quality textbooks to fulfill the education needs of telecommunication and network engineering, and to support the development of specialized undergraduate and graduate curricula in this regard. The Series also supports research in this field and helps prepare its scholars for global challenges that lay ahead. The Series have published innovative textbooks in areas of network science and engineering where textbooks have been rare. It is producing high-quality volumes featuring innovative presentation media, interactive content, and online resources for students and professors.

Book proposals are solicited in all topics of telecommunication and network engineering including, but not limited to: network architecture and protocols; traffic engineering; network design, dimensioning, modeling, measurements, and analytics; network management and softwarization; cybersecurity; synchronization and control; applications of artificial intelligence in telecommunications and networks; applications of data science in telecommunications and networks; network availability, reliability, protection, recovery and restoration; wireless communication systems; cellular technologies and networks (through 5G, 6G, and beyond); satellite and space communications and networks; optical communications and networks; heterogeneous networks; broadband access and free-space optical communications; MSO/cable networks; storage networks; optical interconnects; and data centers; social networks; transmission media and systems; switching and routing (from legacy to today's paradigms); network applications and services; telecom economics and business; telecom regulation and policies; standards and standardization; and laboratories.

Proposals of interest shall be for textbooks that can be used to develop university courses, either in full or in part. They should include recent advances in the field while capturing whatever fundamentals that are necessary for students to understand the bases of the topic and appreciate its evolution trends. Books in this series will provide high-quality illustrations, examples, end-of-chapters' problems/exercises and case studies.

This series is indexed in Scopus. For further information and to submit proposals, please contact the Series Editor, Dr. Tarek S. El-Bawab, telbawab@ieee.org; or Mary James, Executive Editor at Springer, mary.james@springer.com.

Shruthi Koratagere Anantha Kumar

Wireless Broadband Access to Bridge the Digital Divide

Springer

Shruthi Koratagere Anantha Kumar
Research & Innovation Associate
StrathSDR, Electronic and Electrical Engineering
University of Strathclyde
Glasgow, UK

ISSN 2524-4345 ISSN 2524-4353 (electronic)
Textbooks in Telecommunication Engineering
ISBN 978-3-032-09260-1 ISBN 978-3-032-09261-8 (eBook)
https://doi.org/10.1007/978-3-032-09261-8

This Springer imprint is published by the registered company Springer Nature Switzerland AG
The registered company address is: Gewerbestrasse 11, 6330 Cham, Switzerland

To my family ...

Note from the Series Editor

This Series has its roots in the Telecommunication Engineering Education (TEE) movement in the USA, which I had the honor of leading from 2008 to 2015. The movement was a result of remarkable impacts telecom networks had already started to have on society, and many more that were looming on the horizon at that time. Major changes in the theoretical foundations of this field of engineering, its design strategies, and its practical applications were unfolding at the same time. The industries we had known as the Telecom, Cable, and Data-Communication industries through the twentieth century were going through technical innovations and were reshuffling, repositioning themselves in the market, and converging, in preparation for the global economy of the twenty-first century.

The TEE movement was prompted by a personal experience I went through as I moved from industry to academia. *Telecom engineering*, or *Network Engineering*, since the network is centric to the evolution telecom has been through for three to four decades so far, became a distinct multidisciplinary field of study. This field blends ingredients from electrical engineering, computer engineering, computer science, and other disciplines, but may not be taught merely as a branch of any one of them, per se. It became evident that society would be better served by adding specialized undergraduate/graduate courses and programs in telecom/network engineering to our higher-education system, especially in the USA where this was not the norm.

The story of the TEE movement has been told and also discussed within broader contexts in the literature (e.g., *IEEE Communications Magazine, Vol. 53, No. 11, pp. 35–9, November 2015*). The movement had accomplished its main goal of adding telecom/network engineering to ABET's portfolio of recognized engineering disciplines. ABET is the Accreditation Board of Engineering and Technology, Inc., which accredits engineering programs in the USA and more than 40 other countries. A lot of work had to be done to capitalize on that milestone, however. The technical literature of this discipline was rich in scholarly research papers, but not as rich in textbooks designed to support the development of new university courses and curricula. These textbooks are needed to facilitate rigorous academic training for a new generation of network engineers. The idea of this Series came up subsequently to help the telecom/ networking community overcome this shortcoming. Since its inception, Springer's TTE (*Textbooks in Telecommunication Engineering*) has been doing this job. The proximity of the acronyms herein–between TEE and TTE–is coincidental, but worth noting!

I am proud of the inroads TTE has made in publishing quality textbooks and in fulfilling the goals and objectives the Series has had since its beginning. By the end of 2024, we approach the milestone of having 35 books, including 20 published and about 15 in the making (whether in the writing, reviewing, or production phases). More than 50 authors have contributed to this Series so far and they come from about 20 different countries and cultures all over the world.

The topics covered by this Series so far scan the entire landscape of telecom and network engineering, including wireless, optical, satellite; access and core; devices, systems, and networks; space-aerial-terrestrial integration; queuing theory and traffic engineering; network management; Internet and Internet of Things (IoT); signal processing; applied electromagnetics; and economy/business aspects of telecom. The Series includes a cellulargenerations primer and has a Wi-Fi primer project under progress. It covers the most cutting-edge topics with textbooks that are carefully designed to strike the right balance between the need to create new courses thereto today and the fact that progress is imminent in each of these topics, and in the short term. These include books that discuss the roles of edge and cloud computing; of blockchains, and of artificial intelligence and machine learning in network engineering. Special topics such as under-water communications, railways telecom, networked navigation systems, semantic communications, video analytics, and broadband to combat the digital divide are also covered by TTE. I have tried my best to make this Series a top de facto global home for quality telecom/network engineering textbooks, and am hoping that many in our community can start to consider it as such now. The fruits of the hard work put by many people into fulfilling this goal are now handy to pick by our community.

I want to thank my wife Shahira and my daughter Nadine for their support of my work in this Series over nearly ten years. I have put a lot of time into this task and they have understood how this work is important to me. I owe a big deal to all the authors of this Series. Words cannot describe my appreciation of their contributions and talent. I would like to also express my thanks to Springer in general, and to Mary James in particular. Mary has been my main partner throughout this journey, and I couldn't have thought of a better partner than her. Finally, I thank all my colleagues in the field of telecom/network engineering and our readership for their continued support of the mission and the endeavor of this great project.

Tarek S. El-Bawab, PhD
April 2026

Preface

In today's world, wireless broadband has become essential, and humans have benefited from the advent of digital connectivity; however, at the same time, more than 40% of the world's population is yet to be connected to the internet. This has led to a digital divide between people with access to digital connectivity and those without access. The digital divide persists where viable business models are lacking. Addressing this gap requires a multifaceted approach considering technology, economics, policy, and infrastructure. This textbook explores key technologies, theories, and the broader ecosystem shaping the digital divide challenge. This book explores strategies to close the digital connectivity gap in rural areas by examining broadband access, wireless technologies, feasibility studies, key initiatives, and the barriers that hinder progress. It also delves into how technological solutions can align with policies and regulations to support effective integration. Solutions emerge from a blend of disciplines, including engineering, economics, network modeling, and policy analysis. By engaging with diverse perspectives, one can better evaluate technology applications and network feasibility to identify the most effective solutions. This book is designed to provide an in-depth investigation of the subject and is organized into ten chapters.

- **Chapter 1** sets the scene and examines the global and national digital divide, the socioeconomic impact of digital exclusion, the role of broadband in modern society, and broadband access technologies.
- **Chapter 2** examines strategies adopted by telecom providers to bridge the digital divide, including the integration of space, air, and ground networks (SAGIN) and supply-side measures to lower costs and enhance service quality. It then shifts to demand-side approaches aimed at promoting digital adoption and user awareness. Finally, it explores economic and environmental frameworks to evaluate the feasibility and long-term sustainability of broadband technologies.
- **Chapter 3** explores one of the ground networks–5G broadband access and its roles in bridging the digital divide by examining the state-of-the-art, worldwide deployments, and economic and environmental frameworks.
- **Chapter 4** explores the space networks–satellite broadband access and its roles in bridging the digital divide by examining the state-of-the-art, worldwide deployments, and economic and environmental frameworks.
- **Chapter 5** explores the air layer technologies–UAV, NTFP, and HAPS broadband access and its roles in bridging the digital divide by examining the state-of-the-art, worldwide deployments, and economic and environmental frameworks.
- **Chapter 6** explores the Wi-Fi broadband access networks and their roles in bridging the digital divide by examining the state-of-the-art, worldwide deployments, and economic and environmental frameworks.
- **Chapter 7** explores the future networks that have the potential of driving innovation in the telecommunication industry for the next 10–15 years and their role in bridging the digital divide.

- **Chapter 8** examines meaningful connectivity as a foundation for an inclusive digital economy. It traces the path from technological research to commercialization, emphasizing the role of trials, and highlights initiatives and pilot programs that address the usage and access gap and support equitable access.
- **Chapter 9** examines how policies, regulations, and standards shape the scalability and adoption of technologies aimed at bridging the digital divide. It also addresses governance challenges in developing scalable solutions and rolling out broadband access initiatives at scale.
- **Chapter 10** focuses on the role of different stakeholders in bridging the digital divide, focusing on the access and usage gap. Furthermore, this chapter outlines potential research topics and questions to ponder to dive deeper into addressing the digital divide challenge.

Who is This Book for?

The primary motivation behind this book is to provide postgraduate students and professionals with a comprehensive, accessible resource on wireless broadband access to bridge the digital divide. It offers foundational insights explored by several people in academia, industry, government, policymakers, and regulatory bodies. I believe that anyone working in this field should have a solid foundational understanding before advancing to research or specialized topics. Beyond education, this book promotes an integrated approach to addressing the global digital divide–bringing together technology, business models, and governance. It offers both a knowledge base and a guidepost for those pursuing effective connectivity solutions in underserved regions.

Hyderabad, India
April 2026

Shruthi Koratagere Anantha Kumar

Acknowledgements In writing this book, I have been fortunate to be assisted by technical experts in many of the sub-disciplines that belong to the telecommunication domain. Some of my colleagues and friends like Robert W. Stewart, David Crawford, Tawachi Nyasulu, Louise Crockett, H. Nwana Sama, Steve Day, Julie Bradford, Hassan Osman, Edward J. Oughton, Bonface Osoro, Sudhir Dixit, Josef Noll, Gerard Parr, Kaplak Gude, Rajesh Kumar Pathak, and Narendra Magra. This book could not have been written without their guidance. I am deeply indebted to these individuals who took their time to provide me with direction and help organize chapters in their areas of expertise.

I would also like to acknowledge the various organizations and projects I had the opportunity to engage with, like University of Strathclyde, Neutral Wireless, Real Wireless, Cenerva, UK India Future Network Initiative, IEEE Future Networks, and Wireless World Research Forum, which helped me identify key directions and gaps crucial to shaping the research for this book.

I would like to thank Prof. Tarek El-Bawab, my editor at Springer, for his unwavering support and guidance throughout this project. I am deeply appreciative of the opportunity to write this book, which he graciously provided. His support and guidance were instrumental in shaping this work and bringing it to fruition. His encouragement, insights, belief, patience, and expertise in all phases of this project were invaluable to me. I would also like to thank all the staff at Springer Nature, especially Mary James, Brian Halm, and others, who consistently went above and beyond to answer my queries and provide assistance in the best possible way. They were always supportive during this journey.

As always, the greatest source of support systems in life often comes from one's colleagues, friends, and family. I must thank my husband, Sachin, and my son, Shlok. They were super excited about this journey and made it possible for me to finish this book. This book is dedicated to them. To Sachin, your belief in me and your endless patience during late-night hours of writing mean more to me than words can express. Your unwavering support and encouragement kept me going through the challenges and triumphs of this creative process. I always enjoy our discussions on the industry needs, and you always bring a fresh perspective to the topics. You always looked out for me and supported me in all possible ways. To my child, Shlok, thank you for your understanding when I was immersed in the writing world. Your smiles and hugs were the best motivation a writer could ask for. To my parents, as always, who encouraged me to pursue my dreams and asked, "Is it done yet"? at just the right times and were always there to support me with everything within their capabilities for me to finish this book. To my brother, Satya, who would hear me out and help me understand what the reader would want to learn from this book. To my parents-in-law, your belief in my dreams and your constant encouragement have been a guiding light on this journey. To my family, your unwavering example of perseverance and dedication inspires me every day. To my friends, your words of encouragement, kindness, understanding, and steadfast belief in my work have meant the world to me. I would especially like to thank Sachin, Satya, Niharika, Gokul, Abirami, and Chandana for listening to my regular updates throughout the journey. I sincerely appreciate the time and effort you all invested in this journey, and your interest and support during the writing process have been incredibly motivating and encouraging.

Lastly, to all those whose names may not appear here but whose love and support have touched my life, thank you. Writing this book has been a labor of love, and I'm deeply grateful to everyone who has been part of this journey.

With deepest gratitude,

Shruthi Koratagere Anantha Kumar

Competing Interests The author has no competing interests to declare that are relevant to the content of this manuscript.

Contents

Part II Multilayered Technological Solutions

Acronyms

2G	2nd generation
3G	3rd generation
3GPP	3rd generation partnership project
4G	4th generation
5G	5th generation
5G.MIL	5G military applications
5G-HOSTS-SAT	5G hybrid open satellite-terrestrial system
5G-PPP	5G Infrastructure Public Private Partnership
6G	6th generation
AAU	Active antenna unit
ACMA	Australian Communications and Media Authority
ACP	Affordable Connectivity Program
AfDB	African Development Bank
A-GPS	Assisted global positioning system
AI	Artificial intelligence
AM	Amplitude modulation
AOCS	Attitude and orientation control systems
AP	Access point
APEC	Asia-Pacific Economic Cooperation
APT	Asia-Pacific Telecommunity
AR	Augmented reality
ARPU	Average revenue per user
ASK	Amplitude shift keying
ATC	Air traffic controller
B2B	Business-to-business
B5G	Beyond 5G
BAT	Buoyant airborne turbines
BBU	Baseband unit
BEAD	Broadband Equity, Access, and Deployment
BEREC	Body of European Regulators for Electronic Communications
BMC	Business model canvas
BPSK	Binary phase shift keying
BS	Base station
BSS	Broadcasting satellite services
BVLOS	Beyond visual line of sight
CaaS	Connectivity-as-a-service
CAGR	Compound annual growth rate
CAPEX	Capital expenditure
CBRS	Citizens' broadcast radio services
CCK	Complementary code keying
CDF	Cumulative distribution function
CEBF	Connecting Europe Broadband Fund

CEF	Connecting Europe Facility
CFmMIMO	Cell-free massive MIMO
CNPC	Control and non-payload communications
COVID	Coronavirus disease
CRAN	Cloud radio access networks
CRTC	Canadian Radio-Television and Telecommunications Commission
CSC	Common service centers
CSI	Channel state information
CSP	Cloud service provider
CSR	Corporate social responsibility
CubeSats	Cube satellite
D2D	Device-to-device
DaaS	Data-as-a-service
DAS	Distributed antenna systems
DDoS	Distributed denial-of-service
DE4A	Digital Economy for Africa
DFS	Dynamic frequency selection
DL	Downlink
DNS	Domain name system
DoD	Department of Defense
DoS	Denial-of-service
DSA	Dynamic spectrum access
DSL	Digital subscriber line
DSP	Digital signal processing
DSSS	Direct sequence spread spectrum
DTA	Digital Transformation with Africa
DVI	Digital Village Initiative
EAC	East African Community
EHT	Extremely high throughput
EIRP	Equivalent isotropic radiated power
EM	Electromagnetic
eMBB	Enhanced mobile broadband
EMERGE	Enabling mobile and enhanced resilient ground-to-space ecosystem
EOL	End-of-life
ESA	European Space Agency
ETSI	European Telecommunications Standards Institute
EU	European Union
FCC	Federal communications commission
FHSS	Frequency hopping spread spectrum
FL	Federated learning
FM	Frequency modulation
FPGA	Field programmable gate array
FSA	Fixed spectrum allocation
FSK	Frequency shift keying
FSO	Free space optical
FSPL	Free-space path losses
FSS	Fixed satellite service
FWA	Fixed wireless access
GCC	Gulf Cooperation Council
GCS	Ground control stations
GDP	Gross domestic product
GDPR	General data protection regulation
GenAI	Generative AI

GEO	Geostationary earth orbit
GHG	Greenhouse gas
GLONASS	Global Navigation Satellite System
GNSS	Global Navigation Satellite System
GP	Gram Panchayat
GPS	Global Positioning System
GPU	Graphic processing unit
GSMA	Global System for Mobile Communications Association
GW	Gateway
GWL	Gateway link
HAPS	High altitude platform stations
HD	High definition
HF	High frequency
HIBS	High altitude IMT base stations
HTA	Heavier-than-air
HTS	High throughput satellite
IC	Integrated circuit
ICANN	Internet Corporation for Assigned Names and Numbers
ICT	Information and communication technologies
IEC	International Electrotechnical Commission
IEEE	Institute of Electrical and Electronics Engineers
IETF	Internet Engineering Task Force
IMT	International Mobile Telecommunications
InP	Infrastructure provider
IoT	Internet of things
IP	Internet protocol
IPTV	Internet protocol television
IR	Infrared
IRS	Intelligent reflective surfaces
ISAC	Integrated sensing and communication
ISED	Innovation, Science, and Economic Development
ISL	Intersatellite link
ISM	Industrial, Scientific, and Medical
ISO	International Organization for Standardization
ISP	Internet service provider
ISS	International Space Station
IT	Information technology
ITU	International Telecommunication Union
KPI	Key performance indicator
KVI	Key value indicators
LAN	Local area networks
LCA	Life cycle assessment
LED	Light emitting diode
LEO	Low earth orbit
Li-Fi	Light fidelity
LoRa	Long range
LoRaWAN	Long range wide area network
LOS	Line of sight
LSA	Licensed spectrum access
LTA	Lighter-than-air
LTE	Long term evolution
LTE-A	Long term evolution–advanced
M2M	Machine to machine

MAC	Medium access control
MAN	Metropolitan area network
MBSP	Mobile Black Spot Program
MCIT	Ministry of Communications and Information Technology
MDB	Multilateral Development Bank
MEC	Multi-access edge computing
MEO	Medium earth orbit
MFA	Multi-factor authentication
MIMO	Multiple-input multiple-output
ML	Machine learning
MLO	Multilink operations
mMIMO	Massive MIMO
mMTC	Massive machine type communications
mmWave	Millimeter wave
MNO	Mobile network operator
MSS	Mobile satellite services
MU-MIMO	Multi user MIMO
NANP	North American Numbering Plan
NASA	National Aeronautics and Space Administration
NB-IoT	Narrowband IoT
NBN	National Broadband Network
NDLP	National Digital Literacy Programme
NFV	Network function virtualisation
NGMN	Next generation mobile networks
NGO	Non-government organization
NHN	Neutral host network
NIST	National Institute of Standards and Technology
NLOS	Non-line of sight
NP	Nondeterministic polynomial time
NPN	Non-public networks
NPV	Net present value
NR	New radio
NREN	National Research and Education Networks
NSA	Non-standalone
NTFP	Network tethered flying platform
NTN	Non-terrestrial networks
O&M	Operations and maintenance
Ofcom	Office of Communications
OFDM	Orthogonal frequency-division multiplexing
OLPC	One Laptop Per Child
OPEX	Operational expenditure
O-RAN	Open radio access networks
OSI	Open system interconnection
OSS	Operations support system
OTFS	Orthogonal time frequency space
OTT	Over-the-top
PFD	Power flux density
PHY	Physical
PM	Phase modulation
PMGDISHA	Pradhan Mantri Gramin Digital Saksharta Abhiyan
PM-WANI	Pradhan Mantri Wi-Fi Access Network Interface
PNBL	Programa Nacional de Banda Larga
PNT	Position, navigation, and timing

PoP	Point of presence
PPP	Public-private partnership
PSK	Phase shift keying
PTMP	Point-to-multipoint
PTP	Point-to-point
PU	Primary users
QAM	Quadrature amplitude modulation
QCL	Quantum cascade laser
QKD	Quantum key distribution
QoE	Quality of experience
QoS	Quality of service
QPSK	Quadrature phase shift keying
R&D	Research and development
RAN	Radio access network
RANGE	Remote area Access Network for 5th GEneration
RCP	Regional Connectivity Program
RedCap	Reduced capability
RF	Radio frequency
RIS	Reconfigurable intelligent surfaces
RMIO	Radio mobile infrastructure operator
ROI	Return on investment
RR	Radio regulations
RSS	Reconfigurable selective surfaces
RU	Resource unit
SA	Standalone
SaaP	Service-as-a-platform
SADC	Southern African Development Community
SAGIN	Space-air-ground integrated network
SANSA	Shared access terrestrial-satellite backhaul network enabled by smart antennas
SAS	Spectrum access system
SATis5	Satellite-terrestrial integration in the 5G
SDG	Sustainable development goal
SDN	Software-defined networking
SDR	Software defined radio
SDS	Software-defined satellite
SFC	Service function chaining
SG	Study group
SIM	Subscriber identity module
SINR	Signal interference to noise ratio
SLA	Service level agreement
SmallSats	Small spacecraft
SME	Small medium enterprise
SNR	Signal to noise ratio
SoC	System on chip
SON	Self-organising network
SOS	Save our souls
SRE	Smart radio environment
SSID	Service set identifier
STEM	Science, technology, engineering, and mathematics
SU	Secondary users
SWOT	Strengths, weaknesses, opportunities and threats
Tbps	Terabit per second

TCO	Total cost of ownership
TDD	Time division duplex
TDRA	Telecommunications and Digital Government Regulatory Authority
TDS	Time-domain spectroscopy
TEA	Techno-economic assessment
THz	Terahertz
TRAI	Telecom Regulatory Authority of India
TRL	Technology readiness level
TSDSI	Telecommunications Standards Development Society, India
TSP	Telecom service provider
TT&C	Telemetry, tracking and control
TV	Television
TVWS	Television white space
TWT	Target wake time
UAE	United Arab Emirates
UAV	Unmanned aerial vehicle
UE	User equipment
UGV	Unmanned ground vehicle
UHF	Ultra high frequency
UK	United Kingdom
UL	Uplink
UN	United Nations
UNDP	UN Development Programme
UNESCO	United Nations Educational, Scientific and Cultural Organization
UNICEF	United Nations Children's Fund
URLLC	Ultra reliable low latency communications
USA	United States of America
USO	Universal service obligation
USOF	Universal Service Obligation Funds
USSR	Union of Soviet Socialist Republics
V2X	Vehicle to everything
VHTS	Very high throughput satellites
VITAL	Virtualized hybrid satellite-terrestrial systems for resilient and flexible future networks
VLEO	Very low earth orbit
VLOS	Visual line-of-sight
VNC	Value network configuration
VNF	Virtual network function
VoIP	Voice over internet protocol
VoLTE	Voice over LTE
VR	Virtual reality
VSAT	Very small aperture terminals
WAN	Wide area network
WEF	World Economic Forum
WEP	Wired equivalent privac
WG	Working group
Wi-Fi	Wireless fidelity
WiMAX	Worldwide interoperability for microwave access
WLAN	Wireless local area networks
WNI	Wi-Fi network interface
WPA	Wi-Fi protected access
WRC	World Radio Conference
WSP	Wireless service provider

WTO	World Trade Organization
XL-mMIMO	Extra large massive MIMO
Zigbee	Zonal Intercommunication Global-Standard

List of Figures

List of Tables

Part I
Digital Divide and Strategies

In today's world, the telecommunication sector is driving innovation across all sectors. Mapping the data on worldwide coverage data and analyzing the coverage and usage trends helps in understanding the gap in global connectivity. Part I of the book offers an in-depth exploration of the global connectivity gap by defining what a digital divide is and examining root causes, socioeconomic impacts, and disparities. This underscores the urgency of connecting underserved populations by emphasizing strategic approaches to bridge the digital divide–ranging from innovative broadband access and emerging business models to scalable solutions that ensure long-term impact. There is a need to address the access, affordability, and usage gap, which play a crucial role in bridging the digital divide.

Part I also explores the strategies in bridging the gap that focus on going beyond the technological solutions, such as space-air-and-ground networks (SAGIN), infrastructure sharing, spectrum sharing, and network optimization (technological integration and network planning). For enhancing user adoption of enhanced digital services, there is a need to focus on factors such as affordability, digital literacy, digital awareness, digital skills, and user confidence. Furthermore, this part also investigates the framework for qualitative and quantitative assessments to model the feasibility of technologies to extend the networks in remote and rural areas. Additionally, it is essential to examine the proposed technological solutions for affordability and environmental impact, to ensure that they are not only viable but sustainable. Together, these insights contribute to shaping a long-term business model that balances innovation with affordability, inclusivity, and ecological responsibility.

Therefore, Part I provides an overview of the digital divide gap and the strategies to bridge this gap. The concepts and frameworks explored in Part I are applied to technologies discussed in Part II.

Broadband Access and Digital Divide

1

The widespread use of the Internet and digital services has greatly benefited humanity since the advent of telecommunication services, enhancing vital areas such as information access, communication, education, economic opportunity, healthcare, empowerment, efficiency, creativity, innovation, and environmental sustainability [1]. These services offered could be either using wired or wireless technologies. Internet broadband has been perceived as one of the essential needs of human life and contributes significantly to the overall growth of the national gross domestic product (GDP) [2]. Internet penetration has been increasing from 6% in the year 2000 to 65.7% in 2023. The data traffic has been growing exponentially over the years, requiring the operators to upgrade their infrastructure to meet the demand. Broadband access is now seen as a key enabler for social and economic progress, and there is a need to build a digitally inclusive society for all [3].

1.1 Broadband Access

Let us take a step back and begin with understanding *what broadband is* and *why it is important*.

Broadband technology refers to high-speed, high-bandwidth connections that offer higher capacity compared to a dial-up telephone connection. Broadband services are used in use cases ranging from business to personal applications and from small to large-scale businesses. There has been a growing demand for reliable and high data rates for these applications, which is driving the evolution of wireless and wired technologies. There are still around 34% of people who remain unconnected and lack access to the Internet, and this scenario is commonly referred to as the digital divide. The growth of telecommunication services in regions with a digital divide is dependent on several factors beyond technology, such as language, affordability, landscape and terrain, digital awareness, return on investment (ROI), business models, policies, and regulations [4].

The telecommunications network can be classified into two major categories: terrestrial networks and non-terrestrial networks (NTN).

- The terrestrial network utilizes wired or wireless technologies that operate through land-based infrastructure.
- On the other hand, NTN technologies depend on space-based or high-altitude platforms to deliver connectivity. Both types of telecommunication networks are vital in today's interconnected world.

The key technologies discussed in this book are shown in Fig. 1.1, which include terrestrial networks such as 4th generation (4G), 5th generation (5G), 6th generation (6G), wireless fidelity (Wi-Fi), light fidelity (Li-Fi), long-range Wi-Fi, and fiber, and NTN technologies such as low earth orbit (LEO), medium earth orbit (MEO), and geostationary earth orbit (GEO) satellites, high altitude platform stations (HAPS) (and the ones that support International Mobile Telecommunications (IMT) base station (BS) are called high altitude IMT base stations (HIBS)), unmanned aerial vehicles (UAVs), hot air balloons, network tethered flying platforms (NTFPs), and future networks. This book draws on a wide range of sources, including research papers, white papers, industry reports, projects, and test trials, to explore how each of these technologies will help overcome the global digital divide. While the content in this book covers a wide range of topics, the content is designed to be approachable for readers without specialized backgrounds. Key ideas for the areas covered in this book are presented using nonspecialist vocabulary, with subjects and phrases defined as needed.

S. Koratagere Anantha Kumar, *Wireless Broadband Access to Bridge the Digital Divide*,
Textbooks in Telecommunication Engineering, https://doi.org/10.1007/978-3-032-09261-8_1

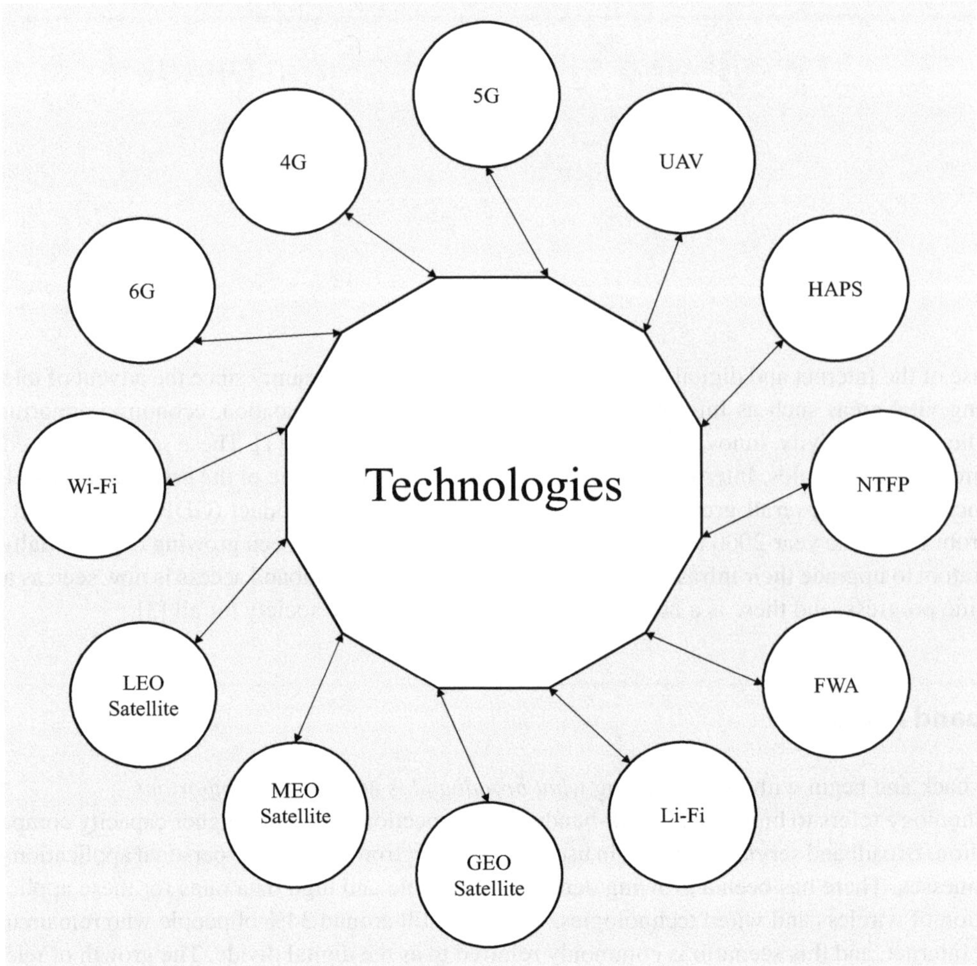

Fig. 1.1 Broadband access technologies

1.1.1 Evolution of Telecommunication Industry

Figure 1.2 shows the development of the telecommunication industry throughout history and highlights significant turning points and milestones, including the invention of the electric telegraph in 1838 and the emergence of ultra-high-speed technologies in 2019 and beyond [5, 6]. The technological advancements in the telecommunications sector have undergone a protracted and constant state of change since their inception to meet the demands of end users and the dependent industries. As of today, there is ongoing research in several directions that will become commercial in the coming years, and some of these are discussed in depth in this book.

The evolution of these technologies has enabled high-speed, reliable, and secure digital broadband access. Today, digital broadband connection is an essential infrastructure that enables people and organizations to interact, communicate, and access information online via the Internet. There is an exponentially growing global need for reliable, secure, and ultra-high-speed Internet services, fueled by the expansion of coverage of the different telecommunication technologies. This surge is crucial for social advancement, economic expansion, and technological innovation worldwide. Almost 5.3 billion individuals on Earth are connected with the help of a wide range of telecommunication technologies. Each technology has a different data rate, coverage, and reliability. For example, 5G supports a peak data speed of 20 Gbps download and 500 Mbps upload [7].

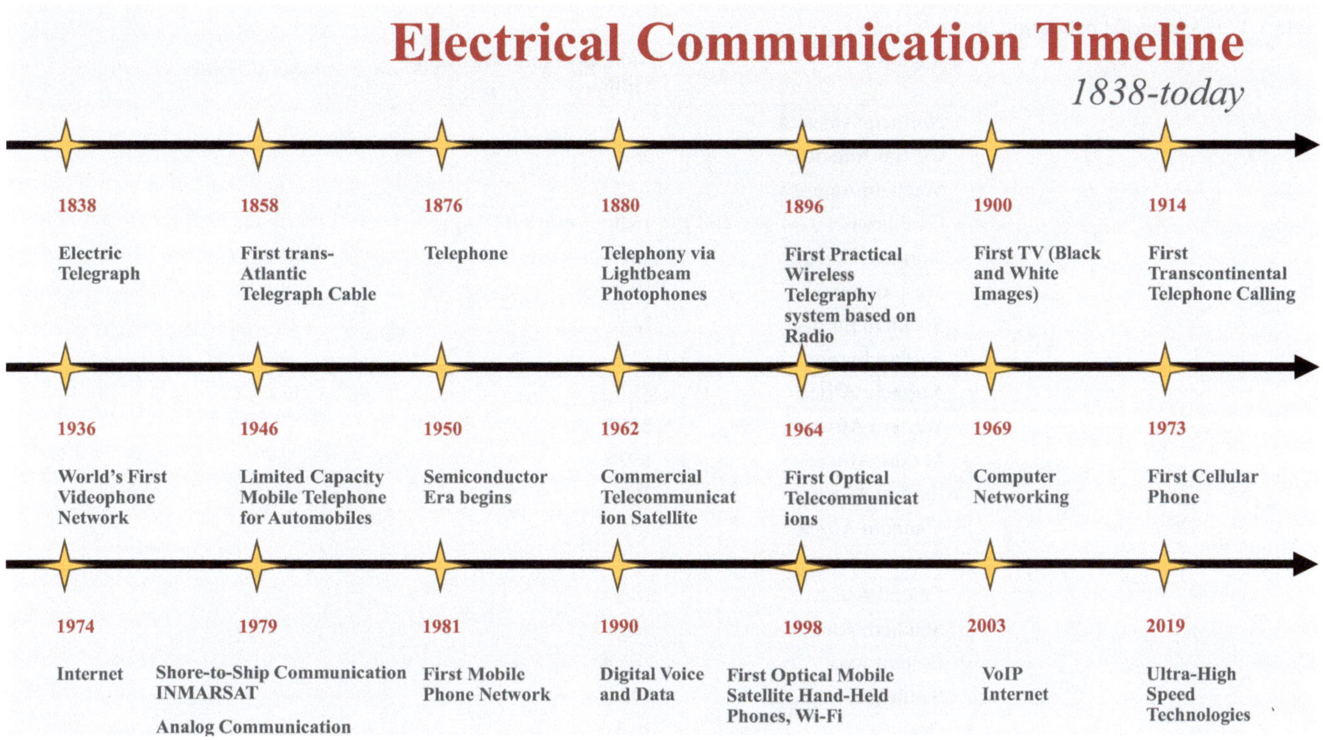

Fig. 1.2 Telecommunication industry timeline

1.1.2 Worldwide Broadband Coverage

Several specialized firms aggregate and analyze data across key factors, including broadband performance, number of cells and towers, site location, coverage, frequency of operations, adoption trends, revenue trends, deployment rates, and emerging technologies. Additionally, data on broadband adoption, penetration, data rates, technological maturity, and standards compliance worldwide are gathered and analyzed by international organizations, for example, International Telecommunication Union (ITU), to frame international policies and regulations as well as help in assessing the infrastructure needs of each country and deploying a data-driven broadband technology to meet the requirements [8, 9].

As per the January 2025 data released by ITU and Datareportal on "*Digital 2024 Global Overview*," only 5.56 billion out of 8.2 billion people are connected to the Internet [9]. There are 5.24 billion social media users in the world. The average user uses the Internet for 6 hours and 38 min daily. The Internet adoption rate as a percentage of the total population is highest in Northern Europe at 97.6%, followed by North America at 96.8%, Western Europe at 94.5%, Southern Europe at 90.2%, and Eastern Europe at 88.4%. Meanwhile, some other regions report Internet adoption rates around 70+%, such as Southern America, Central America, Southern Africa, Central Asia, Western Asia, Eastern Asia, South-eastern Asia, and Oceania. In contrast, regions in Northern Africa and Southern Asia have an Internet adoption rate of over 50%, while regions in Western Africa have 42.3%, Middle Africa has 32.1%, and Eastern Africa has 26.7%.

1.1.3 Digital Divide

The term "digital divide" describes a gap between geographic areas where there are distinct disparities among those who can and cannot access digital services online. A recent report by ITU and the World Bank shows that there is room for improvement in digital connectivity, both in terms of coverage and capacity [10]. As of today, there are around 34% of people around the world who are yet to be connected to the Internet. Similarly, among those who are connected to the Internet, millions of users lack reliable capacity and data rates; that is, they lack meaningful connectivity [11]. Table 1.1 shows the statistics on the worldwide distribution of unconnected populations [9, 12]. Figure 1.3 shows the worldwide Internet adoption rate.

Table 1.1 Unconnected population across the globe

Region	Number of individuals (in Millions)
Northern America	12
Central America	38.5
Southern America	77.2
Caribbean	13.3
Northern Europe	2.6
Western Europe	10.8
Southern Europe	14.8
Eastern Europe	33.5
Northern Africa	85.9
Western Africa	256.8
Middle Africa	139.8
Eastern Africa	360
Southern Africa	18.7
Western Asia	71.7
Central Asia	18.8
Southern Asia	988.7
Eastern Asia	385.8
South-Eastern Asia	196.4
Oceania	10.3

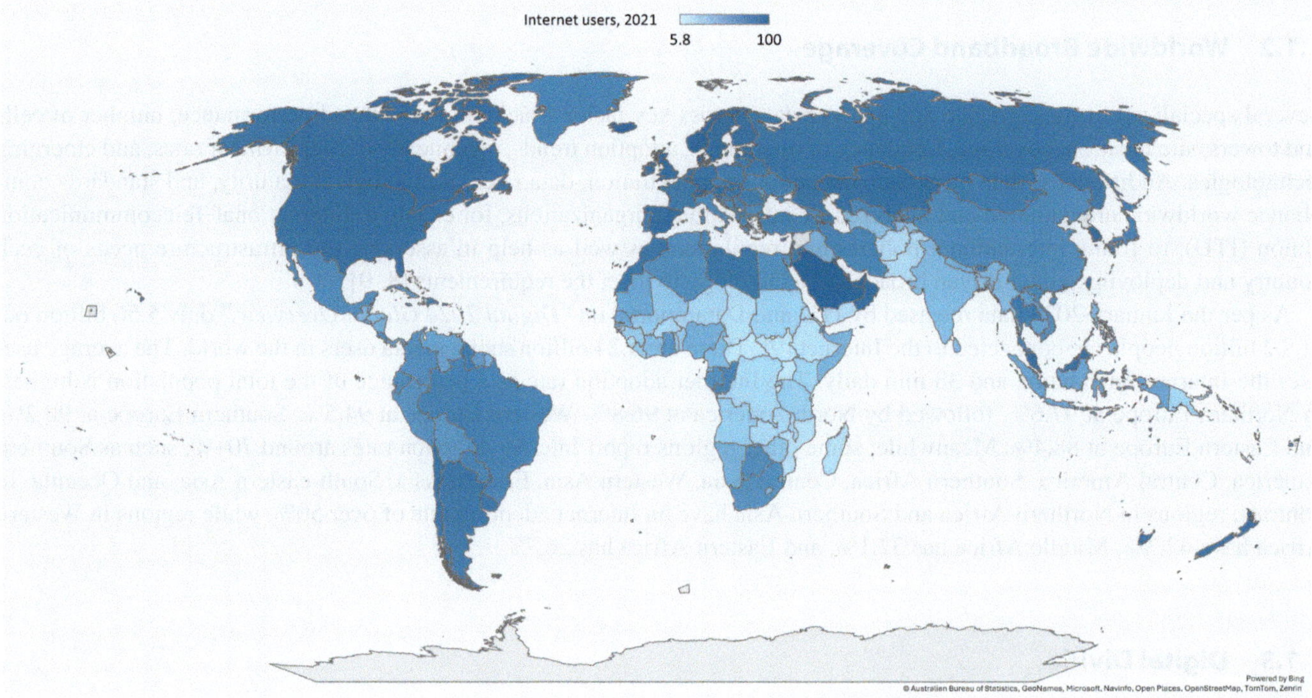

Fig. 1.3 Worldwide Internet adoption rate

1.1.4 Statistics on Digital Divide

Globally, the top ten countries with the lowest levels of Internet adoption ($\leq$20% adoption rate), ranked highest to lowest, include North Korea, the Central African Republic, Burundi, South Sudan, Niger, Yemen, Afghanistan, Ethiopia, Burkina Faso, and Madagascar. Furthermore, these countries also face challenges in other sectors, such as access to electricity, basic drinking water, and basic sanitation services [9]. These operational characteristics should also be considered while developing technology solutions for these places. Additionally, in countries such as Burundi, Eritrea, Somalia, Afghanistan, Guinea, Mozambique, Timor-Leste, and Comoros, the average price per Mbps (fixed broadband) against average monthly income is over 50%, according to research [12].

These statistics demonstrate the disparity in Internet quality and accessibility between countries and territories. Studies have shown that connecting the unconnected to digital services would help the GDP, for example, connecting an extra 10% of schools can improve children's education time and also increase the GDP by around 1.1% [13]. Therefore, connecting the unconnected would contribute toward the economic upliftment of the country, generate newer revenue streams for the people, and provide employment opportunities as well as social well-being for individuals.

The issues of the digital divide must be addressed in both developed and developing nations. However, the issue is more acute in developing nations, where affordability and priority are primary drivers for digital adoption. In contrast, developed countries face an affordability challenge stemming from the availability of advanced technologies, leading to elevated costs for services. Often, the marginalized communities lack the digital skills to use the services confidently in the underserved regions. Likewise to the aforementioned, the World Bank is concerned with comprehending how global digital connectivity affects social inclusion and economic growth, ultimately impacting the overall global development. It also focuses on the digital divide and the equitable usage of broadband infrastructure access [14].

1.2 Why Is There a Digital Divide?

As our dependency on digital technology grows, there is a dire need to bridge this gap and minimize the digital divide. A recent report by ITU and the World Bank justifies that there is room for improvement in digital connectivity, both in terms of coverage and capacity as well as reliability [10, 15]. We do not yet know how to provide connectivity to 100% of the earth and 100% population around the globe since it is a multi-faceted challenge in terms of accessibility, availability, and affordability [16, 17].

The coronavirus disease (COVID)-19 epidemic demonstrated a real digital connectivity gap that must be addressed in both rural and urban locations. People with weak or no Internet access were the most affected by the epidemic. Contributing factors to this situation include limited access to devices, inadequate broadband connectivity, and insufficient technological skills. These factors prevented many students from participating in virtual classes, hindering interpersonal communication, and restricting adults from working remotely [18, 19]. Therefore, there is a need to understand the key reasons leading to this digital divide.

The digital divide has several ramifications affecting revenue, social well-being, and opportunities for advancement. Figure 1.4 shows the key reasons for the digital divide, such as prohibitively high infrastructure rollout costs and low ROI resulting in a scarcity of Internet services in rural and sparsely inhabited areas, and low adoption rates and data consumption per user [20]. These barriers can be categorized into two main categories–coverage and usage gaps, which together result in the digital divide as shown in Table 1.2.

Several key factors contribute to the digital divide, including:

- **Affordability**: People in places with a digital divide have lower per capita income, and the cost of providing telecommunication services in those areas is high, making the service unaffordable for many [21]. To address this issue, governments all over the world are subsidizing broadband access in rural regions and offering discounts for underserved communities, but the approach is not economically sustainable yet [22].
- **Access to Infrastructure**: The broadband operators (terrestrial or NTN) are not able to deploy the required information and communication technologies (ICT) infrastructure due to challenges in terms of climate, geographical terrain, population distribution, network traffic generated, access, power supply, spectrum, network cost, profitability, and population distribution [21].
- **Cultural and Social Barriers**: An essential barrier to overcoming the digital divide is the lack of content and resources available in local languages in the region that is underserved [23]. Only 1.5 billion people out of the 8+ billion people know

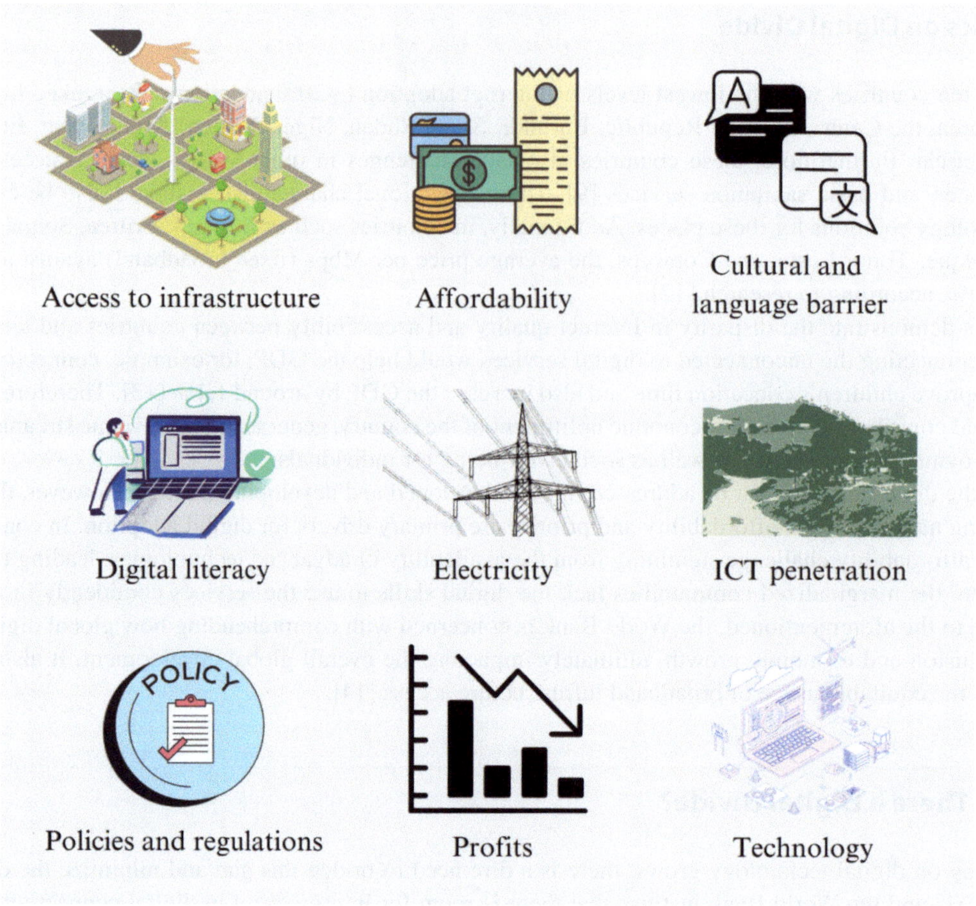

Fig. 1.4 Barriers to global digital divide

English; however, the majority of the ICT content is only available in English, and this is a major barrier. In a slightly different context, women are not allowed to be educated and use ICT services [24].

- **Digital Literacy**: This relates to the users' lack of appropriate technological knowledge necessary to operate digital services, such as accessing services via smartphones, laptops, or other gadgets, which is preventing them from reaping the benefits offered by digital connectivity [22]. People will demand better services once they realize the benefits of rural digital solutions and how they will enhance their lives.

Table 1.2 Coverage and usage gap

Barrier	Usage gap	Coverage gap
Access to infrastructure	✗	✓
Affordability	✓	✗
Cultural and language	✓	✗
Digital literacy	✓	✗
Electricity	✓	✓
ICT penetration	✓	✓
Policies	✓	✓
Profitability	✗	✓
Technology	✗	✓

- **Electricity**: Intermittent or no power supply is another major challenge in providing telecommunication services in many areas. Around 660 million people worldwide, even today, are without access to electricity, and many people have intermittent power supply [25]. Solutions should prioritize ensuring an uninterrupted power supply while encouraging the usage of renewable energy sources; an approach that would help to minimize the overall carbon footprint.
- **ICT Penetration**: Countries with developed economies often exhibit more advanced ICT infrastructure and greater rates of Internet usage, but developing nations may lag owing to a lack of funding, political will, or technological know-how [26, 27].
- **Policies**: The existing telecommunication policies need to be supportive to encourage telecom service providers (TSPs) to deploy telecommunication networks in underserved regions and complement the national-level infrastructure [28]. To enable broader deployment by local operators, policy reforms must prioritize spectrum and infrastructure sharing, streamlined market entry conditions, and fair competition regulations. Additionally, targeted subsidies, alignment with net-zero targets, climate action strategies, and support for open-source technologies are essential to foster innovation and sustainability in the telecom sector.
- **Profitability**: Due to the capital-intensive nature of the telecommunication sector, the operators are not keen on deploying the network in areas with low or no profits. The cost of deploying the network is exorbitant in underserved communities, coupled with the low demand and poor ROI [29]. To attract unconnected users, there is a need for tiered service pricing and newer business models.
- **Technology**: In underserved areas, the technology key performance indicators (KPIs) includes high coverage, high data rates per user, high energy efficiency, high spectral efficiency, spectrum sharing, ease of legal paperwork, support for all use cases using a single network, and seamless roaming capabilities [21]. An appropriate technological solution should begin with assessing regional demand, followed by designing and optimizing a supply model that leverages a mix of last-mile access and backhaul technologies, all while operating within defined cost constraints.

1.2.1 Socioeconomic Impacts of Digital Divide

Limited or no access to internet connectivity and associated technologies significantly deepens the digital divide, profoundly affecting individuals' daily lives and contributing to broader societal disparities. It affects the development across personal, professional, economic, educational, and social dimensions. The digital divide is causing a wide disparity among the people who are connected and those who aren't. It is thus important to understand the socioeconomic impacts of the digital divide and why it is crucial to bridge the gap. The lack of access to the Internet and related technologies leads to adverse scenarios, such as:

- **Education**: Around 2.2 billion children and young adults under the age of 25 lack access to the Internet at home, according to studies by the ITU and United Nations Children's Fund (UNICEF) [30]. When schools and the education sector implemented the policy of remote learning through the use of the Internet and associated technologies during the COVID-19 pandemic. Those without access to the Internet can be cut off from the opportunity to develop their skills, leading to educational gaps, such as falling short of the minimum competency levels in reading and math, and missing employment opportunities and other skills, which affect their personal and societal well-being.
- **Economy**: When Small medium enterprises (SMEs) and entrepreneurs have limited broadband access, they face roadblocks in terms of scaling their operations, reaching a wider customer base, and offering limited financial transaction modes. Furthermore, the GDP of a nation is directly proportional to the digital connectivity and its adoption. However, limited access to these technologies and low investments in digital services and digital literacy of the population inhibit the country's economic growth.
- **Social**: People can share knowledge, opinions, and experiences with friends, family, and a global audience thanks to digital technologies that enable social interactions at personal, professional, and business levels. People who lack access to these digital technologies may miss out on these opportunities and experience feelings of isolation since they are unable to communicate in the same way as someone who has access to the internet. A person with limited to no digital literacy skills is unable to reap the benefits offered by digital connectivity, especially older adults and marginalized individuals are impacted by this.
- **Society**: From a societal viewpoint, a community with no or limited Internet access faces challenges in accessing essential digital services, such as e-governance, digital banking, digital investments, emergency response, e-healthcare, booking

services, and skill development programs. Next, in today's healthcare system, broadband access plays a vital role, from booking an online consultation to enabling precision remote surgery. Yet, the absence of reliable connectivity creates a critical gap, depriving communities of timely medical support, digital care pathways, and life-saving technologies.

- **Culture:** Digital exclusions prevent the conservation and preservation of the culture, traditions, and knowledge of different communities, especially in rural and indigenous areas. Communities without access to the Internet would not be able to preserve their original traditions, literature, music, and art to pass them to the younger and future generations. Furthermore, the artists, creators, and local entrepreneurs have few or no tools to showcase and monetize their work to the global market.

As digital ecosystems become central to governance and innovation, bridging this divide is not just a technical necessity; it's a socioeconomic imperative that demands collaborative, sustainable solutions.

1.2.2 Funds for Bridging the Gap

The digital divide problem is also one of the main focuses of United Nations (UN) sustainable development goal (SDG) 9: developing resilient infrastructure, supporting inclusive and sustainable industrialization, and stimulating innovation [31]. Achieving these targets in the telecommunications industry requires building broadband infrastructure to bridge the digital divide, and empowering individuals to benefit from those technologies to enhance their quality of life, improve economic prospects, and unlock new opportunities for growth and participation

In recent years, an array of worldwide pilot projects and investments have been concentrating on harnessing a variety of broadband access technologies to close the digital divide, enhance connectivity, and promote digital inclusion, as shown in Fig. 1.5. These funds act targeted interventions to address the telecom market failures in underserved areas and for unconnected citizens. These funds help to bridge the digital gap through regulatory measures by stepping in where competition among industry players falls short, and there exist commercial limitations to address the usage gap, to improve the network adoption and usage among consumers.

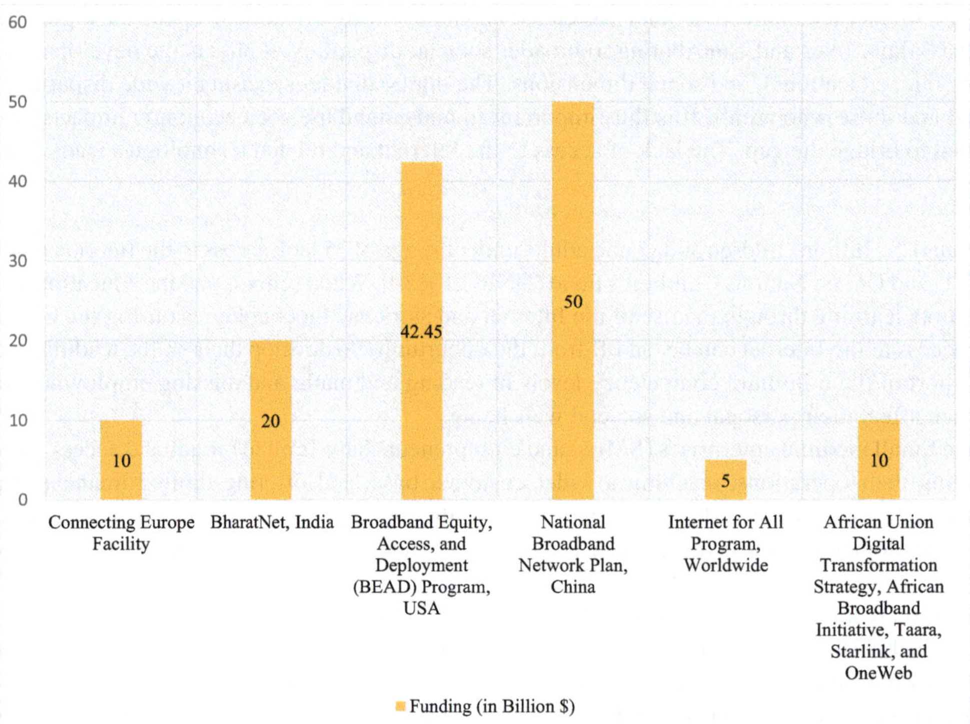

Fig. 1.5 Worldwide investments aimed to bridge the digital divide

- **Connecting Europe Facility (CEF)**: This project aims to bridge Europe's digital divide by developing broadband infrastructure using broadband technologies, such as fiber optic, and minimize the digital disparities [32].
- **BharatNet**: This initiative in India aims to provide high-speed broadband connectivity to all users by laying fiber-optic point of presence (PoP) until the Gram Panchayat (GP), which consists of a cluster of villages [33].
- **Broadband Equity, Access, and Deployment (BEAD) Program**: The project in the USA aims to reduce digital disparities, expand broadband infrastructure, and offer affordable Internet services [34].
- **National Broadband Network Plan**: This Chinese initiative to enhance broadband infrastructure across all regions aims to minimize digital disparity in access and usage across the country [35].
- **Internet for All Program**: In Brazil, this project aims to provide broadband services in rural areas using terrestrial and NTN broadband technologies [36, 37].
- **Africa Initiatives**: Numerous programs in Africa are working closely to bridge the digital divide by advancing digital inclusion programs and deploying infrastructure using terrestrial and NTN technologies. The network aims to offer reliable, affordable, and secure Internet services for all [38].

1.2.3 Stakeholders

Stakeholders are individuals, organizations, or firms that can impact or be impacted by the outcome of the project or program. The primary stakeholders are the ones who, if withdrawn from the project/program, would significantly cause a failure. Understanding the stakeholders, their objectives, and their effect (positive, negative, or neutral) is crucial at the outset of the project to minimize risks and to get further insight into the roles that the stakeholders have performed; this analysis ought to be documented. Key stakeholders involved in bridging the digital divide who may have an impact on or be impacted include TSPs, regulatory bodies, communities, standardization bodies, suppliers and vendors, governments, investors, and customers or end users, as shown in Fig. 1.6. The foremost responsibilities of stakeholders in bridging the digital gap are as follows:

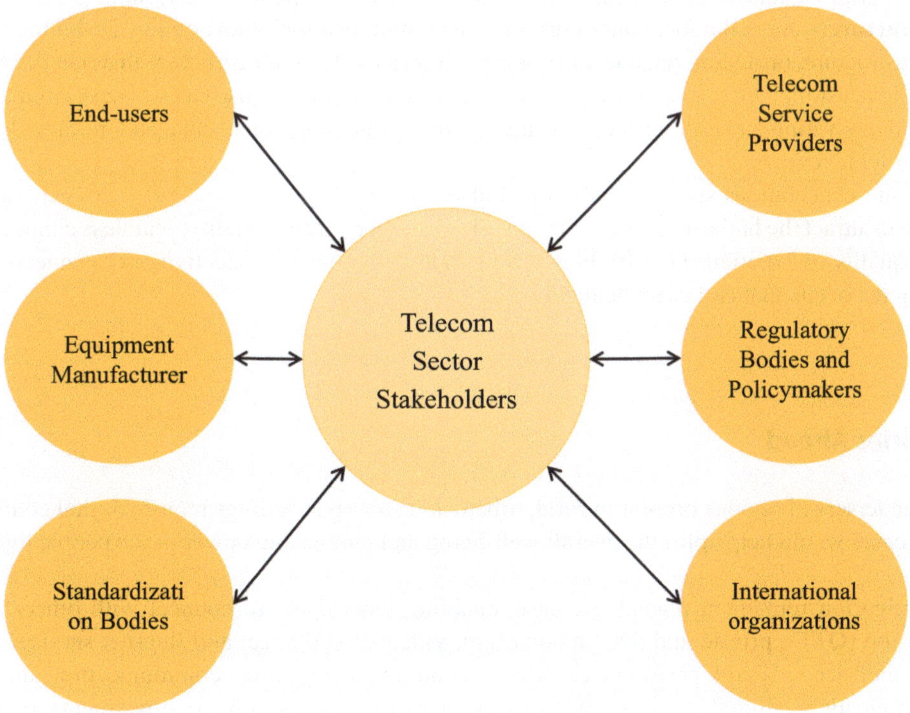

Fig. 1.6 Stakeholders in the telecommunication sectors

- **TSPs** include firms who operate as mobile network operators (MNOs), internet service providers (ISPs), cloud service providers (CSPs), infrastructure providers (InPs), satellite service providers, and wireless service providers (WSPs). These stakeholders are responsible for deploying and operating broadband network technologies across numerous markets, as well as competing with one another to provide services to end users.
- **Telecommunications Regulatory Bodies** play an important role in regulating the telecom sector, whose roles involve assuring fair competition, high-quality service, and customer protection. Each nation has its own regulatory structure and authority to help ensure that the telecom business runs smoothly, fairly, and effectively. Regulators protect the interests of consumers while fostering innovation. Some of the regulatory bodies across the world include ITU, Federal communications commission (FCC) in the United States of America (USA), Body of European Regulators for Electronic Communications (BEREC) in the European Union (EU), Telecom Regulatory Authority of India (TRAI) in India, Australian Communications and Media Authority (ACMA) in Australia, Canadian Radio-Television and Telecommunications Commission (CRTC) in Canada, Office of Communications (Ofcom) in the United Kingdom (UK), and Telecommunications and Digital Government Regulatory Authority (TDRA) in the United Arab Emirates (UAE).
- **International Organizations** play a pivotal role in harmonizing the global telecommunication sector. Their efforts focus on reducing manufacturing and service costs, promoting spectrum harmonization, and fostering innovation. Notable organizations such as ITU, World Trade Organization (WTO), European Telecommunications Standards Institute (ETSI), World Bank, Institute of Electrical and Electronics Engineers (IEEE), Asia-Pacific Telecommunity (APT), and Global System for Mobile Communications Association (GSMA) contribute significantly to these advancements. By enabling the use of standardized spectrum bands, they help minimize interference, particularly in cross-border communications, while driving innovation and ensuring seamless global connectivity.
- **Standardization Bodies** are organizations that develop, coordinate, and promote international standards for the evolution of telecommunication technologies. These standards ensure that technologies are compatible and interoperable across the globe to promote international technology harmonization, communication, and trade. These organizations collaborate to ensure that telecommunications systems and equipment can operate smoothly across national borders, laying the foundation for the global communications network. Notable organizations contributing toward standards in the telecommunication domain include International Organization for Standardization (ISO), International Electrotechnical Commission (IEC), IEEE, ETSI, Internet Engineering Task Force (IETF), 3rd generation partnership project (3GPP), 5G Infrastructure Public Private Partnership (5G-PPP), and Telecommunications Standards Development Society, India (TSDSI).
- **Equipment Manufacturers** are at the forefront of driving telecommunication innovation by designing and manufacturing state-of-the-art infrastructure, providing reliable and quality of service (QoS) infrastructure that handles the massive traffic. They could also design custom hardware and software to meet specific industrial use cases. These companies play a pivotal role in designing and innovating end user devices that are capable of enabling voice, data, and multimedia communications across diverse geographic regions.
- **End Users**' needs and expectations spur innovation and also drive the competition. End users generate revenue for the TSPs, who compete to attract the highest number of potential subscribers. Better quality, seamless connectivity, low latency, security, enhanced quality of experience (QoE), high throughput, high device connectivity per connection, and ubiquitous coverage are among the needs that end users demand.

1.2.4 Opportunities Ahead

Several use cases in underserved regions present in rural, urban, and suburban settings require digital connectivity, as shown in Fig. 1.7. These use cases would help uplift the overall well-being and purchasing power of the people living in underserved regions.

Personal communication remains a pivotal use case, enabling individuals to connect with others and access digital services and over-the-top (OTT); private and live broadcasting; video surveillance; and Internet services [26, 29]. **Remote workers** benefit from enhanced network performance, which enhances collaboration, communication, and productivity [39]. For **smart villages** applications, digital connectivity supports smart lighting, e-banking, e-payments, e-governance, digital assistance for the elderly, emergency support, and disaster management [40]. Enhanced digital infrastructure in rural areas can boost their economies, enhance global trade, and improve living standards.

The world is increasingly embracing **mobile financial services**. The digital economy is driven by financial corporations and fintech companies, offering swift and portable solutions made possible by wireless broadband access. **Smart agriculture**

Fig. 1.7 Use cases for digital connectivity

and farming require broadband networks to support the extensive connectivity and monitoring use cases, including the deployment of internet of things (IoT) devices for various sensing applications [41]. **Remote health** monitoring and surgical procedures are crucial for delivering healthcare in rural areas [42]. E-healthcare systems, especially during the pandemic, helped alleviate the strain on traditional healthcare facilities. **Online education** and remote learning are made possible by high-speed digital networks, allowing students and adults to participate in virtual classrooms from anywhere in the world [23].

The **transportation sector** leverages use cases, such as live traffic monitoring, real-time fleet management, and vehicle to everything (V2X) communication. Transportation powered by digital connectivity enhances the safety and efficiency of autonomous vehicles [43]. **Local businesses** and **large enterprises** can establish an online presence with reliable Internet connectivity, allowing them to compete globally [29, 44]. Major companies can also tap into local markets, which would be challenging without Internet access.

1.3 Wireless Broadband to Bridge the Digital Divide

Broadband technologies have transformed global connectivity through mobile or wireless communication technologies. Since 1980, the telecommunications sector has been transforming to enhance several KPIs, such as maximizing data rates (uplink and downlink), coverage, mobility, energy efficiency, spectrum efficiency, and number of users and devices; reducing latency; supporting numerous applications, open-source software, and vendor-independent hardware; and limiting network congestion.

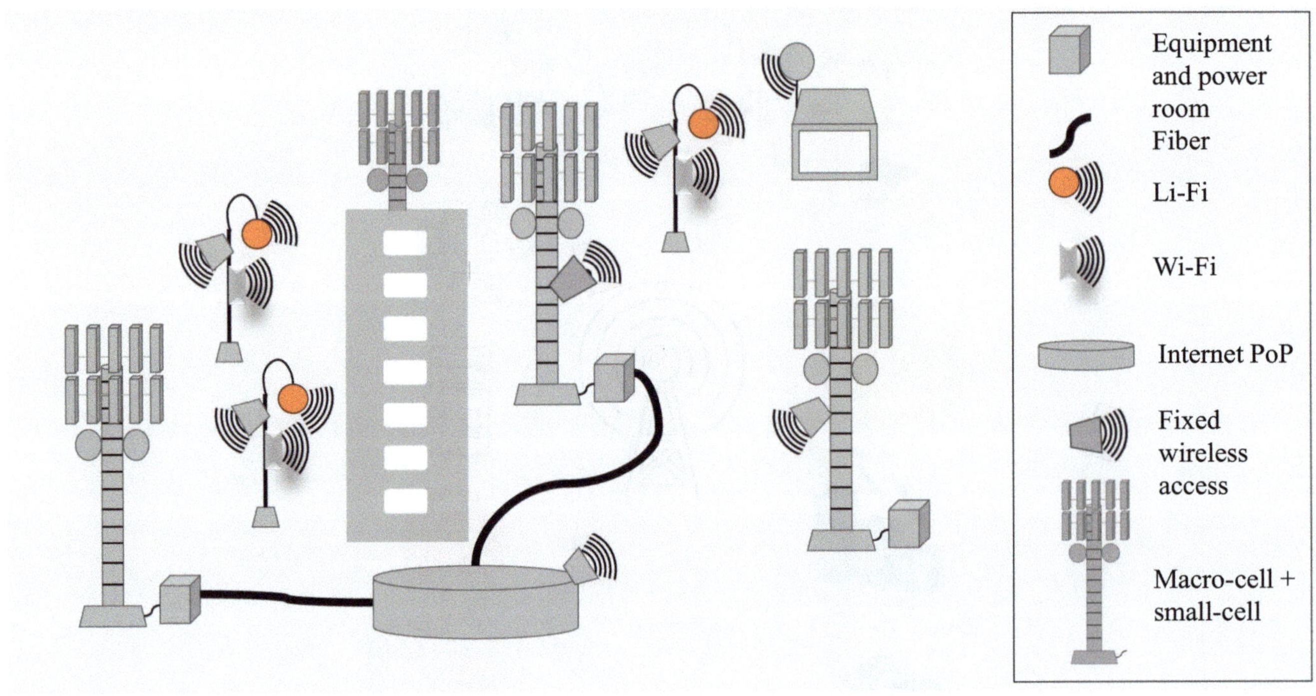

Fig. 1.8 Terrestrial broadband technologies

1.3.1 Terrestrial Broadband Technologies

Terrestrial broadband technologies provide faster data rates and offer shorter latency and greater reliability, at lower costs than NTN alternatives. The architecture of terrestrial wireless broadband access is shown in Fig. 1.8. Terrestrial networks include both wired and wireless communication systems.

The wired communication networks encompass a range of backhaul technologies such as fiber, copper cables, subsea cables, Ethernet, distributed antenna systems (DAS), wired telephone networks, wired television networks, wired relay networks, and coaxial cables. Wired communication networks are prized for their reliability, stability, consistent QoS, and resilience toward weather conditions, making them ideal as the backbone for the telecommunication industry. However, it is not feasible to deploy them everywhere, mainly because of high installation costs, complex cabling requirements, and ongoing cable maintenance challenges. Hence, wireless broadband networks have been gaining popularity in providing broadband services to end users at an affordable price.

Wireless broadband communication networks complement the wired network, usually form a part of the last- or middle-mile networks, and include technologies such as cellular communication systems (2nd generation (2G), 3rd generation (3G), 4G, 5G, and 6G), Wi-Fi, Li-Fi, worldwide interoperability for microwave access (WiMAX), fixed wireless access (FWA), wireless relay networks, Bluetooth, Global Positioning System (GPS), long range wide area network (LoRaWAN), Zonal Intercommunication Global-Standard (Zigbee), microwave, and IoT. The coverage area and QoS for any technology depends on several factors, including frequency of operation, power levels, interference tolerance, signal reuse framework, spectrum available, telecommunication regulations, clutter, and path of propagation. Each successive generation of wireless communication networks offers a significantly improved performance in terms of higher data rates, enhanced energy efficiency, reduced latency, and faster deployment times. Additionally, newer technologies focus on minimizing deployment costs while incorporating sophisticated computational capabilities to optimize overall network functionality and reliability.

Wireless networks also face several challenges, including data security, energy efficiency, higher packet loss compared to wired networks, signal propagation challenges, signal sensitivity to weather conditions, power consumption for wide-scale network deployment, interference with other signals operating in the same or neighboring frequency bands, spectrum allocation, cell location optimization, and resource redundancy. To address these limitations, wireless networks must incorporate advanced encoding, modulation, and security features. Despite certain trade-offs, wireless technologies are widely adopted,

Table 1.3 Evolution of cellular communication technologies

Feature	1G	2G	3G	4G	5G	6G
Timeframe	1980s	1990s	2000s	2010s	2020s	2030 (expected)
Technology	Analog communication	Digital communication	Mobile broadband	IP-based networking	Multi-band (low, mid, and high)	AI-powered, THz spectrum
Data Speed	–	64 Kbps	0.2–2 Mbps	0.1–1 Gbps	1–10 Gbps	<1 Tbps
Features	Basic voice calls with limited call quality	Better voice quality, enhanced security using encryption	Enhanced data transfer rates, multimedia capabilities	Lower latency, improved network stability, higher bandwidth	More reliable connections, energy efficiency, enhanced capacity	Near-zero latency, massive connectivity for advanced IoT, sustainability-focused designs
Limitations	No data services, prone to interference, scalability challenges, and low security	Very slow data rates for Internet	Higher latency compared to later generations	Needs significant infrastructure investment	Limited coverage, high implementation costs, challenges with mmWave signal penetration	Still in research and development phases

mainly because of ease of network deployment, minimal cabling requirements, support for user mobility, remote diagnostics capabilities, customized coverage, and compatibility with diverse end user devices and versatile equipment.

1.3.1.1 Cellular Communication Systems

A cellular communication system is a technology that allows mobile devices to communicate with each other using high frequency (HF) electromagnetic (EM) waves [45]. The EM waves transmitted from a mobile device are transmitted to the nearest BS, from where the signal is routed to the destination device using positioning technologies such as enhanced cell ID, GPS, IP routing, observed time difference of arrival, assisted global positioning system (A-GPS), and new radio (NR) positioning [46, 47]. Table 1.3 shows the different cellular communication systems [6, 48]. Among these, the promising technologies to bridge the digital divide are 4G, 5G, and, in the future, 6G, as described briefly:

- **4G** cellular communication systems were designed to overcome the 3G issues using technologies such as long term evolution (LTE) and WiMAX. 4G supports 0.1 - 1 Gbps, digital communication, voice calls, text, multimedia, data, streaming video, voice over LTE (VoLTE) calls, voice over internet protocol (VoIP), IP services, faster broadband services, low latency, interoperability, digital broadcasting, and trustworthy transmission, and robust data encryption [6]. Additionally, while 4G offers significantly improved data rates compared to 3G, the rapid growth in data usage and the proliferation of connected devices have pushed the limits of 4G networks, leading to the development of the next generation of mobile communication systems.
- **5G** cellular communication systems use orthogonal frequency-division multiplexing (OFDM), a modulation technique for digital signals across multiple channels to reduce interference and support both narrowband and wideband applications [49]. The network has peak data rates of 10 Gbps per user, latency of less than 1 ms, faster download and upload speeds, supports over a million IoT devices simultaneously, allows multiple devices on the same network, provides high-quality Internet connectivity, and enables device-to-device (D2D) communication that supports machine to machine (M2M) communication and IoT applications. 5G supports an array of technologies, which would be discussed in Chap. 3. The key features of 5G are very high speed, versatile applications, remote maintenance, energy efficiency, mobility up to 500 kmph, easy scalability, support for neutral hosts and spectrum sharing, resource allocation as per demand, isolation of slices, 5G features, easy setup, remote monitoring and maintenance of the network, and cloud computing [6, 50]. Despite its advanced features, 5G faces challenges, including the need for significant infrastructure investment, managing interference in the higher frequency bands, and ensuring security and privacy in a highly connected environment. Additionally, widespread deployment of 5G is still ongoing, with many regions, particularly those with a digital divide, yet to fully benefit from its capabilities.
- The development of **6G** cellular communication system and its enabling technologies are anticipated to address the limitations and expand the capabilities of 5G [51, 52]. 6G aims to deliver unprecedented data rates, projected to reach up to 1 Tbps, and offer ultra-low latency as low as 0.1 ms, pushing the boundaries of digital communication even further. 6G will

introduce novel features such as ubiquitous connectivity, enabling high-speed, high-capacity communication; thus, playing a crucial role in narrowing the digital divide in underserved regions. 6G is expected to bridge the digital divide by providing seamless connectivity as well as access to essential services in remote and underserved areas, including education, healthcare, and public safety. Despite its ambitious scope, the challenges for 6G include the need for new spectrum allocation, particularly in the sub-Terahertz (THz) and THz bands; the development of energy-efficient technologies to manage the increased power consumption; and the creation of robust 6G infrastructure. Nevertheless, the potential of 6G to enable a hyper-connected, intelligent society represents a significant leap forward in the evolution of global communication systems.

1.3.1.2 Alternative Wireless Broadband Solutions

Non-cellular terrestrial broadband technologies offer unique advantages in delivering reliable and high-speed communication for applications ranging from wireless backhaul for remote and rural areas to last-mile connectivity in dense urban areas. These technologies complement the traditional wired backhaul broadband networks by mitigating key challenges related to deployment costs and scalability.

- **Wi-Fi** is a family of IEEE 802.11 that is used for wireless local area networks (LAN) access for connecting devices to Internet. Wi-Fi operates on different frequency bands. They are used most for last-mile connectivity and occasionally as long-distance backhaul for terrestrial networks. The speeds supported by the Wi-Fi vary with the generation of Wi-Fi technology. For example, IEEE 802.11b supports up to 11 kbps, while IEEE 802.11ax supports 10 Gbps. The standards have been continuously evolving to support higher spectral efficiency and bandwidth, increased data rates, a higher number of connected devices, lower power consumption, and multiple-input multiple-output (MIMO). Wi-Fi competes with cellular communication networks by offering reliable indoor connectivity and public hotspots in remote, rural, and urban regions [53]. Several studies have shown that around 80% of data traffic is generated indoors, which shows that it has the potential to bridge the digital divide [54].
- **WiMAX** is a wireless metropolitan area network that operates using IEEE 802.16 standards to provide wide area coverage and high-speed data rates [55]. It can support up to 75 Mbps per user data rates and easily scale to accommodate a high number of users. It is a cost-effective solution when it is difficult to deploy wired networks. However, the key challenges with WiMAX are interference, handover challenges, limited device adaptability and market penetration, and security challenges. This technology is generally used as a middle-mile wireless backhaul technology and supports applications such as telephone, VoIP, smart grids and metering, last-mile broadband access for rural connectivity, and broadcasting services [56].
- **FWA** is a technology that operates using licensed bands of 4G and 5G and unlicensed bands, to provide wireless connectivity between two points [57]. FWA links can be either a point-to-point (PTP) or point-to-multipoint (PTMP), to provide higher coverage and ultra-high speed data rates at a lower cost than fiber backhaul connectivity. It is useful in areas with no or poor coverage, helps in reducing the digital divide, and provides a reliable service. When FWA uses licensed bands, spectrum planning is necessary to avoid interference on the incumbent operator's network and maintain QoS for end users. FWA is a key competitor for fiber, cable digital subscriber line (DSL), and copper lines in suburban and rural areas, as well as in emerging markets where there are no legacy backhaul technologies [58]. The main challenges with FWA include the sensitivity of the signal to weather, prohibitively high initial investment, interference from other users, security issues, line of sight (LOS) to cell towers, limited radio frequency bands, and increased latency compared to fiber backhaul technologies.
- **Free Space Optical (FSO)** communication is a wireless technology that uses light to transmit data over free space [59]. The studies have highlighted the potential of using light in FSO for backhaul applications in metropolitan area network (MAN), remote satellite signaling backhaul, disaster management, short-time high broadcast events, temporary high-demand broadcasting events such as Internet-based Internet protocol television (IPTV), VoIP, live broadcasting, and IoT [60]. The setup of free space optical (FSO) is easier than the guided medium optical communication via fiber, making it a popular choice for backhaul for rural areas [61]; however, FSO is easily impacted by LOS visibility and weather conditions such as temperature, atmospheric water content, and rain. The advantages of FSO are cost efficiency, ease of setup and maintenance, and integration with existing technologies [62].
- **Microwave** frequency bands ranging between 1 to 100 GHz within the radio frequency communication bands are used for carrying out LOS PTP communications, and they are usually used as a backhaul link [63, 64]. Microwave frequencies allow high-capacity wireless data to be transmitted between two towers. Microwave backhaul communication allows HF reuse in the networks and is popular in applications including satellite communication, radar systems, PTP links, and radio

astronomy, backhaul links in remote and rural areas [65]. Microwave backhaul operates in the frequency range of 6-42 GHz, offering data rates up to several Gbps [66]. PTP microwave backhaul offers long-distance transmission of data and is cost-effective in areas where laying cables is difficult [67, 68]. Microwave backhaul faces issues such as lack of LOS, sensitivity to climatic conditions, interference from other technologies operating on the same frequency bands, data security, and complex network maintenance requirements [69, 70].

1.3.2 Non-terrestrial Broadband Technologies

Figure 1.9 shows the non-terrestrial broadband technologies such as LEO, MEO, and GEO satellites; HIBS, HAPS, UAVs, and NTFP. It provides information regarding the distance from the Earth's surface for different NTN technologies. In the olden days, there was a need for separate devices that could transmit and receive from the satellite; however, with the advancement of NTN as part of 3GPP standards, today, there are commercial end user devices that support NTN and terrestrial communication using a single device. Affordable satellite-based portable device standards would potentially assist in bridging the global digital divide and support various IoT and narrowband IoT (NB-IoT) applications. In the future, non-terrestrial broadband technologies are expected to support higher-speed data rates and larger bandwidth capacity than a terrestrial network.

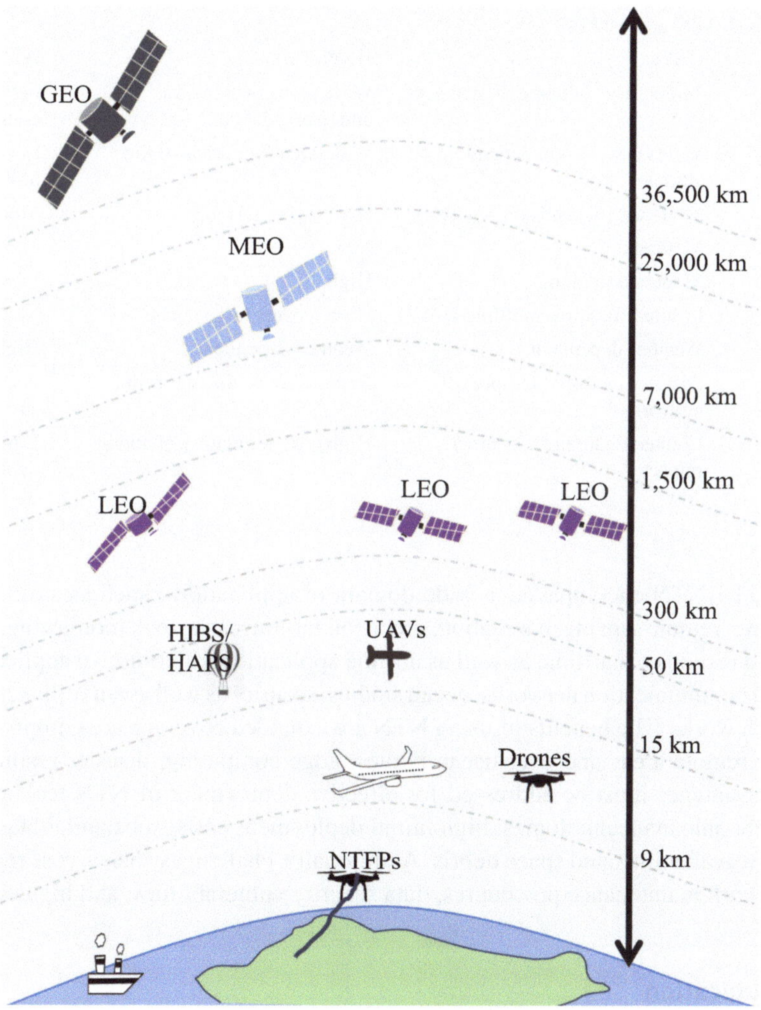

Fig. 1.9 Non-terrestrial broadband technologies

Table 1.4 Comparison of LEO, MEO, and GEO Satellites

Attribute	LEO	MEO	GEO
Altitude	200–2,000 km above Earth	2,000–35,786 km above Earth	35,786 km above Earth
Orbit Period	90–120 min (fast orbit)	2–12 h	24 h (matches Earth's rotation)
Coverage Area	Small; requires many satellites for global coverage	Moderate; fewer satellites needed than LEO	Large; a single satellite covers 1/3 of Earth
Latency	Very low (20–40 ms)	Moderate (50–150 ms)	High (600 ms)
Applications	Real-time applications: IoT, online gaming, video conferencing	Navigation, broadband, regional coverage	Broadcasting (TV, radio), weather monitoring, fixed broadband
Cost	Lower launch cost; frequent replacements due to shorter lifespan	Moderate cost; satellites last longer	High initial cost; long lifespan (15 years)
Advantages	Low latency for real-time use cases; better signal quality	Fewer satellites for coverage; supports regional and global networks	Continuous coverage of fixed areas; reliable for broadcasting
Challenges	High number of satellites required; frequent deorbiting and replacements	Trade-off between cost and coverage; slightly higher latency than LEO	High latency; expensive to launch and maintain

Table 1.5 Comparison of UAVs, HAPS, and NTFPs

Attribute	UAVs	HAPS	NTFPs
Purpose	Temporary cellular connectivity	Wide-area coverage for underserved areas	Persistent tethered communication platforms
Range	Varies (several km depending on altitude)	Wide-area (hundreds of km)	Localized but tethered to ground units
Bandwidth	Moderate (depends on payload)	High (up to 10 Gbps)	Moderate to high
Latency	Moderate	Low	Low
Deployment Cost	Moderate to high	High	Low
Scalability	Limited (requires multiple UAVs)	High (covers large areas)	Moderate (localized coverage)
Reliability	Weather-dependent	Weather-dependent	High (constant power supply)
Use Cases	Disaster recovery, temporary networks	Remote areas, disaster relief	Persistent communication, surveillance
Challenges	Limited endurance, weather impact	High cost, regulatory challenges	Limited mobility, tether dependency

The use cases supported by NTN encompasses a wide domain of applications, such as save our souls (SOS), two-way communication, broadcasting, remote sensing, navigation, precision farming, livestock monitoring, weather monitoring, asset tracking, disaster survey and response, maritime as well as airtime applications, automotive applications, backhaul networks, fleet management, universal communication networks, ocean studies, security as well as surveillance, global roaming, versatile networks, and other research works. The benefits of using NTN are extended coverage area, improved network performance, cost-effective solutions for remote areas and over-ocean bodies, edge computing, disaster resilience, and future-proofing. However, several key disadvantages must be addressed for effective deployment of NTN technologies. These include the development of cost-efficient antenna technologies, high initial deployment costs, substantial latency, sensitivity to weather conditions, limited spectrum availability, and space debris. Additionally, challenges span across regulatory and technological challenges, demanding network maintenance procedures, data security vulnerabilities, and high energy consumption.

1.3.2.1 Satellite Communication

A communications satellite is an artificial satellite that creates a communication channel from a source transmitted to relays and amplifies radio telecommunication signals via a transponder to a receiver at different locations on Earth [71]. Satellite communication is one of the key enablers for regions with poor Internet connectivity or areas where deploying terrestrial networks is not possible [72], and is discussed in detail in Chap. 4. The coverage regions of the satellite can be electronically steered to accommodate any changes in demand. There are several types of satellites in orbit serving diverse functions. The three main kinds of satellites are LEO, MEO, and GEO, of which 8,377 active satellites are orbiting the Earth, constituting 84% LEOs, 3% MEO, and 12% GEO [73]. Table 1.4 shows the comparison of LEO, MEO, and GEO satellites.

1.3.2.2 Alternative NTN Technologies

HAPS, UAV, and NTFP technologies are becoming popular worldwide to address the digital divide challenge in regions where it is hard to deploy terrestrial networks. These technologies are deployed at the stratosphere layer of the Earth's atmosphere to provide high data rates with much lower latency compared to LEO satellites as well. These three technologies operate at different heights and are explored in detail in Chap. 5. The ongoing research in these technologies explores the design where they would support IMT band BS to provide cellular connectivity for a large area, but not as large as a satellite's coverage area. The key applications that these technologies support by providing services include telecommunication, earth observation, military surveillance, precision agriculture, delivery services, disaster relief, emergency telecommunication networks, and environmental monitoring. The advantages of using the technologies include ease of deployment, lower investment cost, rapid deployment, integration with terrestrial networks, high maneuverability, and efficiency. Several challenges need to be addressed before the large-scale commercialization of these technologies is feasible. For example, power challenges, the high cost of maintaining the unit up in the air, cost-effectiveness, interference from terrestrial networks, and frequent take-off and landing. Table 1.5 shows the comparison of these air layer technologies.

1.4 Summary

In the modern world, broadband access is essential, but a substantial proportion of the population remains unconnected because of impediments to digital connectivity. The socioeconomic impacts of the digital divide and opportunities ahead highlight the need to address this issue. There are funds that "intervene" to address telecommunication market failures and aim to bridge the digital divide. Due to the high deployment costs of wired broadband solutions, this book focuses on wireless broadband access technologies as a means of bridging the worldwide digital divide, focusing on affordability and accessibility. This book offers guidelines for developing solutions that leverage both technological and non-technological factors to bridge the digital divide–accessibility, availability, and affordability.

Exercises

1.1 What is broadband access, and why is it considered essential for bridging the digital divide?

1.2 What do worldwide data and coverage trends reveal about the gender divide? Explore why the gender divide exists in digital connectivity.

1.3 Which regions face the most significant challenges in achieving broadband connectivity, and why?

1.4 Why is it important to analyze both terrestrial and non-terrestrial technologies when addressing the digital divide?

1.5 How does broadband access influence the broader goal of bridging the global digital divide?

1.6 The government of a developing country plans to expand broadband access to underserved rural areas. Currently, only 25% of the rural population, totaling 12 million people, have access to broadband Internet. The government aims to increase this coverage to 80% within the next five years.

(a) Calculate how many more people need to gain broadband access to achieve the government's goal.
(b) If the cost of providing broadband access is $30 per person, estimate the total funding required to achieve the target.

1.7 A global organization aims to achieve broadband coverage for 90% of the world's population within the next decade. Currently:

- The global population is estimated to be 8 billion people.
- 65% of the global population currently has broadband access.
- The cost of expanding broadband infrastructure is estimated at $200 per person for urban areas and $500 per person for rural areas.
- 30% of the global population resides in rural areas.

(a) Calculate the total number of people who currently do not have broadband access.
(b) Determine how many more people need to gain access to achieve the 90% coverage goal.
(c) Estimate the total cost of achieving 90% coverage if 50% of those who need access live in urban areas, and the other 50% live in rural areas.

1.8 For problem 1.7, if the cost of providing broadband access in rural areas increases by 20%, how would this affect overall project costs for global coverage?

1.9 Develop a case study highlighting the key challenges contributing to the digital divide in terms of access to essential resources such as electricity, infrastructure development, and security issues like theft and tampering.

1.10 How do ground, air, and space layer broadband technologies compare in terms of coverage, cost, and scalability in underserved regions?

References

1. A. Ramade, A. Serrano-Santoyo, M. Cabrera-Flores, R. Conte, The new digital divide: The confluence of broadband penetration, sustainable development, technology adoption and community participation. Inf. Technol. Dev. **18**, 345–353 (2012)
2. S. Amiri, B. Reif, Internet penetration and its correlation to gross domestic product: an analysis of the Nordic countries. Int. J. Bus. Humanit. Technol. **3**, 50–60 (2013)
3. A.G. Wilhelm,*Digital Nation: Toward an Inclusive Information Society* (MIT Press, 2006)
4. ITU. The affordability of ICT services 2022. Cited in Apr 2023
5. M. El-Sayed, J. Jaffe, A view of telecommunications network evolution. IEEE Commun. Mag. **40**(12), 74–81 (2002)
6. G. O'Regan, A short history of telecommunications, in *A Brief History of Computing* (Springer, 2021), pp. 227–236
7. A.A. El-Saleh, M.A. Al Jahdhami, A. Alhammadi, Z.A. Shamsan, I. Shayea, W.H. Hassan, Measurements and analyses of 4G/5G mobile broadband networks: an overview and a case study. Wirel. Commun. Mobile Comput. **2023**(1), 6205689 (2023)
8. International Telecommunication Union (ITU). Global connectivity report 2022, (2022). Accessed 25 Mar 2025
9. International Telecommunication Union (ITU). ITU data hub (2025). Accessed 25 Mar 2025
10. ITU. Global connectivity report 2022. Cited in 15 Jan 2022
11. International Telecommunication Union (ITU). Universal and meaningful connectivity (umc). Accessed 17 Feb 2025
12. DataReportal. Digital 2024 global overview report (2024). Accessed 25 Mar 2025
13. World Economic Forum. 2.5 billion people lack Internet access: How connectivity can unlock their potential (2024). Accessed 25 Mar 2025
14. World Bank. Digital development (2025). Accessed 25 Mar 2025
15. Organisation for Economic Co-operation and Development (OECD). Closing broadband connectivity divides for all (2025). Accessed 11 July 2025
16. Cenerva. Universal service-dr. charles jenne. Accessed 13 Jun 2025
17. D.C. Gómez. The three levels of the digital divide: barriers in access, use and utility of internet among young people in Spain. Interações: Sociedade e as novas modernidades **34**, 64–91 (2018)
18. K.N. Kuteesa, C.U. Akpuokwe, C.A. Udeh, Theoretical perspectives on digital divide and ict access: comparative study of rural communities in Africa and the United States. Comput. Sci. IT Res. J. **5**(4), 839–849 (2024)
19. V. Korovkin, A. Park, E. Kaganer, Towards conceptualization and quantification of the digital divide. Inf. Commun. Soc. **26**(11), 2268–2303 (2023)
20. J. Lähteenmäki, The evolution paths of neutral host businesses: antecedents, strategies, and business models. Telecommun. Policy **45**(10), 102201 (2021)
21. M. Harvey, D.P. Hastings, G. Chowdhury, Understanding the costs and challenges of the digital divide through uk council services. J. Inf. Sci. **49**(5), 1153–1167 (2023)
22. J.A. Hoyos Muñoz, D.C. Valencia, Trends and challenges of digital divide and digital inclusion: a bibliometric analysis. J. Inf. Sci. 01655515221148366 (2023)
23. A. Afzal, S. Khan, S. Daud, Z. Ahmad, A. Butt, Addressing the digital divide: access and use of technology in education. J. Soc. Sci. Rev. **3**(2), 883–895 (2023)

24. A. Acilar, Ø. Sæbø, Towards understanding the gender digital divide: a systematic literature review. Global knowl. Mem. Commun. **72**(3), 233–249 (2023)
25. World Bank. Progress on basic energy access reverses for first time in a decade. Accessed 26 Mar 2025
26. A. Caragliu, C.F. Del Bo, Smart cities and the urban digital divide. npj Urban Sustain. **3**(1), 43 (2023)
27. M. Hasan Laskar, Examining the emergence of digital society and the digital divide in India: a comparative evaluation between urban and rural areas. Front. Sociol. **8**, 1145221 (2023)
28. H. Gruber, J. Hätönen, P. Koutroumpis, Broadband access in the EU: an assessment of future economic benefits. Telecommun. Policy **38**, 1046–1058 (2014)
29. S.K.A. Kumar, G.V. Ihita, S. Chaudhari, P. Arumugam, A survey on rural internet connectivity in India, in *2022 14th International Conference on COMmunication Systems & NETworkS (COMSNETS)* (IEEE, 2022), pp. 911–916
30. International Telecommunication Union and UNICEF. How many children and young people have internet access at home? Accessed 13 Jun 2025
31. United Nations Sustainable Development. Take action for the sustainable development goals. Cited in 23 Aug 2022
32. European Commission. Connecting Europe Facility-European Commission
33. BharatNet. Bharatnet project
34. BroadbandUSA. Broadband equity access and deployment program
35. China Ministry of Industry and Information Technology. National broadband network plan: Enhancing connectivity across China, 2024. The specific URL might have changed. For the latest information, please refer to the official Ministry website or contact the relevant government office
36. Freedom House. Brazil: Freedom on the net 2022 country report
37. ITU. Brazil's new strategy aims for internet in all schools
38. Broadband Commission. Working group on broadband for all: A digital moonshot for Africa
39. K. Li, P. Xirui, J. Song, B. Hong, J. Wang, The application of augmented reality (AR) in remote work and education (2024), arXiv:2404.10579
40. K. Bokun, J. Nazarko, Smart villages concept–a bibliometric analysis and state-of-the-art literature review. Prog. Plan. **175**, 100765 (2023)
41. J. Zhao, D. Liu, R. Huang, A review of climate-smart agriculture: recent advancements, challenges, and future directions. Sustainability **15**(4), 3404 (2023)
42. S. Subha, T.M. Inbamalar, C.R. Komala, L.R. Suresh, S. Boopathi, K. Alaskar, A remote health care monitoring system using internet of medical things (iomt), in *2023 3rd International Conference on Innovative Practices in Technology and Management (ICIPTM)* (IEEE, 2023), pp. 1–6
43. C. Liu, L. Ke, Cloud assisted internet of things intelligent transportation system and the traffic control system in the smart city. J. Control Decis. **10**(2), 174–187 (2023)
44. P. Vassilakopoulou, E. Hustad, Bridging digital divides: a literature review and research agenda for information systems research. Inf. Syst. Front. **25**(3), 955–969 (2023)
45. Y. Akaiwa, *Introduction to Digital Mobile Communication* (John Wiley & Sons, 2015)
46. P.K. Enge, The global positioning system: Signals, measurements, and performance. Int. J. Wirel. Inf. Netw. **1**, 83–105 (1994)
47. G. Yanying, A. Lo, I. Niemegeers, A survey of indoor positioning systems for wireless personal networks. IEEE Commun. Surv. Tutor. **11**(1), 13–32 (2009)
48. W. Jiang, B. Han, *Cellular Communication Networks and Standards: The Evolution from 1G to 6G* (Springer Nature, 2024)
49. A. Kumar, M. Gupta, A review on activities of fifth generation mobile communication system. Alex. Eng. J. **57**(2), 1125–1135 (2018)
50. 5G-PPP. 5G-PPP
51. N. Chukhno, O. Chukhno, S. Pizzi, A. Molinaro, A. Iera, G. Araniti, Approaching 6g use case requirements with multicasting. IEEE Commun. Mag. **61**(5), 144–150 (2023)
52. G. Xylouris, N. Nomikos, A. Kalafatelis, A. Giannopoulos, S. Spantideas, P. Trakadas, Sailing into the future: technologies, challenges, and opportunities for maritime communication networks in the 6g era. Front. Commun. Netw. **5**, 1439529 (2024)
53. Q. Chen, G. Yu, H. Shan, A. Maaref, G.Y. Li, A. Huang, Cellular meets WiFi: traffic offloading or resource sharing? IEEE Trans. Wirel. Commun. **15**(5), 3354–3367 (2016)
54. B. Feng, C. Zhang, J. Liu, Y. Fang, D2D communications-assisted traffic offloading in integrated cellular-WiFi networks. IEEE Internet Things J. **6**(5), 8670–8680 (2019)
55. S.A. Ahson, M. Ilyas, *WiMAX: Standards and Security* (CRC Press, 2018)
56. S. Ahmadi, An overview of next-generation mobile WiMAX technology. IEEE Commun. Mag. **47**(6), 84–98 (2009)
57. M. Hashemi, M. Coldrey, M. Johansson, S. Petersson, Integrated access and backhaul in fixed wireless access systems, in *86th Vehicular Technology Conference (VTC-Fall)* (IEEE, 2017), pp. 1–5
58. E.-M. Oproiu, I. Gimiga, I. Marghescu, 5G fixed wireless access-mobile operator perspective, in *International Conference on Communications (COMM)* (IEEE, 2018), pp. 357–360
59. A.K. Majumdar, Fundamentals of free-space optical communications systems, optical channels, characterization, and network/access technology. Opt. Wirel. Commun. Broadband Glob. Internet Connect **186**, 55–116 (2019)
60. H. Kaushal, V.K. Jain, S. Kar, *Free Space Optical Communication*, vol. 60. (Springer, 2017)
61. A.K. Majumdar. Free space optical communications to connect the unconnected globally in remote places. Technology issues relevant to implementing systems, in *Trends in Wireless Communication and Information Security: Proceedings of EWCIS*, vol. 2020 (2021), pp. 3–7
62. V.W.S. Chan, Free-space optical communications. J. Lightwave Technol. **24**(12), 4750–4762 (2006)
63. V. Camarchia, M. Pirola, R. Quaglia, *Electronics for Microwave Backhaul* (Artech House, 2016)
64. X. Huang, Y.J. Guo, A. Zhang, V. Dyadyuk, A multi-gigabit microwave backhaul. IEEE Commun. Mag. **50**(3), 122–129 (2012)
65. Y. Li, A. Cai, G. Qiao, L. Shi, S.K. Bose, G. Shen, Multi-objective topology planning for microwave-based wireless backhaul networks. IEEE Access **4**, 5742–5754 (2016)
66. F. Yaghoubi, J. Chen, A. Rostami, L. Wosinska, Mitigation of rain impact on microwave backhaul networks, in *2016 IEEE International Conference on Communications Workshops (ICC)* (IEEE, 2016), pp. 134–139
67. T. Naveh, Mobile backhaul: fiber vs. microwave. Ceragon White Paper **1**, 1–11 (2009)

68. N. Nepomuceno, Network optimization for wireless microwave backhaul. Ph.D thesis, Université Nice Sophia Antipolis, 2010
69. S. Little, Is microwave backhaul up to the 4g task? IEEE Microw. Mag. **10**(5), 67–74 (2009)
70. M.M. Ahamed, S. Faruque, 5G backhaul: requirements, challenges, and emerging technologies, in *Broadband Communications Networks: Recent Advances and Lessons from Practice*, vol. 43 (2018)
71. F. Vatalaro, G.E. Corazza, C. Caini, C. Ferrarelli, Analysis of LEO, MEO, and GEO global mobile satellite systems in the presence of interference and fading. IEEE J. Sel. Areas Commun. **13**(2), 291–300 (1995)
72. H. Al-Hraishawi, H. Chougrani, S. Kisseleff, E. Lagunas, S. Chatzinotas, A survey on nongeostationary satellite systems: the communication perspective. IEEE Commun. Surv. Tutor. **25**(1), 101–132 (2022)
73. Nanoavionics. How many satellites are in space? (2023)

Strategies to Address the Digital Divide 2

Globally, several strategies are being tested and trialed to improve the affordability, accessibility, and availability in the telecommunication sector. Many initiatives and subsidies have been launched to assist stakeholders in bridging the digital divide. These efforts aim to mitigate the challenges posed by the telecommunication industry, which requires significant upfront financial resources and generates earnings subsequently [1]. The strategies need to address both the access gap, such as integration of technologies, spectrum sharing, infrastructure sharing, technology integration, and futuristic network design [2–4]; and the usage gap, such as digital literacy, affordability, and priorities. It is anticipated that these strategies will reduce the total cost of the network and make it a viable business model in underserved regions [5].

2.1 SAGIN

The advancement in the telecommunication sector has led to the development of an innovative vertical homogeneous network architecture known as space-air-ground integrated network (SAGIN), designed to integrate communication systems across different altitude layers seamlessly [2, 6]. This concept is developed to integrate space, aerial, and ground layers to provide ubiquitous and seamless broadband connectivity across the globe. The key features offered by this network would be high-speed Internet and seamless connectivity, specifically targeting underserved communities across both urban and rural geographies. SAGIN aims to leverage the strengths of different technologies to offer a unified, hybrid communication network with high reliability and backup even if any one layer fails; enhance flexibility and scalability to meet the growing demand; improve user mobility experience; and support emerging technologies for remote and rural areas and build an inclusive community.

What is SAGIN? Fig. 2.1 illustrates the architecture for SAGIN, where the three network segments: the space, air, and ground layer, can either operate independently or cooperatively based on the use case [2, 7, 8]. This multi-dimensional architecture is expected to provide ubiquitous coverage, seamless connectivity, and support futuristic applications, such as emergency response, intelligent transportation, ecological remote sensing, rural broadband access, and bridge the digital divide. The space segment consists of GEOs, MEOs, and LEOs satellites to provide wide-area coverage and high-speed Internet in regions beyond the reach of terrestrial networks. Next, the air segment consists of aerial platforms that encompass UAVs, HAPS/HIBS, balloons, and drones that operate in the stratosphere to support applications such as relay, disaster recovery, on-demand coverage, and fill coverage gaps. Finally, the ground segment consists of cellular networks, fiber-optic infrastructure, Wi-Fi, Li-Fi, and other terrestrial networks that provide high-speed connectivity to urban and rural areas. Because of its size and monetary cost, ground network manages a significant portion of communication traffic [7].

Why is SAGIN gaining popularity, when terrestrial networks handle most of the traffic? This is mainly because the current terrestrial networks are not capable of providing ubiquitous, seamless, and global connectivity, especially in remote and rural regions, and it is incapable of withstanding natural disasters [9]. While depending only on NTN solutions, such as satellites, is not sufficient, as the latency offered is higher than that of terrestrial networks. Hence, there comes into picture the aerial networks, which offer the benefits of both terrestrial wireless networks and satellites. As a result, there is a need to integrate the different layers of technologies to cater to the multi-faceted requirements for emerging applications.

Then, the next question is how would SAGIN be implemented, and what are the key driving technologies? To implement SAGIN, there is a need to focus on an integrated network architecture that supports interoperability, efficient resource management, edge computing, as well as AI technologies. There is a need to focus on efficient spectrum allocation across the layers, dynamic traffic routing, and AI-driven network optimization to ensure smooth interfacing and integration across layers. This architecture is expected to lower the overall network cost and improve the network flexibility to offer a wide

S. Koratagere Anantha Kumar, *Wireless Broadband Access to Bridge the Digital Divide*,
Textbooks in Telecommunication Engineering, https://doi.org/10.1007/978-3-032-09261-8_2

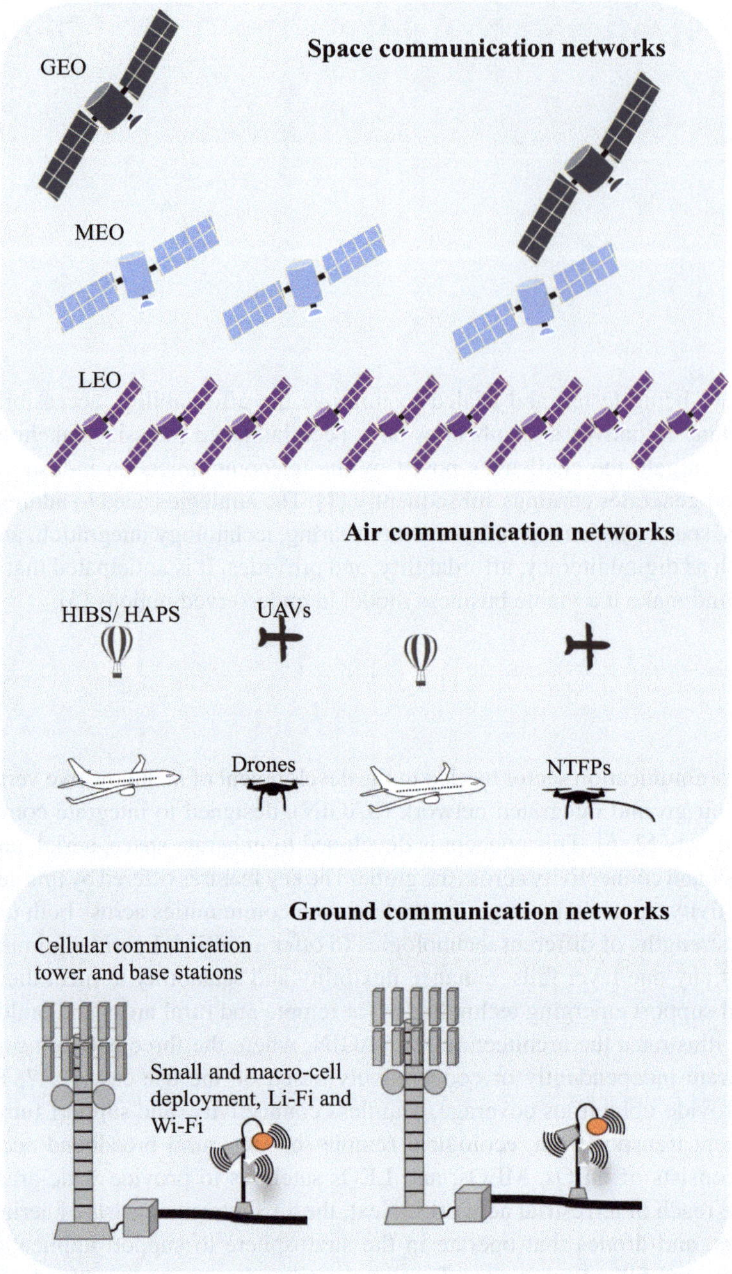

Fig. 2.1 SAGIN architecture

range of services globally. The key technologies that are driving research for SAGIN and related use cases are flexible access and routing methods, efficient resource scheduling, UAV trajectory, and secure communication [9]. Furthermore, there is a potential scope for integration with edge computing, AI, deep space communication, quantum computation, blockchain, and machine learning (ML).

SAGIN is part of future networks and partly in 6G networks [2]. Several critical barriers must be addressed for the implementation and widespread adoption of SAGIN. These include coordinating the integration of space, air, and ground networks across protocols, hardware, and software systems; efficiently allocating resources while mitigating latency constraints across heterogeneous technologies; managing the high deployment costs associated with expanding space and terrestrial networks;

navigating spectrum complexity across multiple layers to minimize inter-layer interference; acquiring and synchronizing user location information across network layers; developing advanced mobility management protocols and algorithms; and designing energy-efficient architectures that span the entire multilayer network [6, 10].

2.2 Infrastructure Sharing

Another popular strategy employed by the operators for addressing the demand is telecommunication infrastructure sharing. Enabling multiple operators to share the physical assets helps in maximizing resource utilization, reducing overall cost toward the networks, and accelerating deployment [4, 11]. Infrastructure sharing focuses on the collaboration of more than one operator in utilizing the same physical resources, sites, or technologies instead of deploying them individually. This approach is governed strictly by the regulators to ensure fair competition and a balanced and inclusive environment, and to prevent monopoly in the market. Several studies have highlighted the benefits of various infrastructure sharing strategies as explained in [12–14]. The four major infrastructure sharing strategies in the telecommunication ecosystem include:

- Long time ago, the ***No Sharing*** strategy was used widely, where each operator builds, runs, and maintains their network. In this method, the operator has complete control over the network's hardware, software, and infrastructure, as shown in Fig. 2.2. The network's operators prefer to supervise the network and install a network as per their demand estimations and licensed spectrum bands [15]. Over time, the operators noticed how costly it was to set up and maintain their networks and remain profitable, which encouraged operators to consider other infrastructure deployment strategies. Infrastructure that employs the *No Sharing* approach is capital intensive, underutilized, consumes a long time at each stage, and impedes technological development with slower rollouts. The capital-intensive nature of this strategy would add financial pressure and can gradually weaken an operator's ability to maintain exclusive control over network assets, prompting a shift toward collaborative models and shared ownership.

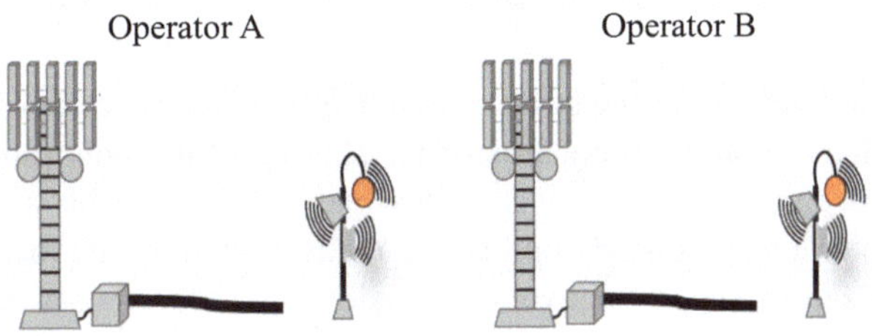

Fig. 2.2 No sharing strategy

- ***Passive Sharing*** is a form of infrastructure sharing strategy that involves sharing non-electronic components of network deployment with other operators. It involves sharing of physical supporting infrastructure such as backhaul, telecommunication sites, ducts, masts, towers, equipment rooms, and related power supplies, air conditioning, and security systems as shown in Fig. 2.3. *Passive Sharing* offers benefits in terms of reduced duplication, ready-to-deploy site location and facilities, ease of network rollouts, increased resource utilization, support network expansion at a reduced cost, and lowered capital expenditure (CAPEX) and operational expenditure (OPEX).

There are two main types of *Passive Sharing*:

- **Backhaul sharing**: This involves sharing only the backhaul transport network from the site where the base stations are deployed.
- **Site sharing**: This involves sharing a physical site, tower, power supply, equipment room, and backhaul at the site location where the base stations are deployed.

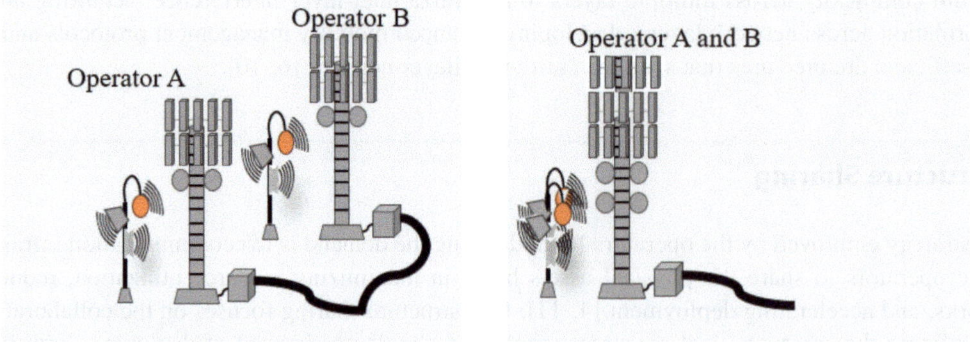

(a) Backhaul sharing (b) Site sharing

Fig. 2.3 Passive sharing strategy: **a** Backhaul sharing, **b** Site sharing

Passive Sharing allows the operators to maintain their competitiveness and still retain control over the network equipment, key performance indicators, and coverage. *Passive Sharing* is widely used around the world by telecommunication operators. As laying fiber is both an expensive and time-consuming procedure, the operators can further save costs by sharing the backhaul [16]. Many studies have shown that *Passive Sharing* reduces CAPEX and OPEX by 16–35% [17]. Each country has laid out policies and regulations to encourage *Passive Sharing* and regulate the competition among the operators. This strategy is widely popular in countries such as the United Kingdom, the United States of America, the European Union, Brazil, India, China, and Canada [4].

The drawbacks of *Passive Sharing* include low decision-making power on the location of the telecommunication site or site degradation parameters, limited potential savings as the network equipment as well as spectrum are not shared, limited benefits if operators lack their sites, a costly negotiation process, and a lengthy negotiation process as the discussion takes place per site.

- ***Active Sharing*** is a form of infrastructure sharing strategy that involves sharing electronic and non-electronic components of network deployment with other operators. It consists of the sharing of not only physical supporting infrastructure but also radio access network (RAN), antennas, BSs, feeder cables, and base transceiver stations, as shown in Fig. 2.4 [18, 19]. *Active Sharing* offers benefits in terms of reduced duplication, ready-to-deploy site location and facilities, ease of network rollouts, increased resource utilization, support network expansion at a reduced cost, and lower CAPEX and OPEX costs compared to the other two sharing strategies. *Active Sharing* is being adapted in different countries and actively explored in developing countries.

  Negotiating for *Active Sharing* takes time, as it involves sharing over a wide range of sites. *Active Sharing* strategy is preceded by a mutual agreement among the service providers or operators (generally, two national or higher level), and these arrangements are governed by the regulatory frameworks and policy guidelines [20]. In such scenarios, generally, the regulator does not permit the mutual sharing of spectrum between them [21]. *Active Sharing* is expected to result in a cost savings of CAPEX by 33–35% and OPEX by 25–33% [21].
  The key benefits offered by *Active Sharing* include a higher reduction in overall costs, optimized resource utilization, enhanced services, accelerated rollouts, scalable infrastructure, reduced service pricing, increased profits to the operators, and higher revenue to the governments. It benefits the industries that rely on telecommunication services for their operations. Generally, regulators place constraints on all operators coming together in *Active Sharing* to maintain market competition and avoid collusion. There is a need for a common consensus between the operators regarding the control over the network resources, network complexity, pricing plans, QoS, and network upgrades [22].
- The final form of the infrastructure sharing strategy is ***neutral host network (NHN)***, which involves end-to-end network sharing that includes spectrum, core networks, RAN, BSs, backhaul, telecommunication sites, gateways, ducts, masts, towers, equipment rooms, power supplies, air conditioning, feeder cable, base transceiver station, and security systems [19]. In this deployment type, multiple operators coexist on a single physical network and are separated via network virtualization. Dedicated virtual resources on the common physical infrastructure would be assigned to all operators, based on their service level agreement (SLA), across the front mile to the final mile of the network. In contrast to the sharing techniques that were

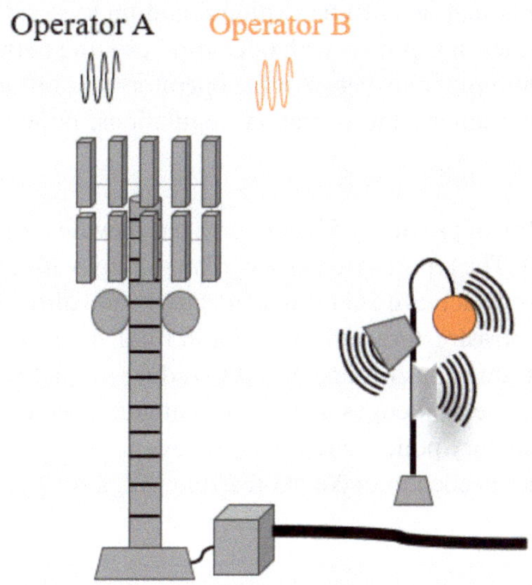

Fig. 2.4 Active sharing strategy

previously discussed, the potential operators would only have a network arrangement with the *NHN* infrastructure operator [21, 23]. A single operator, a group of operators collaborating, or a third-party operator can all deploy the *NHN* and lease the capacity to other potential operators who are interested in providing service in the region of interest, and are known as InP. Each tenant can customize the KPIs, which is required to cater to their end users in terms of SLA with the InP [24]. Figure 2.5 shows *NHN* architecture, which could transform the telecommunication network competition into a service rather than an infrastructure-based one. TSPs can solely focus on providing service, and InPs can focus on building, operating, and maintaining the network. This model would encourage new entrants to enter the capital-intensive market and complement the services provided by the large operators. The key benefits of this strategy are cost efficiency and extreme suitability for remote, rural, and dense urban areas, attracting new entrants, maximizing resource utilization, energy efficiency, spectrum efficiency, and other factors that would encourage wide-scale deployment of the services [25].

This approach is widely explored in 5G, satellite, and advanced Wi-Fi technologies since they require dense deployment, and this may be achieved with greater affordability by employing *NHN*. Given the novelty of this kind of infrastructure

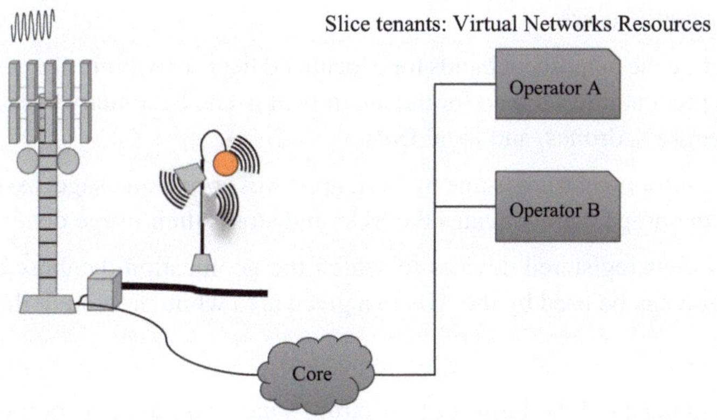

Fig. 2.5 Neutral host networks strategy

sharing approach, further study is needed to comprehend how it will affect the businesses of telecommunications and its dependent sectors. The main concerns that need to be explored and understood with regards to *NHN* are technological complexities, security, data breach issues, integration with operators' existing networks, pricing strategies of each operator, trust among the operators, timeline commitments between the operators and infrastructure provider, spectrum usage, legal issues, dispute management, collusion among the operators, regulations, policy impacts, and differentiation among the operators [4, 25].

Different studies analyze the feasibility of providing 5G networks nationwide. For example, the requirements for deploying a nationwide 5G network in the UK [26]. The report assumes an urban setting, with 2×2 sharing between the operators would be feasible. The policy must be inclusive to achieve 5G rural connectivity and allow 1×4 sharing between operators, i.e., the four national MNOs sharing common infrastructure and spectrum in rural areas. The report also suggested using alternative solutions for backhaul and shared spectrum bands for operations to reduce costs. Similarly, to bring 5G connectivity to India, the white paper by TRAI [27] discusses the challenges in terms of network specifications, architecture, spectrum, backhaul, small cell deployment, investment, regulatory policy, and use cases. As pointed out by the research, newer business models, higher investments, and deployments are needed to make 5G feasible [27, 28].

2.3 Spectrum Sharing

The next strategy is spectrum sharing. It complements the existing long-term spectrum licensing when the required spectrum band for operation is not available [29]. Spectrum sharing is a technology that allows SUs to coexist on the same frequency channels as a licensed or incumbent user to maximize spectral band usage and spectrum efficiency [30]. Spectrum sharing is enabled by programmable components–software-defined radios (SDRs), embedded systems, field programmable gate array (FPGA) computers; and hardware components–system on chip (SoC) [31]. SDR has been studied for some time; however, it is being used quite extensively in the telecommunication industry with the most recent developments in technology.

The spectrum databases shield the incumbents from secondary users' interference and also enable registered secondary users (SU) to access the database to determine what frequencies are accessible for their use in a certain region. The spectrum database also maintains a log of the spectrum usage by the SUs using a spectrum sensing mechanism, which involves utilizing a spectrum analyzer to plot the amplitude of an input signal vs frequency regularly to check if users are present across different bands [32]. By possibly reallocating underutilized bands to secondary users, spectrum sensing aids in maximizing spectral usage. Future decisions on spectrum policy and spectrum access methods may benefit from using real-time spectrum sensing [32, 33]. The development of SDR technology has made this possible for future radio networks. By facilitating dynamic spectrum sharing, the spectrum database optimizes the benefits of spectrum utilization and allows for real-time, efficient spectrum use [31]. Figure 2.6 shows the generic architecture for spectrum sharing technology [34].

The key terminologies are:

- **Incumbent or primary users (PU)**: They often hold long-term spectrum licenses, and their coverage may be state- or nationwide, depending on the nation's spectrum regulations. Incumbents could be operators such as the military, MNOs, ISPs, TV broadcasters, and satellite users.
- **SUs**: These licensed users utilize the incumbent bands for a localized license to minimize costs and are not allowed to operate over a certain power level to prevent interference for the incumbent users. Secondary users could be InPs, micro-operators, private networks, mobile operators, drones, and local ISPs.
- **Geolocation Database**: It monitors spectrum usage by SUs, approves spectrum usage, assigns frequencies to SUs, informs the national regulator about the usage, authenticates the SUs, and stores their usage details.
- **Spectrum manager**: This allows registered devices to search the geolocation database and determine what frequencies and the corresponding duration can be used by the SUs in a given area while protecting the signals of the incumbents from interference.

There is a growing need to understand the impact of spectrum sharing on the telecommunications industry to formulate policies and frameworks for the wide-scale usage of this technology. The spectrum bands that support sharing and technical requirements–such as power levels, coverage regions, transmission levels, guard bands, location, duration of usage, resource allocations, bandwidth, and interference levels–need to be standardized through the implementation of spectrum access system

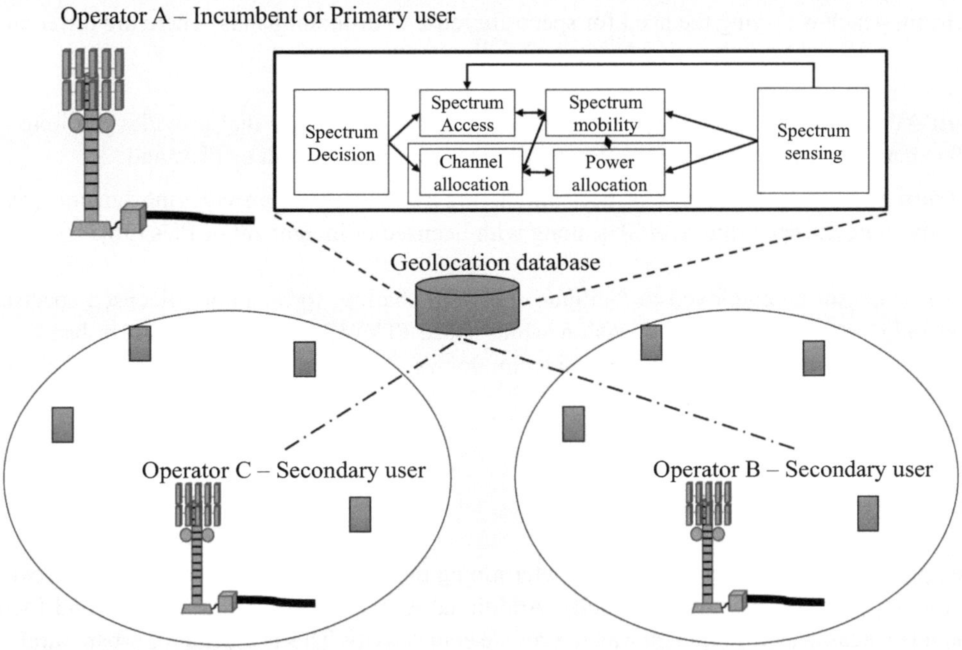

Fig. 2.6 Generic architecture for spectrum sharing technology

Table 2.1 Comparison of shared spectrum access technologies

Feature	LSA	TVWS	CBRS (USA)	Shared access license (UK)
Frequency range	Varies by country; EU: 2.3 GHz	470–790 MHz	3550–3700 MHz	3800–4200 MHz (n77)
License type	Licensed (long-term)	License-exempt/ lightly licensed	Tiered access (PAL, GAA)	Low and medium-power shared licensing
User tiers	Incumbents + Secondary users	Secondary users (underutilizes TV bands)	3 tiers: Incumbents, PAL, GAA	Licensed users with technical limitations
Spectrum coordination	Central database (regulator or license administrator)	Geolocation database	SAS + ESC system	Licensed under Ofcom with location-specific terms
QoS guarantee	Moderate to High (depending on framework)	No QoS guarantee	Moderate (PAL users), Low (GAA)	Depends on license type
Coverage range	Regional or nationwide	Long range (sub-GHz propagation)	Localized to regional (up to 150 MHz bandwidth)	Low-power: 50m radius, Medium-power: rural-wide
Use cases	Industrial verticals, private 5G, rural broadband backhaul	Rural broadband backhaul	Industry 4.0, indoor, seaports, rural broadband backhaul	Local indoor, rural mobile, private wireless
Primary objective	Localized licensing with QoS	Utilize vacant TV bands in rural areas	Efficient use of mid-band spectrum with flexibility	Simplified access to spectrum for private networks

(SAS). The key aspects that a spectrum sharing framework needs to focus on are the duration of spectrum access, the spectrum rights and access limits, access fees, terms, dispute management, security, and incumbent as well as SUs protection [35]. A few issues remain unanswered, including how spectrum sharing influences current business models, how operators' collaboration impacts spectrum misuse and its effects on long-term spectrum demand, and compatibility with existing devices in the market.

Spectrum sharing is becoming necessary in the telecommunications industry since the airwaves are becoming crowded and congested with the growing demand for numerous use cases. The demand for wireless connectivity use cases is fueled by the available spectrum, which is driving the need for spectrum reuse in different bands. There are different methods to share the spectrum:

- **Fixed Spectrum Allocation (FSA)** is a type of spectrum sharing technology that provides the long-term capability to opportunistically share the spectrum with SUs along with licensed or incumbent or PUs, and
- **Dynamic Spectrum Access (DSA)** is a type of spectrum sharing technology that provides the dynamic (real-time) capability to opportunistically share the spectrum with SUs along with licensed or incumbent or PUs [36].

Numerous frameworks can be employed to facilitate spectrum sharing; for example, licensed spectrum access (LSA), citizens' broadcast radio services (CBRS), television white space (TVWS), and shared access bands [37–40]. Table 2.1 shows the overview of some popular spectrum sharing technologies. Spectrum sharing is also being explored as part of 6G communication systems as well [41].

2.4 Network Optimization Strategies

Network resource optimization plays a crucial role in determining the coverage, QoS, and associated cost of a network, and helps utilize the resources to their maximum capacity. Additionally, it also improves scalability and focuses on providing coverage in the intended areas in a cost-efficient way to provide connectivity. This is crucial for urban, rural, and remote region deployments in a resource-constrained environment. For example, urban deployments are resource-constrained in terms of space to lay cables and masts, whereas rural and remote regions have resource constraints in terms of cost and infrastructure that can be deployed.

2.4.1 Technological Integration

The integration of multiple technologies is vital in the telecommunications sector. A multitude of technologies work in tandem to support both end users who are consumers of such technologies and older applications [42]. Because upgrading to the latest technology is typically quite expensive, operators would prefer to maximize the utilization of existing infrastructure as long as it meets the demand in terms of QoS with minor upgrades, and handles the network congestion. Operators would benefit from a well-planned technological integration to better grasp the complex process of optimizing the system architecture such that newer technologies are backward compatible with previous technologies, yet capable of satisfying the demands of current users. The complex optimization process needs to consider several factors such as resource planning, CAPEX, compliance with policies, regulations, and standards, operations and maintenance (O&M) cost, developing support across different domains of telecommunication businesses, impact on current and legacy technologies, affordability issues, and scalability of technology.

The main advantages of well-planned technological integration are to maximize QoS for the users, meet the user demands, provide seamless connectivity, assess the risk associated with the newer technologies, analyze their growth strategy in real-time, provide smoother O&M support, enhance security, minimize the budget for brownfield deployments, and improve user experience. These benefits will help the operators leverage the insights from the technological integration to assess the newer technology rollout timelines, challenges, expected revenues, cost minimization, and network performance efficiency.

As of today, there is a growing need to convert physical resources to virtual/cloud resources to improve their utilization. 5G and multi-access edge computing (MEC) rely on virtualization and the cloud for providing KPIs as per 3GPP standards [43, 44]. The operators around the world are integrating their existing broadband technologies, especially with Wi-Fi, as it supports data offloading and is operating on the unlicensed spectrum band. However, the challenge lies in managing the interference of large-scale Wi-Fi networks, user authentication, handover complexity, mobility management, and billing integration [45, 46].

2.4.2 Network Planning

Network planning involves strategic decision-making on different factors related to the network deployment and operations, such as the location of the telecommunication sites, coverage area, frequency bands of operation, network topology, network architecture, servers required, demand-supply assessment, selection of appropriate software and hardware components, MEC deployment, MIMO technology, beamforming, network capacity, traffic handling algorithms, operations management, and predictive maintenance [47]. Network planning plays a significant role in improving network coverage, QoS, QoE, energy efficiency, cost savings, and other factors that would make a technology affordable, scalable, sustainable, and reliable [48].

The different strategies and technologies are dependent on efficient network planning for affordable services and resource maximization, such as:

- **Demand assessment** helps in assessing the demand of the end users that the network needs to support. The network deployed should be capable of supporting future upgrades to next-generation technologies and also be cost-effective. This would help in **designing a network architecture** that would meet the required demand, considering constraints in terms of existing infrastructure and cost. This helps in performing network dimensioning to determine requirements in terms of hardware, software, physical resources, power, spectrum, and equipment. **Network traffic engineering** involves predicting the peak congestion requirements for each cell site to avoid congested routes. It includes analyzing user behavior to forecast and reroute crowded routes, determining the shortest distance path, and handling users during peak hours.
- Next step is the **actual deployment** of the network, after the procurement of the components from the previous step. The deployment involves a detailed project implementation plan and considers all the risks that would either slow or hinder the deployment.
- Once the deployment is complete, there is a need to **test, validate, and document** the network behavior and performance against the benchmark performance in terms of QoS, vulnerability to data breach, and security standards. The log of the network performance at different stages would help in assessing the network health by the TSPs.
- **O&M** is a critical factor that is capital intensive since it handles all the repairs, network breakdowns, and network issues; monitors the software and hardware upgrades for the network; and performs security updates to the latest version available. This step also involves periodically assessing whether the existing infrastructure would be able to support the desired applications or not. The **survivability** of the network involves a continuation of services, data flow, and other operations even when certain components are broken or undergoing network upgrades. The network downtime should be minimized to lower the revenue loss for the operators as well as businesses dependent on digital connectivity. Studies by the United Nations (UN) have shown that a network shutdown for a few hours in a year in 46 countries resulted in a loss of $20.5 billion [49].
- As the network's expansion progresses, there will be a greater need to **train** staff to diagnose errors during installation, and O&M. continuation of services, data flow, and other operations even when certain components are broken or undergoing network upgrades. The network downtime should be minimized to lower the revenue loss for the operators as well as businesses dependent on digital connectivity.
- In today's world, strategic decision-making is significantly dependent on the data of consumer needs and their usage behavior [50]. In order to strategically locate their businesses and optimize their earnings, the networks are becoming reliable and efficient in gathering vast amounts of data. This data is then processed using big data algorithms, deep learning, AI, and other data analysis technologies. Enhanced network reliability and site placement to optimize coverage, QoS, and QoE are made possible by the **data-driven design**, which also helps with cost savings, better traffic management, data security, and energy efficiency.

2.5 Economic Modeling

The deployment of newer technologies depends on several factors such as technological advancements, demand-supply balance, equipment manufacturing, policies enforced by the regulatory body, and the cost of various components and their ROI. For 5G and beyond technologies, the cost of deployment has been increasing with technological advancement, which is driving the need to explore newer business cases and revenue models. The roles and duties of stakeholders in the telecommunications sector must be aligned with business models to determine the ROI and evaluate affordability and willingness-to-pay. Every organization that capitalizes on technological advancements needs to revamp its business models based on the deployment context, which includes remote, rural, suburban, urban, and dense urban areas.

Telecommunication network economics is the study of technological advancements within the telecommunications sector and the economic elements of service provision [51, 52]. In the past, the technical and financial facets were considered separately. However, with dynamic advancements in the telecommunication sector, there is a need to explore both these aspects simultaneously. Hence, in this section, the discussion revolves around the collaborative approach to exploring technical and financial facets.

2.5.1 Scenario Planning

Scenario planning aids in the formulation of a broad spectrum of future probable scenarios while taking into consideration present trends and uncertainty. Scenario planning will assist in eliminating judgmental ambiguity in corporate decisions by forecasting the behavior of various stakeholders in uncertain situations. As Schoemaker [53] states, foreseeing the future requires the knowledge of:

- Things we know we know.
- Things we know we do not know.
- Things we do not know we do not know.

This model was developed for military purposes in the 1950s to simulate war scenarios and assess the most strategic move by considering basic, political, economic, and technological trends. These results help in narrowing down the future likely scenarios for a product or solution. As per Schoemaker, the generic process involved in scenario planning is derived from [53] and is given below:

Step 1: Define the time frame and scope of the analysis.
Step 2: Identify the major stakeholders.
Step 3: Identify basic trends.
Step 4: Identify key uncertainties.
Step 5 to 8: Scenario construction and assessment.

This approach takes an extensive amount of time and money, but it is well-known because it helps with risk management, executive and stakeholder decision-making, assessing opinions on the future course of the company, allocating funds in case of emergency, readiness for risk, and projecting potential future events that could impact the business [54]. The main obstacles with this strategy are the trustworthiness of data used to estimate future potential elements influencing firms, as well as the ever-evolving business market around us [55]. The four main types of scenario planning are quantitative scenarios, operational scenarios, normative scenarios, and strategic management scenarios.

Scenario planning is crucial for addressing the digital divide as it enables stakeholders to account for the uncertainties in the initiatives and potential solutions to be employed. It also helps in assessing the risk involved in developing adaptive strategies to ensure digital inclusivity for all.

2.5.2 Strategic Planning

A business needs to understand the demand for its product or service in the target market, as well as other potential markets where the demand is positive for its business. The business model helps in understanding the various aspects of the viability of commercial success in terms of services offered, target customers, pricing model, infrastructure, organization matters, trading, operations, and pricing [56, 57]. The businesses need to closely study the market and rethink their existing business strategies to suit the market trend; otherwise, they might lose business to their competitors. Strategic planning is a structured process used to define the goals, priorities, and outline the actions required to achieve the goals. Strategic planning involves developing business models that are stable during market fluctuation, and includes strengths, weaknesses, opportunities and threats (SWOT) analysis, business model canvas (BMC), and value network configuration (VNC). There are different types of business models depending on the organization's target customers, such as retailer, manufacturer, fee-for-service, subscription, profit-sharing, freemium, bundling, franchise, pay-as-you-go, brokerage, and affiliate.

2.5.2.1 SWOT Analysis

A SWOT analysis helps identify a company's strengths, weaknesses, opportunities, and threats, providing valuable insights into its business landscape [58]. It helps in the strategic decision-making process to examine the internal possibilities and the external difficulties. Additionally, SWOT analysis aids in determining the benefits and drawbacks of an option, as well as the likelihood of success. A SWOT analysis assists in making an unbiased assessment of the firm's market position, challenges to ongoing business prospects, and insights by identifying the gaps and market uncertainties that might threaten the business [59]. The template of SWOT analysis is Table 2.2. This allows stakeholders the capacity to assess their inventory and develop strategies to assist them in harnessing their strengths and opportunities while mitigating weaknesses and threats.

Table 2.2 SWOT analysis template

Strengths	Weaknesses
Opportunities	Threats

2.5.2.2 Business Model Canvas

BMC is a strategic management tool that uses a visual framework to create, describe, and analyze business models [60]. Table 2.3 shows that the template is made up of nine major components: customer segments, value propositions, channels, customer relationships, revenue streams, key resources, key activities, key partnerships, and cost structure. Alexander Osterwalder proposed the nine-block model [61], which enables the business organization to think in terms of business models rather than just products. Businesses that put these holistic aspects on a single canvas may receive a comprehensive perspective of their operations, discover opportunities for improvement, and stimulate innovation [62]. This method encourages team members to collaborate, making it simpler to evolve and modify company plans in response to market changes. Finally, the BMC provides a framework for aligning a company's actions with its strategic objectives. The Canvas excels at defining a current line of business to address issues such as:

(a) What does product/market fit mean for this business?
(b) Where has the company been focusing, and is it still relevant to "a"?
(c) What are the key revenue, cost, and profit drivers, and how to improve those?

The core beliefs of BMC and SWOT analysis should assist in gaining a better understanding of the organization's business models. It should assist in strengthening the business, and if there is any doubt in any of these, the steps should be repeated until a clear image is obtained. Maintaining records and continuing to enhance SWOT analysis and the BMC allows for improved data-driven decision-making.

Table 2.3 Business model canvas

Key partners	Key activities	Value proposition	Customer relationships	Customer segments
	Key resources		Channels	
Cost			Revenue streams	

2.5.3 Value Chains and Value Network Configuration

Value network models examine services in which an organization works as a middleman to create value that generates income for them, rather than typical business beliefs [63]. Business generates value in the form of money, knowledge, reputation, and loyalty to the deliverables. Value networks are quite complex, and they include the study of different actors and their technical roles to understand the flow of products, revenues, and services of a traditional value chain. With time, knowledge is becoming highly dependent on the data and information, which generates significant revenue for the businesses [64]. Value chains describe the activities involved in the entire process from the beginning to the end of a product until it reaches a customer. This helps businesses understand their differentiation, optimize costs, and improve the overall quality of service for the end users. The value chain also helps the industry understand the key stakeholders, interdependencies, and value offered by their services to increase their value. This model also helps them identify their main competitors and the differentiation their solutions, products, and services offer compared to others [65].

VNC comprises interconnected business actors and technological components [66, 67], that help businesses analyze the most profitable value they can offer. Business strategies are interrelated to produce value and bring ambiguity in the business roles to be played by the actors, who are sometimes interchangeable. Value networks assist in understanding the roles played by actors across many business organizations [67]. The concepts discussed in this section can be applied to various telecommunication technologies to understand their business case for areas with a digital divide. An interesting aspect of the digital divide is that each area has its unique challenges, which leads to the conclusion that no one solution fits all. The network economics model would be unique for villages within the same country as well.

For example, [68] examines the concept of scenario planning for 5G micro-operators, and suggests three general 5G business models along with the corresponding value ecosystems that each offers in terms of scalability, adaptability, and sustainability. We must comprehend the potential outcomes of the suggested approach since it differs from the conventional MNO business model for local services. In [69], the author explores the SWOT analysis, VNC models, and scenario planning for 5G network slicing supporting NHN as a solution to bridge the digital divide.

2.5.4 Technical and Stochastic Modeling

After analyzing the economic models, the next step is to determine which technologies are necessary to satisfy user demand. This will help determine the total cost of constructing and operating the networks. The technical parameters that determine the network performance are constrained by the cost, which determines the techno-economic feasibility of any deployment.

Table 2.4 describes the key technical parameters required for designing a network. These parameters together determine the performance of the communication links, QoS and QoE offered by the technology, latency, and data rates, and KPIs offered by the network. The link budget for a communication link determines whether it is successful in transmitting the intended data or not. It provides a detailed analysis of power and resource constraints and accounts for losses and gains at each stage of the transmission path.

By creating a stochastic model based on the network's technical specifications, we can substantially improve the simulator used to understand and forecast signal propagation in the target area. This model becomes crucial for precisely predicting network performance in rural locations, where topography and environmental variables can have a substantial impact on signal strength and coverage. Signal fading, interference, and multipath effects, all of which are frequently unanticipated in rural areas, are examples of the unpredictability and randomness present in the propagation environment that the simulator can better account for by using the stochastic model.

2.5.4.1 Link Budget

A link budget is a qualitative calculation that helps in assessing whether a communication link will be successful and is mathematically modeled as:

$$P_r = P_t + G_t - L_t - L_f - L_{prop} + G_r - L_r \tag{2.1}$$

where P_r is the received power in dBm, P_t is the transmitted power in dBm, G_t is the transmitter antenna gain in dBi, L_t is the transmitted losses in dB, L_f is the free-space path losses (FSPL) in dB, L_{prop} is the propagation losses in dB, which depend on the environment (rural, suburban, urban, dense urban), antenna mispointing, fading margin, polarization mismatch, and interference, G_r is the receiver antenna gain in dBi, and L_r is the receiver losses in dB. Link budgets are essential for

Table 2.4 Technical parameters for network deployment

Category	Parameters
RAN	Frequency bands, bandwidth, duplex mode, transmit power, antenna height, type, modulation scheme, carrier aggregation, TDD UL/DL configuration, MIMO configuration
Core network	Core architecture, interface protocol, latency requirements, network slicing configuration, user plane and control plane separation, QoS profiles and bearer
Transport and backhaul	Backhaul type, bandwidth capacity, latency and jitter, redundancy and failover configuration, synchronization, interface
Site parameters	Geographical coordination, site elevation, environmental conditions, power supply type, battery backups, tower type and height
Spectrum and licensing	Licensing frequency blocks, spectrum efficiency, interference management techniques, regulatory compliance metrics, emission limits
Network optimization parameters	Handover parameters, load balance settings, coverage and capacity KPIs, call drop thresholds, data throughput thresholds
Security and management	Encryption standards, authentication mechanism, network monitoring tools, OSS/BSS integration, fault and performance alarms
Service and applications	Voice services, data services, multicast and broadcast capabilities, IoT network configuration

system design, performance evaluation, and regulatory compliance. Both terrestrial network and NTN technologies provide an exponential increase in data rate throughput. However, this depends on factors including adaptive modulation and enhanced spectral efficiency through encoding techniques [70].

2.5.4.2 Channel Encoding

Channel encoding in wireless communication helps to protect data from errors during transmission by adding redundant bits to the data before transmission, which helps the receiver detect any errors introduced during transmission. This mainly happens due to noise, interference, or other impairments in the communication channel. Encoding techniques play a crucial role in delivering and ensuring reliable transmission of data over any distance [71–73]. Table 2.5 shows some of the common encoding techniques [74].

Table 2.5 Channel encoding techniques

Technology	Channel coding techniques
5G Networks	LDPC, Polar Codes
Satellite Networks	RS Codes, Convolution Codes, Turbo Codes
6G Networks	Polar Codes, LDPC, ML-based Decoding
Wi-Fi	Convolution Codes, LDPC
HAPS	Polar Codes, LDPC
HIBS	Turbo Codes, RS Codes

2.5.4.3 Channel Modulation

Channel modulation is a process of converting digital data into useful signals, which is suitable for transmission by varying certain properties such as amplitude, frequency, or phase, to encode the information [71]. The data would be transmitted efficiently and reliably using channel modulation techniques. Some of the basic analog channel modulation techniques include amplitude modulation (AM), frequency modulation (FM), and phase modulation (PM); however, these are very sensitive to noise and have poor transmission efficiency. To overcome the shortcomings of analog channel modulation, digital modulations were introduced, which include amplitude shift keying (ASK), frequency shift keying (FSK), phase shift keying (PSK), binary phase shift keying (BPSK), quadrature phase shift keying (QPSK), quadrature amplitude modulation (QAM), orthogonal frequency-division multiplexing (OFDM), direct sequence spread spectrum (DSSS), and frequency hopping spread spectrum (FHSS) [70]. Digital modulation is more robust compared to analog modulation and helps in high performance,

Table 2.6 Modulation techniques and their benefits across technologies

Technology	Modulation techniques	Benefits
5G Networks	OFDM, QAM (various levels)	High spectral efficiency, robustness against multipath; higher data rates through variable QAM levels
Satellite Networks	PSK, APSK, OFDM	PSK for long-distance robustness, APSK for resilience to non-linear distortion, OFDM for handling multipath and improving throughput
6G Networks	OTFS, Index Modulation, THz Modulation	OTFS provide robustness to Doppler and fading, index modulation enhances spectral efficiency, and THz modulation enables ultra-high-speed data
Wi-Fi	PSK (various levels), OFDM	PSK offer noise robustness; OFDM supports high data rates and resistance to multipath effects
HIBS and HAPS	PSK (various levels), OFDM	Ensures robustness and high throughput in high-altitude communications via PSK and OFDM

error detection, and higher transmission data rates. These schemes are adapted to suit the requirements of each technology in terms of data rate, bandwidth, power efficiency, spectral efficiency, operational constraints, weather conditions, and noise. Channel modulation is a vast subject, and Table 2.6 discusses some of the prominent techniques for each technology [70]. For example, 6G uses a wide range of modulation techniques such as orthogonal time frequency space (OTFS) to be resilient to Doppler shift and multipath fading, index modulation for improved spectral efficiency, and Terahertz (THz) modulation for extremely high data rates.

Both channel modulation and channel encoding are essential for ensuring reliable and efficient communication across various wireless network systems. Stochastic modeling is used to quantitatively model these inputs and gives us a better understanding of realistic propagation QoS offered by the service provider.

2.5.4.4 Stochastic Modeling

Stochastic modeling is a mathematical technique for predicting and analyzing systems with random variability [75]. Unlike deterministic models, which yield a single, fixed output for a given set of inputs, stochastic models use random variables to provide a variety of different outcomes [76]. These models employ probability distributions to explain the possibility of various events occurring, which aids in understanding the range and likelihood of possible outcomes. A stochastic model is a technique for forecasting statistical features of potential outcomes by accounting for random variance in one or more parameters across time, which is usually based on changes found in historical data using established methodologies.

Stochastic modeling in telecommunications is used in various ways to enhance network performance and reliability. For instance, in traffic modeling, stochastic processes such as Poisson processes and Markov chains help predict data packet arrivals and network congestion, enabling better resource allocation and congestion management [76]. In wireless networks, stochastic geometry models analyze the spatial distribution of users and base stations to optimize coverage and minimize interference; predict and mitigate the impact of component failures, ensuring continuous service; and assist in dynamic resource allocation, adjusting bandwidth and processing power based on real-time traffic conditions to maintain quality of service. Overall, stochastic modeling is crucial for designing robust, efficient, and adaptive telecommunication networks [75].

The steps involved in the stochastic modeling are:

- **Stochastic processes**: Identify the process and relevant parameters in the system that exhibit uncertainty.
- **Probability distributions**: Select an appropriate probability distribution function that best models the stochastic behavior of the process.
- **Random variable generator**: Generate a random variable that would help in generating the data to represent the variability in the selected parameters.
- **System modeling**: Develop a system model that represents the structure and behavior of the system and also incorporates uncertainty into the model to simulate a real-world scenario.
- **Monte Carlo simulation**: Analyze the outcomes in different scenarios, but vary different parameters to systematize the performance.

- **Statistical analysis**: Measure the simulation results using statistical measures such as mean, variance, confidence intervals, and cumulative distribution functions (CDFs).
- **Outcome analysis**: To assess the parameters that play a critical role in the system design, perform sensitivity, scenario, and risk analysis.

With the advancement in wireless technologies, there is a need to evaluate the network performance in a heterogeneous scenario where different wireless technologies coexist [77, 78]. Stochastic geometry helps in studying the signal transmission behavior against the distance, considering factors such as error probability, clutter along the transmission path, outage probability, power control, frequency reuse, and signal interference to noise ratio (SINR) factors [79]. Stochastic geometry is a powerful analytical tool that can perform the system-level performance analysis of signal transmission and understand the behavior of the network where there are multiple users [80]. The model could help understand the system performance in deployment conditions ranging from dense urban scenarios to remote rural deployments for any technology of interest [81, 82]. The key factor in modeling lies in the selection of the formulas for link budget, SINR model, and system throughput for the corresponding technology and deployment type [83].

2.5.5 Techno-Economic Assessment Framework

A techno-economic model is a comprehensive tool that can help comprehend and evaluate the business's scale, profitability, and risk [84, 85]. It involves an analysis of factors involving technological, financial, market, and regulatory opportunities as well as risk factors. It is used in many projects and considers almost all of the foreseeable factors that could have an impact on the business, such as market demand, market penetration, competitive landscapes, and inflation, and performs sensitivity analysis [86]. Additionally, it also helps evaluate regulatory requirements and perform risk analysis, ensuring that the business complies with set standards and legal frameworks to avoid any regulatory challenges. The key components of a techno-economic assessment (TEA) include technical evaluation, economic analysis, market assessment, and environmental impact. In reality, techno-economic models are commonly employed in research projects to determine the viability and impact of new technologies [87].

There are different methods to assess the cost involved in the network deployment, which depend on the framework applied [88]. As per the ITU reports on the regulatory accounting guide, there are several reasons why cost estimation plays a crucial role in the telecommunication industry [89, 90], such as:

- For understanding the in-depth technical requirements for deployment and operations, TEA would help in understanding the spectrum needs. This would further help the regulator assess the services offered, the market competition, pricing strategies, and profitability for the public as well as private networks. After a detailed study, the regulator would develop suitable policies and regulations promptly to support the technological advancements.
- There are stringent requirements for TSPs to comply with legal policies, regulatory norms, administrative policies, and government policies to follow fair trade practices. TEA helps in understanding the compliance with these.
- The cost assessment would also help in bridging the digital divide and ensure that tariffs are affordable for the people living in the country. This analysis would also help in understanding the market competition required in different regions to upgrade the telecommunication infrastructure, ease of market entry, and market failure scenarios.
- The assessment would also help to understand the environmental impact by considering factors such as sustainability, resource usage, emissions, net-zero, climate change, and pollution.

Understanding the supply, demand, and cost characteristics allows us to quantitatively analyze the problem and determine the best approach to bridging the digital divide by crafting supportive policies [86]. The regulator aims to understand operators' cost accounting methods in order to develop regulations and policy recommendations, such as price control measures, legal standards, business transparency initiatives, universal service costing, and policy frameworks for traffic flows, wholesale pricing, sharing requirements, data protection, universal service obligations, entry and exit processes, and dispute management; to curb anti-competitive practices [91].

The next step involves asset valuation using different techniques such as historic cost (values at purchased price), indexation (appropriate indices are applied for an asset purchased in historical time to estimate the current valuation), absolute valuation

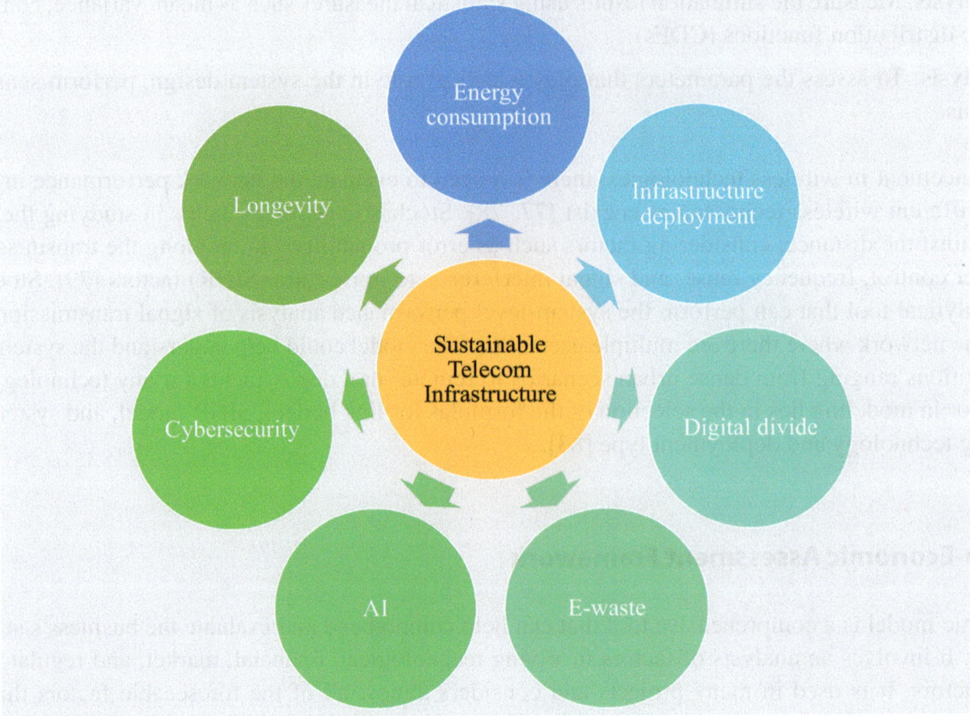

Fig. 2.7 Sustainable telecom infrastructure

(a product of an existing number of assets with current acquisition price), and modern equivalent assets (estimation of a discontinued product). Once the asset prices are estimated, there is a need to understand the cost standards for services and products.

TEA models help in understanding the feasibility of different broadband solutions, in terms of technology and economic benefits. Several studies, both academic and industry, publish research discussing the scope of the technology and various deployment strategies to improve their ROI. TEA is particularly important in the context of rapidly evolving infrastructures such as 5G and beyond, allowing stakeholders to make informed decisions regarding investments and operational strategies [92]. The policymakers and regulators explore the TEA models in framing the policies for the deployment of newer technologies and bridging the digital divide across the world [93]. This integration is vital for optimal network planning, ensuring that investments are both viable and strategically sound.

There are studies that study TEA for rural areas to assess the potential of the technology of interest to bridge the digital divide. For digital divide challenges, technological determinants along with digital literacy, affordability, and value offered in people's lives play a critical role in improving the feasibility [94]. These studies' observations aid stakeholders in creating policies that support improved decision-making while taking into account stakeholder and consumer needs [95]. The wide range of technical alternatives available in today's world emphasizes the necessity of knowing which technology is most suitable for each area of interest because they have different, unique challenges.

The economic implications of telecommunication investments are further illustrated by studies examining the relationship between telecommunications infrastructure and economic growth, such as in Africa [17, 96, 97]. In summary, the techno-economic evaluation of telecommunications technologies is a complex procedure that necessitates the integration of social, economic, and technological factors. The procedures and frameworks created for TEA will be crucial in directing investments and guaranteeing long-term growth in the telecom sector as the industry changes, especially with the introduction of 5G and other emerging technologies in bridging the digital divide.

2.6 Environmental Modeling

There is a growing requirement around the world to consider environmental degradation and climate change, and explore the interrelationships between various factors impacting them [98]. This is very crucial for our planet's health and well-being, and directly impacts the future of our planet. Environmental modeling is critical for policymakers to make informed decisions and for businesses to develop sustainable practices. Several frameworks and standards measure environmental sustainability, and some of the popular frameworks are–**Life cycle assessment (LCA)** to measure the environmental impacts of a product, process, or system across the entire lifespan of the product. **Scenario-based carbon modeling** helps in projecting emissions under different future scenarios such as policy changes, technology upgrades, or market shifts. **Scope 1–3 emissions accounting** for classification under greenhouse gas (GHG) protocol, in which **scope 1** accounts for direct emissions from owned operations, **scope 2** is indirect emissions from purchased electricity, and **scope 3** is from all other indirect emissions.

The environmental modeling section delves into environmental sustainability–energy efficiency, sustainable infrastructure, and carbon emissions, highlighting their interconnectedness and importance in shaping a sustainable future. Environmental modeling includes representations and simulations to consider environmental systems, their interactions, and changes [99]. The modeling can be performed in different ways to account for several variables impacting the climate. The three main types of models are descriptive models, predictive models, and normative models. The latter is used often in evaluating the outcomes of strategies and policy and assists in the decision-making process [100]. Figure 2.7 shows some of the key sustainable infrastructure practices that would help reach the telecommunication sector's climate goals.

By this time, you would be wondering what is a sustainable infrastructure. Sustainable practices and infrastructure would help achieve the demand for the telecommunication industry without damaging the environment [101]. The key benefits offered by deploying sustainable infrastructure and following sustainable practices and business models include minimizing carbon emissions and resource consumption. These objectives are met by using energy-efficient technologies and renewable energy sources, adhering to regulations, inspiring environmentally conscious consumer behavior, building strong community relationships, and promoting innovative sustainable solutions. The goal of sustainable telecom infrastructure practices is to reduce the detrimental impact on the environment while simultaneously enhancing network reliability and efficiency [102].

2.6.1 Energy Consumption

As of 2024, the energy consumption of the telecommunication industry and ICT sector accounts for about 2–3 and 4% of global energy consumption, respectively [103]. According to GSMA research, energy expenditure accounts for 20–40% of OPEX for telecom companies, and these numbers can go higher in countries that rely on diesel, such as Africa and Southeast Asia. Several studies have shown that advanced computing wireless broadband technologies, such as 5G and Wi-Fi 7, would likely require more power to operate over a large-scale deployments compared to the older technologies [104, 105]. Consequently, operators and stakeholders must pursue alternative reliable energy sources that meet ethical standards, economic viability, environmental sustainability, and commercial feasibility.

Several countries across the globe are exploring alternative energy sources that are environmentally friendly and minimize their OPEX toward energy consumption, which would help in lowering their overall total cost of ownership (TCO) [106]. Around 70% of the energy consumption of telecommunication towers is toward RAN [107]. Hence, there is a need to minimize this or increase the usage of green energy solutions per mobile site. Many studies are now investigating alternative sources of energy, particularly green energy derived from natural resources that can be replenished more quickly than they are consumed, while causing little to no harm to the environment. Green energy involves using energy from sources such as solar, wind, tidal, hydroelectric power, and biogas to operate the telecommunication network and infrastructure [108]. This would help them contribute to the climate change challenges.

2.6.2 Carbon Emissions

In this book, we mentioned carbon emissions several times. There is a need to understand what carbon emissions are and why they are an important factor in today's climate change goals [109]. The combination of carbon with oxygen, which are the fourth and third most prevalent elements, respectively, results in carbon emissions. This combination happens when biological material decomposes, something is burned, and air from living beings is inhaled. The excess carbon dioxide, CO_2, generated from these processes is polluting the Earth. Since CO_2 absorbs infrared radiation, it is called a greenhouse gas and helps in keeping the planet warm. Beyond a certain concentration in the atmosphere, CO_2 causes the overall temperature of the Earth to increase and alter our climate. Climate change causes several problems, including rising global temperatures, more intense storms, more droughts, warmer oceans, rising sea levels, rising oceans, extinction of species, scarcity of food, health hazards for all living things, air quality, ecosystem disruption, and poverty [110]. These circumstances might lead to a catastrophe in the lives of humans and other living things and are upsetting the Earth's natural equilibrium.

It has been estimated that the telecommunication industry contributed to around 2% of the world's carbon emissions and generates significant e-waste, which is either discarded, burned, or illegally traded [111]. Today, telecommunication services are focusing on sustainable practices that would help in reducing carbon emissions globally. Studies have shown that by upgrading the existing infrastructure to energy-efficient technologies and promoting digitalization, the operators can reduce the carbon emissions by 15%, in addition to boosting their QoS, performance, and economic growth. Many TSPs are adopting sustainability practices, such as waste management, and renewable energy adoption to achieve the 2030 target on emission reduction and achieve net-zero targets by 2050.

2.7 Usage Gap

Bridging the digital divide is a multi-faceted strategy focusing on ensuring accessibility, affordability, and availability. In the above strategies, we focused on the access and coverage gap. Now, there is a need to explore the strategies employed to bridge the usage gap.

2.7.1 Affordability and Priorities

A significant portion of people who are unconnected lack disposal financial resources to purchase internet connectivity plans, digital services, or even the devices to access the digital content [112]. Studies show that financial constraints significantly limit both internet access and adoption, and without alternative ways to obtain free or affordable devices, individuals miss out on digital connectivity's advantages. Therefore, there is a need to enhance the affordability to bridge the digital divide. Some of the commonly employed methods to encourage usage are:

- Subsidized Internet plans targeted for low-income households.
- Subsidized device rates to access digital services for low-income households and students.
- Reusing public-sector information technology (IT) equipment for underserved population.
- Tiered pricing models to encourage adoption in regions with low income.
- Public Wi-Fi and community Wi-Fi to promote digital adoption of the technology and related services.

Along with the challenge of affordability, there is a need to focus on meeting the priorities of the local communities in underserved regions [16]. Strategies must be context-driven and inclusive in nature to encourage digital adoption, while supporting the priorities of local communities, such as:

- Building a self-sustainable telecommunication infrastructure that supports the other sectors, such as renewable energy, water management systems, and other priorities of local communities.
- Community-led initiatives, such as public Wi-Fi, and local ISPs.
- Culturally relevant programs, such as youth engagement, women-focused initiatives, children-focused education, and tailored content in the local language.
- Economic empowerment through microfinance, mobile banking, e-commerce, remote work opportunities, and e-governance.
- Developing a regulatory framework for encouraging fair and competitive pricing mechanisms to encourage adoption in underserved regions.

- Encouraging spectrum and infrastructure sharing to drive the cost of service provisioning low, which in turn would improve the service affordability.

2.7.2 Digital Literacy and Awareness

There is another category of population, who despite having sufficient income for devices and data services, remain unconnected with digital services due to inadequate digital literacy skills. Hence, aside from cost and affordability, there is another key aspect contributing to the digital divide: consumer digital awareness [113, 114]. Several UN SDGs targets for 2030 focus on improving digital literacy and awareness among users to increase the adoption of digital connectivity-related services. The roles of SDGs and ICT are explored further to understand their impact on each other and how they drive the economy of the nation [115]. Digital connectivity facilitates the exercise of human rights and raises awareness regarding them. The UN focuses mainly on three targets to study the SDGs and ICT relationship that includes ICT-enabled solutions, services, and resources, and SDGs; and ICT for measuring SDGs.

ICT plays a crucial role in 17 UN SDGs for applications worldwide, especially sustainable development goal (SDG) 1, 4, 5, 7, 8, 9, 10, 11, 13, and 16 [116]. The telecommunication infrastructure directly contributes toward SDG 9, that is, industry, innovation, and infrastructure. For example, farmers would be able to access a knowledge base to enhance their farming practices and find solutions to their difficulties, as well as share their expertise with others across the world. Another potential benefit would be access to online educational findings, enabling people to address their local hurdles by using global solutions from our communities with similar challenges. ICT would also help the governing bodies and industries make informed decisions on the needs of the community, surveillance, solutions to cater to their challenges, evidence-based policymaking, and monitoring of critical resources. It will help in the economic, social, and political development of the community.

So far, we have discussed the benefits of digital literacy, but what does it mean? Digital literacy is a necessary skill in today's technologically advanced environment. It signifies an ability to successfully browse, assess, and use digital tools and information. Digital literacy enables people to take full advantage of the digital age by teaching them how to operate computers and mobile devices, as well as how to access and critically analyze online material. Furthermore, it requires being aware of online privacy and security precautions, as well as the ability to communicate and work successfully in digital contexts [117]. In summary, digital literacy is more than just understanding how to utilize gadgets and software; it is about having the knowledge and abilities to flourish in a digitally networked world.

Digital literacy is becoming a fundamental need for human life for individuals of all age groups, which would empower them to navigate digital technology confidently, responsibly, and effectively. Embracing digital literacy is key to unlocking the full potential of the digital age and building a more inclusive and informed society. Digital literacy should be accompanied by policies and regulations to regulate the proper usage of technology, as well as skills to build a progressive world. This would help to build an independent, resilient, and knowledgeable community.

2.8 Summary

This chapter highlighted the importance of focusing on strategies beyond technological deployment that would focus on improving the coverage, in turn, the network availability. For bridging the access and affordability gap, strategies including SAGIN, infrastructure sharing, spectrum sharing, and efficient and optimized network planning are explored. Furthermore, economic and environmental modeling play a vital role in measuring the network performance, associated cost and carbon emissions, business models, ROI, and energy consumption. These parameters would help in building a sustainable network that would meet the requirements as per the demand-supply balance of the community in focus. Part II utilizes these frameworks to develop a model for technologies discussed in Chap. 1, which helps in assessing these metrics. Finally, the strategies employed in bridging the usage gap are discussed and will be explored in depth in Part III of this book.

Exercises

2.1 A telecom operator is tasked with providing network coverage to remote villages with limited infrastructure. The area has rugged terrain, making traditional tower installations challenging.

(a) Evaluate which infrastructure sharing strategy would be most effective in this scenario, when 4 operators would want to provide service in the region.
(b) Discuss how spectrum sharing could be leveraged to enhance connectivity.
(c) Propose a phased deployment plan with emphasis on cost efficiency and scalability.

2.2 A region with newly deployed broadband networks struggles with low adoption rates due to limited digital literacy.

(a) Design a digital literacy campaign targeting students, small business owners, and elderly users.
(b) Propose a collaboration strategy involving local authorities, NGOs, and telecom operators.
(c) Evaluate how improved digital awareness can maximize the benefits of recent infrastructure upgrades.

2.3 There is a need to understand the user uptake for a newer technology. Using a logistic growth model, estimate the number of new internet users in a region with a current population of 5 million, given a maximum user capacity of 10 million and a growth rate of 1.5%. The formula for the logistic growth model is:

$$U(t) = \frac{K}{1 + \frac{K-U_0}{U_0} e^{-rt}}$$

where $U(t)$ is the number of internet at time t, K is the maximum users capacity, U_0 is the initial number of users, r is the growth rate, e is Euler's number 2.718.

Modify the parameters, K, U_0, and r, and examine the impact of the adoption, and plot them on a graph, and draw an inference from it.

2.4 A wireless communication system transmits data between two nodes separated by a distance of 5 km. The system has the parameters as: speed of signal propagation is 3×10^8 m/s, transmission rate is 10 Mbps, packet size is 1500 bytes, processing delay at each node is 2 ms, and queuing delay is 1.5 ms. Calculate the end-to-end latency of a single packet transmission. Use the formula:

$$\text{Propagation Delay} = \frac{\text{Distance}}{\text{Propagation Speed}}$$
$$\text{Transmission Delay} = \frac{\text{Packet Size (in bits)}}{\text{Transmission rate}}$$

2.5 (a) If the frequency is 2.4 GHz and the distance is 1 km, what is the free-space path loss L_{fs}?
$L_{fs} = 20\log_{10}(f) + 20\log_{10}(d) - 147.55$ where f is the frequency (in MHz) and d is the distance (in km).
(b) Given $P_t = 20$, dBm, $G_t = 15$, dBi, $G_r = 10$, dBi, and $L_{fs} = 100$, dB, what is the received power P_r?
(c) If receiver sensitivity is -90, dBm and $P_r = -80$, dBm, what is the link margin?
(d) Including additional losses like cable attenuation L_c or fading margins F_m, how does affect the link?
$L_{total} = L_{fs} + L_c + F_m$.

2.6 Study the differences in deployment costs, latency, and scalability between urban and rural telecommunication projects. Provide quantitative comparisons and analyze how these differences impact network planning.

Understand the revenue model for urban and rural scenarios by considering a population of 3,000 users using cellular communication technologies. Consider the network uptake to be around 40% in the beginning. Use the subscription-based model to study the potential revenue at different average revenue per user (ARPU) values.
($R = \sum_{i=1}^{T=5} 12ZS_i$, where R is the revenue from the users, Z is the ARPU, S_i is the number of subscribers for the year i, and the study is for 5 years period.) Also, ARPU ratio Rural : Urban = 0.3 : 1 to 0.5 : 1

Next, plot the values for different values of ARPU, investment duration, and number of subscribers at year 0.

2.7 Conduct a case study on the energy consumption of satellite communication (LEO, MEO, GEO) versus terrestrial fiber networks. Compare their sustainability metrics, including energy efficiency and carbon footprint.

2.8 Propose a scenario plan for deploying low-cost internet solutions in a flood-prone region. Include key assumptions, risks, and alternative strategies for maintaining connectivity during floods.

2.9 Perform a case study to understand the time taken by the users to switch from 4G to 5G, and study the time taken for growth in ARPU by considering any one country of interest. This is modeled using the diffusion model. The Bass model defines the rate of adoption as:
$f(t) = (p + qF(t))(1 - F(t))$, where $F(t)$: Cumulative fraction of adopters at time t, p: Coefficient of innovation, q: Coefficient of imitation.
The cumulative number of adopters at time t is $A(t) = N \times F(t)$, where N is the total potential market size.

Similarly, the ARPU growth rate is modeled as $Z(t) = Z_{\text{base}} + \Delta Z \cdot f(t)$.

2.10 Design an optimization algorithm to minimize deployment costs of high-altitude platform stations (HAPS) while maximizing their coverage area and bandwidth efficiency. Include mathematical constraints and decision variables.

References

1. J.R. Ram, When sharing isn't always caring: understanding telecom infrastructure sharing in the multilateral context. Available at SSRN 2971852 (2017)
2. J. Liu, Y. Shi, Z.M. Fadlullah, N. Kato, Space-air-ground integrated network: a survey. IEEE Commun. Surv. Tutor. **20**(4), 2714–2741 (2018)
3. E. Yaacoub, M.-S. Alouini, A key 6G challenge and opportunity—connecting the base of the pyramid: a survey on rural connectivity. Proc. IEEE **108**(4), 533–582 (2020)
4. S.K.A. Kumar, E.J. Oughton, Infrastructure sharing strategies for wireless broadband. IEEE Commun. Mag. **61**(7), 46–52 (2023)
5. S. Malisuwan, W. Kaewphanuekrungsi, N. Suriyakrai, Infrastructure sharing in telecommunications: fundamental and analysis. Int. J. Appl. Eng. Res. Dev., ISSN (P) 2250–1584 (2015)
6. H. Cui, J. Zhang, Y. Geng, Z. Xiao, T. Sun, N. Zhang, J. Liu, W. Qihui, X. Cao, Space-air-ground integrated network (SAGIN) for 6G: requirements, architecture and challenges. China Commun. **19**(2), 90–108 (2022)
7. J. Zhao, Y. Zhu, K. Cai, Z. Gao, Z. Han, L. Hanzo, Multiple-antenna aided aeronautical communications in air-ground integrated networks: channel estimation, reliable transmission, and multiple access. IEEE Wirel. Commun. **31**(2), 105–111 (2023)
8. B. Li, Z. Fei, Y. Zhang, UAV communications for 5G and beyond: recent advances and future trends. IEEE Internet Things J. **6**(2), 2241–2263 (2018)
9. K. Yang, Y. Wang, X. Gao, C. Shi, Y. Huang, H. Yuan, M. Shi, Communications in space-air-ground integrated networks: an overview. Space: Sci. Technol. **5**, 0199 (2025)
10. N. Cheng, W. Quan, W. Shi, W. Huaqing, Q. Ye, H. Zhou, W. Zhuang, X. Shen, B. Bai, A comprehensive simulation platform for space-air-ground integrated network. IEEE Wirel. Commun. **27**(1), 178–185 (2020)
11. D.-E. Meddour, T. Rasheed, Y. Gourhant, On the role of infrastructure sharing for mobile network operators in emerging markets. Comput. Netw. **55**(7), 1576–1591 (2011)
12. F.E. Idachaba, Telecommunication cost reduction in Nigeria through infrastructure sharing between operators. Pac. J. Sci. Technol. **11**(1), 272–276 (2010)
13. A. Saif, A. Alashwal, Q. Abdullah, S.H. Alsamhi, A. Ameen, A. Salh, Infrastructure sharing and quality of service for telecommunication companies in Yemen, in *International Congress of Advanced Technology and Engineering (ICOTEN)* (IEEE, 2021), pp. 1–6
14. O.C. Nosiri, C.K. Agubor, E.U. Ekwueme, Telecom infrastructure sharing, a panacea for sustainability, cost and network performance optimization in Nigeria telecom industry. Int. J. Sci. Eng. Res. **6**(8), 621–626 (2015)
15. T. Frisanco, P. Tafertshofer, P. Lurin, R. Ang, Infrastructure sharing and shared operations for mobile network operators from a deployment and operations view, in *NOMS - Network Operations and Management Symposium* (IEEE, 2008), pp. 129–136
16. J. Hou, L. Sun, T. Shu, Y. Xiao, M. Krunz, Economics of strategic network infrastructure sharing: a backup reservation approach. IEEE/ACM Trans. Netw. **29**(2), 665–680 (2020)
17. E.J. Oughton, Policy options for broadband infrastructure strategies: a simulation model for affordable universal broadband in Africa. Telemat. Inform. **76**, 101908 (2023)
18. Sharing Infrastructure for Mobile Networks. Accessed on 04 Apr 2025
19. K. Samdanis, X. Costa-Perez, V. Sciancalepore, From network sharing to multi-tenancy: the 5G network slice broker. IEEE Commun. Mag. **54**(7), 32–39 (2016)
20. K. Amadasun, M. Short, T. Crosbie, Telecommunication infrastructure sharing a remedy for the reduction of network operator cost and environmental pollution, in *2020 IEEE International Conference on Environment and Electrical Engineering and 2020 IEEE Industrial and Commercial Power Systems Europe (EEEIC/I&CPS Europe)* (IEEE, 2020), pp. 1–6
21. S.K.A. Kumar, R.W. Stewart, D. Crawford, S. Chaudhari, Techno-economic study of 5G network slicing to improve rural connectivity in India. IEEE Open J. Commun. Soc. **2**, 2645–2659 (2021)
22. L. Cano, A. Capone, G. Carello, M. Cesana, M. Passacantando, On optimal infrastructure sharing strategies in mobile radio networks. IEEE Trans. Wireless Commun. **16**(5), 3003–3016 (2017)
23. I. Tognisse, A.D. Kora, J. Degila, Infrastructure sharing model to connect the unconnected in rural areas. ITU J. (2021)
24. R. Bajracharya, R. Shrestha, H. Jung, H. Shin, Neutral host technology: the future of mobile network operators. IEEE Access **10**, 99221–99234 (2022)
25. J. Lähteenmäki, The evolution paths of neutral host businesses: antecedents, strategies, and business models. Telecommun. Policy **45**(10), 102201 (2021)
26. S. Friedner, R. House, 5G infrastructure requirements in the UK. *LS telcom UK, Tech. Rep.* (2016)

27. S. Shrivastava, Path leading approach to 5G deployment for the Indian communication service providers. Telecom Bus. Rev. **15**(1), 37 (2022)
28. S.K. Gupta, Trai releases white paper on enabling 5g in india. *TRAI* (2019)
29. J.M. Peha, Approaches to spectrum sharing. IEEE Commun. Mag. **43**(2), 10–12 (2005)
30. S. Bhattarai, J.-M.J. Park, B. Gao, K. Bian, W. Lehr, An overview of dynamic spectrum sharing: ongoing initiatives, challenges, and a roadmap for future research. IEEE Trans. Cogn. Commun. Netw. **2**(2), 110–128 (2016)
31. L. Zhang, M. Xiao, W. Gang, M. Alam, Y.-C. Liang, S. Li, A survey of advanced techniques for spectrum sharing in 5G networks. IEEE Wirel. Commun. **24**(5), 44–51 (2017)
32. M.M. Buddhikot, Understanding dynamic spectrum access: models, taxonomy and challenges, in *2nd IEEE International Symposium on New Frontiers in Dynamic Spectrum Access Networks* (IEEE, 2007), pp. 649–663
33. R. Horvitz, Geo-database management of white space vs. open spectrum. SocArXiv (2013)
34. L. Gavrilovska, D. Denkovski, V. Rakovic, M. Angjelichinoski, Medium access control protocols in cognitive radio networks: overview and general classification. IEEE Commun. Surv. Tutor. **16**(4), 2092–2124 (2014)
35. H. Feng, B. Chen, K. Zhu, Full spectrum sharing in cognitive radio networks toward 5G: a survey. IEEE Access **6**, 15754–15776 (2018)
36. W.S.H.M.W. Ahmad, N.A.M. Radzi, F.S. Samidi, A. Ismail, F. Abdullah, M.Z. Jamaludin, M. Zakaria, 5G technology: towards dynamic spectrum sharing using cognitive radio networks. IEEE Access **8**, 14460–14488 (2020)
37. ETSI. Spectrum sharing frameworks for temporary, dynamic, and flexible spectrum access for local private networks. Accessed on 26 Mar 2025
38. M. Grissa, A.A. Yavuz, B. Hamdaoui, Trustsas: a trustworthy spectrum access system for the 3.5 GHz CBRS band, in *IEEE Conference on Computer Communications* (IEEE, 2019), pp. 1495–1503
39. M. Rahman, A. Saifullah, A comprehensive survey on networking over TV white spaces. Pervasive Mob. Comput. **59**, 101072 (2019)
40. W. Webb, A. Medeisis, L.F. Minervini, Evolved spectrum usage rights: a catalyst for liberal spectrum management reform. Telecommun. Policy **48**(3), 102715 (2024)
41. A. Pärssinen, M.-S. Alouini, M. Berg, T. Kürner, P. Kyösti, M.E. Leinonen, M. Matinmikko-Blue, E. McCune, U. Pfeiffer, P. Wambacq, White paper on RF enabling 6G: opportunities and challenges from technology to spectrum. Technical report, University of Oulu (2021)
42. V. Sathya, S.M. Kala, K. Naidu, Heterogenous networks: from small cells to 5G NR-U. Wirel. Pers. Commun. **128**(4), 2779–2810 (2023)
43. T.Q. Duong, N.-S. Vo, Wireless communications and networks for 5G and beyond. Mob. Netw. Appl. **24**, 443–446 (2019)
44. A. Gautam, P. Thakur, G. Singh, Mobile communications and networks beyond 5G, in *5G and Beyond Wireless Networks* (CRC Press, 2024), pp. 190–206
45. K. Lee, J. Lee, Y. Yi, I. Rhee, S. Chong, Mobile data offloading: how much can WiFi deliver? IEEE/ACM Trans. Netw. **21**(2), 536–550 (2012)
46. H. Zhou, H. Wang, X. Li, V.C.M. Leung, A survey on mobile data offloading technologies. IEEE Access **6**, 5101–5111 (2018)
47. F. Bahlke, O.D. Ramos-Cantor, S. Henneberger, M. Pesavento, Optimized cell planning for network slicing in heterogeneous wireless communication networks. IEEE Commun. Lett. **22**(8), 1676–1679 (2018)
48. B.B. Haile, E. Mutafungwa, J. Hämäläinen, A data-driven multiobjective optimization framework for hyperdense 5G network planning. IEEE Access **8**, 169423–169443 (2020)
49. World Economic Forum. Internet shutdowns: What you need to know. Accessed on 13 Mar 2025
50. A. Fernández-Fernández, E. Coronado, A. Erspamer, G. Samaras, V. Theodorou, S. Siddiqui, Unlocking the path toward intelligent telecom marketplaces for beyond 5g and 6g networks. IEEE Commun. Mag. **61**(3), 28–34 (2023)
51. P. Verma, F. Zhang, *The Economics of Telecommunication Services* (Springer, 2020)
52. A. Gupta, D.O. Stahl, A.B. Whinston, The economics of network management. Commun. ACM **42**(9), 57–63 (1999)
53. P.J.H. Schoemaker, Scenario planning: a tool for strategic thinking. MIT Sloan Manag. Rev. (1995)
54. M. Amer, T.U. Daim, A. Jetter, A review of scenario planning. Futures **46**, 23–40 (2013)
55. G.D. Peterson, G.S. Cumming, S.R. Carpenter, Scenario planning: a tool for conservation in an uncertain world. Conserv. Biol. **17**(2), 358–366 (2003)
56. C. Zott, R. Amit, Business model design: an activity system perspective. Long Range Plan. **43**(2–3), 216–226 (2010)
57. G.A. Steiner, *Strategic Planning* (Simon and Schuster, 2010)
58. G. Houben, K. Lenie, K. Vanhoof, A knowledge-based SWOT-analysis system as an instrument for strategic planning in small and medium sized enterprises. Decis. Support Syst. **26**(2), 125–135 (1999)
59. D.W. Pickton, S. Wright, What's SWOT in strategic analysis? Strat. Chang. **7**(2), 101–109 (1998)
60. A. Murray, V. Scuotto, The business model canvas. Symphonya. Emerg. Issues Manag. 94–109 (2015)
61. A. Osterwalder, Business model generation: a handbook for visionaries, game changers, and challengers (2010)
62. E.S. Türko, Business plan vs business model canvas in entrepreneurship trainings, a comparison of students' perceptions. Asian Soc. Sci. **12**(10), 55–62 (2016)
63. E.A. Zamora, Value chain analysis: a brief review. Asian J. Innov. Policy **5**(2), 116–128 (2016)
64. J. Hellin, M. Meijer, *Guidelines for Value Chain Analysis* (Food and Agriculture Organization of the United Nations (FAO), Rome, Italy, 2006)
65. P.C. Ensign, Value chain analysis and competitive advantage. J. Gen. Manag. **27**(1), 18–42 (2001)
66. V. Allee, Reconfiguring the value network. J. Bus. Strat. **21**(4), 36 (2000)
67. T. Casey, T. Smura, A. Sorri. Value network configurations in wireless local area access, in *9th Conference of Telecommunication, Media and Internet* (IEEE, 2010), pp. 1–9
68. P. Ahokangas, M. Matinmikko-Blue, S. Yrjölä, V. Seppänen, H. Hämmäinen, R. Jurva, M. Latva-Aho, Business models for local 5G micro operators. IEEE Trans. Cogn. Commun. Netw. **5**(3), 730–740 (2019)
69. S.K.A. Kumar, 5G network slicing for rural connectivity: multi-tenancy in wireless networks. *University of Strathclyde* (2023)
70. T.S. Rappaport, *Wireless Communications: Principles and Practice* (Cambridge University Press, 2024)
71. E. Basar, I. Altunbas, Space-time channel modulation. IEEE Trans. Veh. Technol. **66**(8), 7609–7614 (2017)
72. M. Vázquez-Castro, F. Pérez-Fontán, B. Arbesser-Rastburg, Channel modeling for satellite and HAPS system design. Wirel. Commun. Mob. Comput. **2**(3), 285–300 (2002)

73. M. Khammassi, A. Kammoun, M.-S. Alouini, Precoding for high throughput satellite communication systems: a survey. IEEE Commun. Surv. Tutor. (2023)
74. K. Deergha Rao et al., *Channel Coding Techniques for Wireless Communications* (Springer, 2015)
75. F. Baccelli, M. Klein, M. Lebourges, S. Zuyev, Stochastic geometry and architecture of communication networks. Telecommun. Syst. **7**, 209–227 (1997)
76. H. ElSawy, A. Sultan-Salem, M.-S. Alouini, M.Z. Win, Modeling and analysis of cellular networks using stochastic geometry: a tutorial. IEEE Commun. Surv. Tutor. **19**(1), 167–203 (2016)
77. Y. Hmamouche, M. Benjillali, S. Saoudi, H. Yanikomeroglu, M. Di Renzo, New trends in stochastic geometry for wireless networks: a tutorial and survey. Proc. IEEE **109**(7), 1200–1252 (2021)
78. P.C. Pinto, A. Giorgetti, M.Z. Win, M. Chiani, A stochastic geometry approach to coexistence in heterogeneous wireless networks. IEEE J. Sel. Areas Commun. **27**(7), 1268–1282 (2009)
79. X. Yu, C. Li, J. Zhang, K.B. Letaief, *Stochastic Geometry Analysis of Multi-Antenna Wireless Networks* (Springer, 2019)
80. M. Haenggi, J.G. Andrews, F. Baccelli, O. Dousse, M. Franceschetti, Stochastic geometry and random graphs for the analysis and design of wireless networks. IEEE J. Sel. Areas Commun. **27**(7), 1029–1046 (2009)
81. Y. Hmamouche, M. Benjillali, S. Saoudi, On the role of stochastic geometry in sixth generation wireless networks, in *2020 10th International Symposium on Signal, Image, Video and Communications (ISIVC)* (IEEE, 2021), pp. 1–6
82. O.B. Osoro, E.J. Oughton, A techno-economic framework for satellite networks applied to low earth orbit constellations: assessing Starlink. OneWeb and Kuiper. IEEE Access **9**, 141611–141625 (2021)
83. T. Tuovinen, N. Tervo, A. Pärssinen, Analyzing 5G RF system performance and relation to link budget for directive MIMO. IEEE Trans. Antennas Propag. **65**(12), 6636–6645 (2017)
84. M. Kantor, K. Wajda, B. Lannoo, K. Casier, S. Verbrugge, M. Pickavet, L. Wosinska, J. Chen, A. Mitcsenkov, General framework for techno-economic analysis of next generation access networks, in *12th International Conference on Transparent Optical Networks* (IEEE, 2010), pp. 1–4
85. G.S. Murthy, Chapter two - techno-economic assessment, in *Biomass, Biofuels, Biochemicals*, ed. by G.S. Murthy, E. Gnansounou, S.K. Khanal, A. Pandey (Elsevier, 2022), pp. 17–32
86. P.H. Kobos, T.E. Drennen, A.V. Outkin, E.K. Webb, S.M. Paap, S. Wiryadinata, Techno-economic analysis: best practices and assessment tools. Technical report, Sandia National Lab. (SNL-CA), Livermore, CA (United States); Sandia National ... (2020)
87. M. Bromwich, C. Hong, Costs and regulation in the UK telecommunications industry. Manag. Account. Res. **11**(1), 137–165 (2000)
88. R. Roy, Cost engineering: why, what and how? *Decision Engineering Report Series* (2003)
89. ITU-D. Economic policies and methods of determining the costs of services related to national telecommunication/ICT networks. Accessed on 26 Oct 2024
90. International Telecommunication Union (ITU). ITU-D question 4/1 final report 2021. Accessed on 27 Mar 2025
91. L.A. Ims, *Broadband Access Networks: Introduction Strategies and Techno-economic Evaluation* (Springer, 2012)
92. E.J. Oughton, W. Lehr, Surveying 5G techno-economic research to inform the evaluation of 6G wireless technologies. IEEE Access **10**, 25237–25257 (2022)
93. E.J. Oughton, N. Comini, V. Foster, J.W. Hall, Policy choices can help keep 4G and 5G universal broadband affordable. Technol. Forecast. Soc. Chang. **176**, 121409 (2022)
94. V. Krizanovic Cik, D. Zagar, K. Grgic, A framework for optimal techno-economic assessment of broadband access solutions and digital inclusion of rural population in global information society. Univers. Access Inf. Soc. **17**(3), 517–540 (2018)
95. C. Bendicho, Techno-economic assessment in communications: models for access network technologies, in *Future of Information and Communication Conference* (Springer, 2021), pp. 315–341
96. S.H. Lee, J. Levendis, L. Gutierrez, Telecommunications and economic growth: an empirical analysis of sub-Saharan Africa. Appl. Econ. **44**(4), 461–469 (2012)
97. F. Atsu, C. Agyei, W. Phanuel Darbi, S. Adjei-Mensah, The impact of telecommunication revenue on economic growth: evidence from Ghana. Afr. J. Econ. Manag. Stud. **5**(2), 195–208 (2014)
98. M.D. Ahmed, D. Sundaram, Sustainability modelling and reporting: from roadmap to implementation. Decis. Support. Syst. **53**(3), 611–624 (2012)
99. K. Henderson, M. Loreau, A model of sustainable development goals: challenges and opportunities in promoting human well-being and environmental sustainability. Ecol. Model. **475**, 110164 (2023)
100. D.C. Eddy, S. Krishnamurty, I.R. Grosse, J.C. Wileden, K.E. Lewis, A normative decision analysis method for the sustainability-based design of products. J. Eng. Des. **24**(5), 342–362 (2013)
101. A.M.T. Thomé, P.S. Ceryno, A. Scavarda, A. Remmen, Sustainable infrastructure: a review and a research agenda. J. Environ. Manag. **184**, 143–156 (2016)
102. S. Thacker, D. Adshead, M. Fay, S. Hallegatte, M. Harvey, H. Meller, N. O'Regan, J. Rozenberg, G. Watkins, J.W. Hall, Infrastructure for sustainable development. Nat. Sustain. **2**(4), 324–331 (2019)
103. J. Malmodin, N. Lövehagen, P. Bergmark, D. Lundén, ICT sector electricity consumption and greenhouse gas emissions-2020 outcome. Telecommun. Policy **48**(3), 102701 (2024)
104. S. Sangeetha, J. Logeshwaran, M. Faheem, R. Kannadasan, S. Sundararaju, L. Vijayaraja, Smart performance optimization of energy-aware scheduling model for resource sharing in 5G green communication systems. J. Eng. **2024**(2), e12358 (2024)
105. E. Oughton, G. Geraci, M. Polese, V. Shah, D. Bubley, S. Blue, Reviewing wireless broadband technologies in the peak smartphone era: 6G versus Wi-Fi 7 and 8. Telecommun. Policy **48**(6), 102766 (2024)
106. N.R. Deevela, T.C. Kandpal, B. Singh, A review of renewable energy based power supply options for telecom towers. Environ., Dev. Sustain. **26**(2), 2897–2964 (2024)

107. C. Lubritto, Telecommunication power system: energy saving, renewable sources and environmental monitoring. *IntechOpen* (2010)
108. J. Pearson, T. Wagner, J. Delorit, S. Schuldt, Meeting temporary facility energy demand with climate-optimized off-grid energy systems. IEEE Open Access J. Power Energy **7**, 203–211 (2020)
109. B.C. O'neill, M. Dalton, R. Fuchs, L. Jiang, S. Pachauri, K. Zigova, Global demographic trends and future carbon emissions. Proc. Natl. Acad. Sci. **107**(41), 17521–17526 (2010)
110. G. Radonjič, S. Tompa, Carbon footprint calculation in telecommunications companies-the importance and relevance of scope 3 greenhouse gases emissions. Renew. Sustain. Energy Rev. **98**, 361–375 (2018)
111. World Broadband Association. The importance of environmental sustainability in telecom service providers' strategy: World broadband association white paper, 2024. Accessed on 27 Mar 2025
112. D.M. West, Digital divide: improving internet access in the developing world through affordable services and diverse content. Cent. Technol. Innov. Brook. **1**, 30 (2015)
113. D. Bawden et al., Origins and concepts of digital literacy. Digit. Literacies: Concepts, Policies Pract. **30**(2008), 17–32 (2008)
114. I. Koppel, S. Langer, Basic digital literacy-requirements and elements. Revista Práxis Educacional **16**(42), 326–347 (2020)
115. I. Prieto-Egido, T. Sanchez-Chaparro, J. Urquijo-Reguera, Impacts of information and communication technologies on the SDGs: the case of Mayu telecomunicaciones in rural areas of Peru. Inf. Technol. Dev. **29**(1), 103–127 (2023)
116. F.-F. Rothe, L. Van Audenhove, J. Loisen, Digital development, inequalities & the sustainable development goals: what does'leave no-one behind' mean for ICT4D? Inf. Technol. Dev. **29**(1), 9–26 (2023)
117. R. Russell-Bennett, M. Rosenbaum, R. Fisk, M. Raciti, Sdg editorial: improving life on planet earth-a call to action for service research to achieve the sustainable development goals (SDGs). J. Serv. Mark. (2023)

Part II
Multilayered Technological Solutions

Part II would delve deeper into technological solutions that have the potential for bridging the digital divide. Wireless broadband access networks would consist of a range of technologies that would deliver high-speed internet connectivity over long distances where laying a wired infrastructure is not possible or not affordable. The key technologies that are explored in this part include ground networks, air networks, and space networks. These technologies support ultra-high-speed data transmission and a myriad of devices, making them ideal for urban and semi-urban deployments. With a slight modification to the architecture along with their priorities, these technologies can be deployed in rural and remote regions as well.

Part II looks at 5G, satellites, unmanned aerial vehicles (UAVs), network-tethered flying platforms (NTFPs), high-altitude platforms (HAPS) with base stations (HIBS), and Wi-Fi. Though there is a wide range of technological solutions available, there is a need to understand that an extensive analysis of the overall deployment of each technology would determine the feasibility. For an underserved community, there is limited infrastructure, necessitating cost-efficient solutions that are sustainable for the long run. The concepts discussed in Part I are applied to the technologies in Part II, which helps in getting a better picture of the technological suitability for the location of interest.

Finally, Part II also presents the emerging technologies that are termed together as future networks, and includes 6G, antenna technologies, massive MIMO, massive IoT, Li-Fi, THz technologies, reconfigurable intelligent surfaces, and integration with AI/ML. The integration of future emerging technologies with existing technologies would help in enhancing the capabilities of technologies, as well as focusing on the access, affordability, and usage gap. With a long-term vision, future networks play a vital role and are instrumental in enhancing connectivity and promoting digital equality.

3 Ground Networks: 5G

Currently, over 300 operators worldwide have deployed 5G, out of which 72 operators across 131 countries have launched commercial 5G standalone (SA) networks, as of 2025. Several cities have operational 5G deployments with SA and non-standalone (NSA) networks [1, 2]. Emerging players are driving the 5G industry by lowering the cost per gigabyte of service and advancing the technology. The spread of globalization and the ease of entry for new players have fueled widespread 5G deployments worldwide. Despite this momentum, according to studies, traditional approaches for 5G deployments are expensive and unaffordable in underserved regions, making 5G unsuitable and unsustainable for addressing the digital divide [3]. For example, 5G subscriptions in Africa are expected to reach only 180 million by 2029 [4]. However, studies have shown that by employing the strategies discussed in Chap. 2, 5G could become affordable, scalable, and serve the underserved communities [5], which together hold the potential to reshape broadband access for the unconnected. Therefore, there is a need to explore 5G as a solution for bridging the global digital divide by designing innovative 5G business models that address accessibility, availability, and affordability.

3.1 State-of-the-Art of 5G

5G is a wireless technology and promises to deliver ultrafast data in the range of Gbps per user, as low as 1 ms, extremely high spectrum efficiency (3x), hyper-densification (100x), very high connection density (up to 1 billion devices), highly reliable and superior network efficiency (99.9%, 100x), and energy efficiency (10x), compared to the 4G technologies [6]. When user or device density is high, 5G surpasses 4G, providing broadband speeds over 1 Gbps, while accommodating narrowband and wideband services.

Figure 3.1 shows the 5G architecture and the general use cases in different scenarios such as office buildings, seaports, airports, shopping malls, industries, schools, universities, AR/VR, hospitals, retail mobile users, and IoT, transportation and logistic applications [5]. 5G enables remote monitoring and maintenance of networks. 5G core supports cloud radio access networks (CRAN) and virtualization, service-based architecture, control and user plane separation, mobility, and session management, KPI-based SLA, and QoS for new services. 5G also allows loose coupling to support easier interoperability between different devices and standards. Cloud services allow run-time expansion and reduction in the resources and network capacity, which helps better resource utilization to support multiple and additional applications as well [7].

There are two main deployment modes for 5G: SA and NSA, as shown in Fig. 3.2. In NSA mode, 5G RAN operates using the 4G evolved packet core, which manages the control signals. This helps the operators to reuse their existing 4G (LTE and long term evolution - advanced (LTE-A)) infrastructure. However, NSA does not support all functionalities of 5G since the core is mainly designed for 4G functionalities. Whereas a 5G SA includes a 5G RAN operating using a cloud-based 5G core that supports all the intended functionalities offered by 5G and requires significant investment from the operators. Furthermore, 5G SA is more energy efficient than 5G NSA [8].

S. Koratagere Anantha Kumar, *Wireless Broadband Access to Bridge the Digital Divide*,
Textbooks in Telecommunication Engineering, https://doi.org/10.1007/978-3-032-09261-8_3

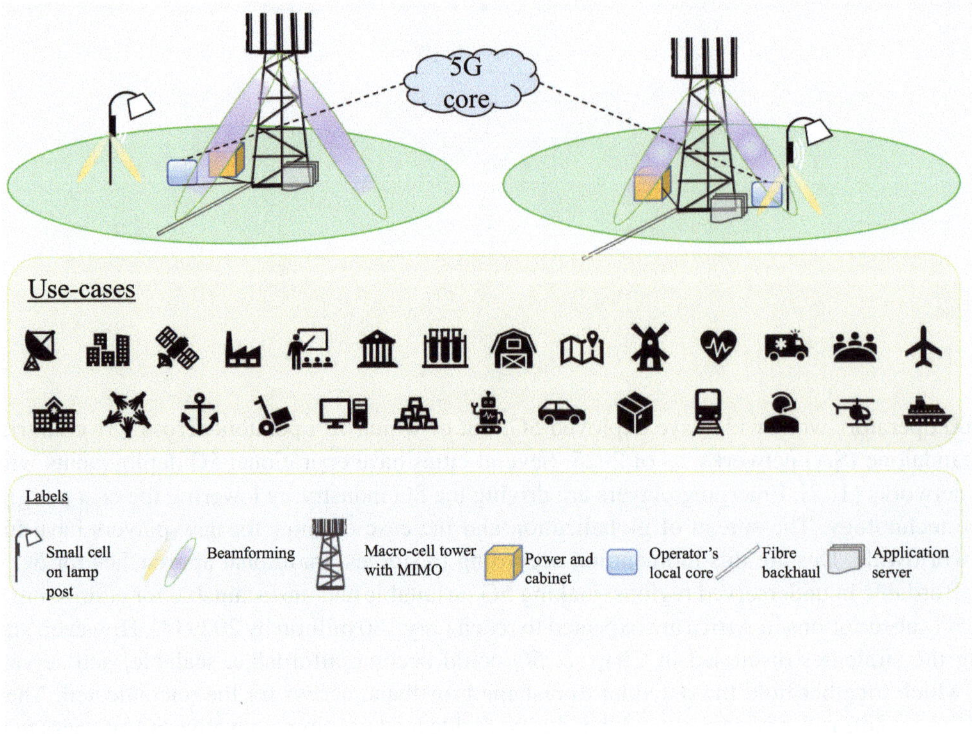

Fig. 3.1 5G architecture and applications

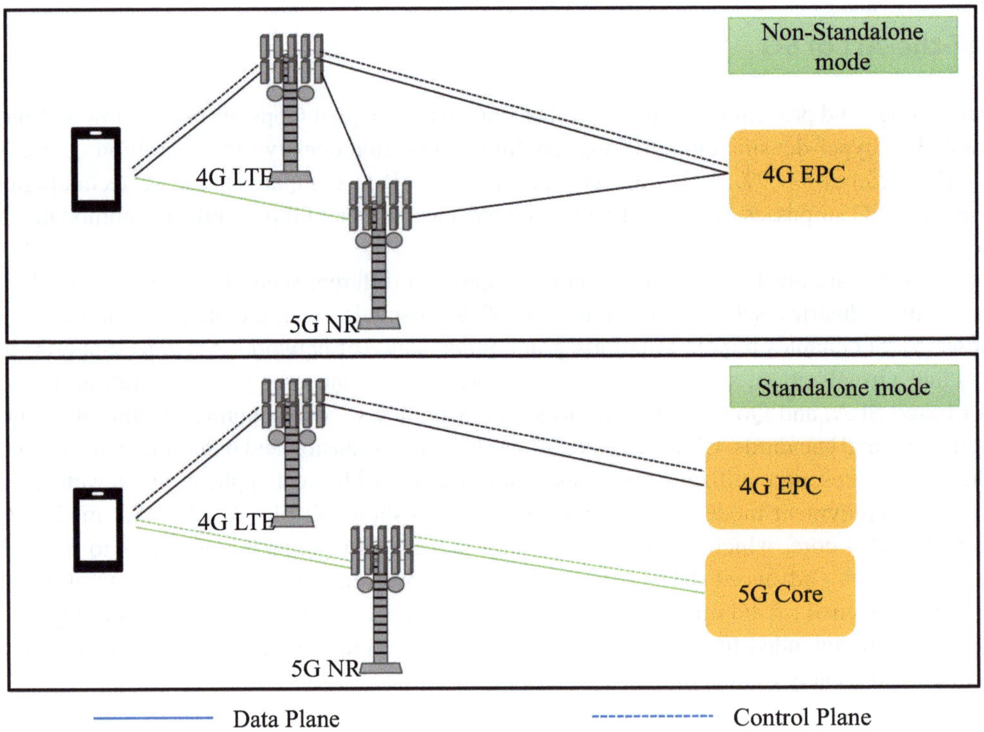

Fig. 3.2 NSA and SA modes of 5G

3.1.1 Spectrum Bands

Figure 3.3 shows the different spectrum bands 5G networks support [9].

- **Low-Band**: This band is in the range of 600–900 MHz, which offers excellent coverage range but data rates around 2-250 Mbps per user, comparable to 700 MHz band supporting 4G. In some countries, this band is also available as part of shared spectrum bands, though only as a pair of 5 MHz bands [10]. Low-band is a popular choice for rural areas and other regions where the digital divide is prominent.
- **Mid-Band**: This uses frequency bands in the range from 1.7 to 4.7 GHz that offer data rates of 100 to 900 Mbps and coverage up to a few kilometers in radius [11]. This band is a popular choice for urban and suburban deployments. Mid-bands are suitable for urban, industrial, suburban, broadcasting, and private network use cases.
- **High-Band Millimeter Wave**: The lower bands of mmWave frequencies from 24 to 47 GHz lie in this category. It offers Gbps speed data rates, which are comparable to wired broadband access but with limited coverage (200 m) [12]. However, mmWave is dependent on the weather and impeded or absorbed by the material it passes through, such as walls, buildings, and foliage. Since the coverage offered by 5G NR is limited, there is a need for dense deployment that increases the cost of the network. Therefore, this band is mostly suitable for ultra-dense environments where crowds of people gather, such as shopping malls, stadiums, concert arenas, live events, and airports. Globally, licensed and deployed 5G mmWave spectrum bands range from 24.5–29.5 GHz [13].

3.1.2 Use Cases and Applications

5G offers lower error rates and enhanced data security compared to earlier generations of cellular communication systems. The three main use cases supported by 5G as part of IMT 2020 are shown in Fig. 3.4:

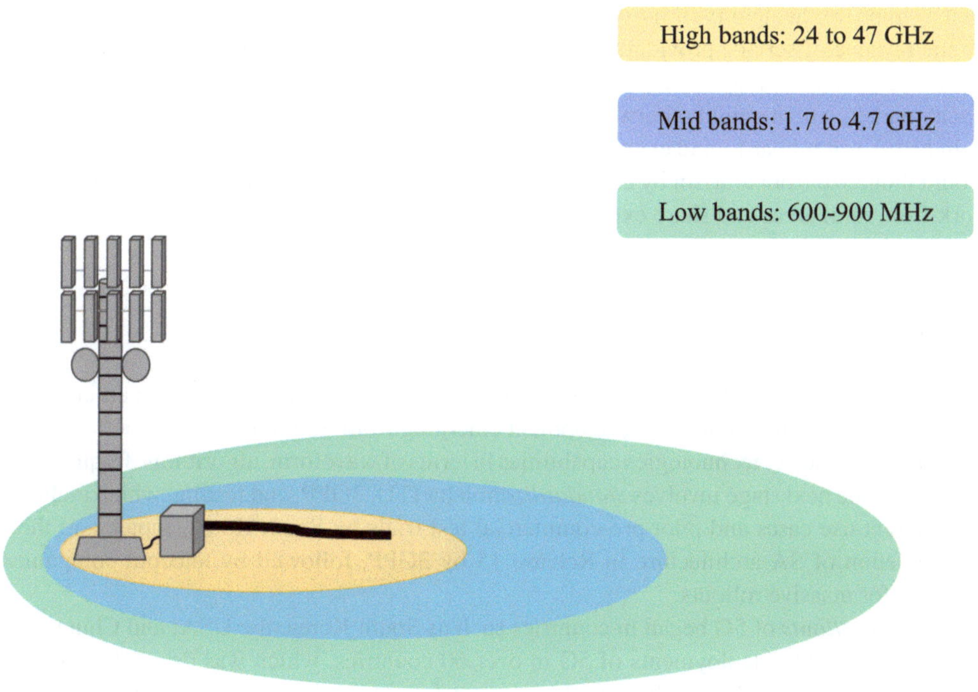

Fig. 3.3 5G spectrum bands coverage

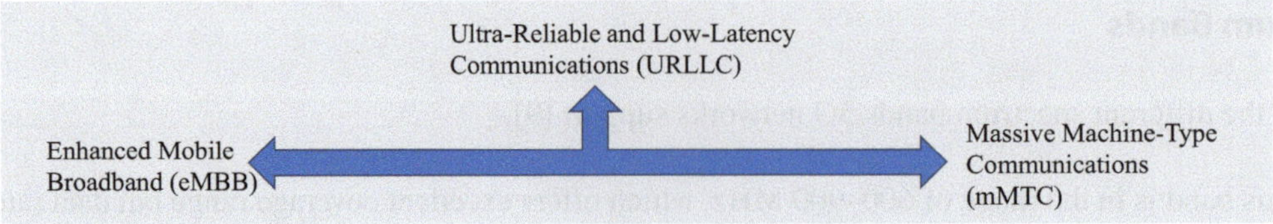

Fig. 3.4 IMT vision 2020 5G use cases

- **eMBB**: Enhanced mobile broadband (eMBB) enables ultrafast broadband on mobile devices for applications that require high bandwidth and low latency. It allows applications such as video streaming, augmented reality, cloud computing, enhanced gaming quality, live streaming of sports, and an ad-hoc network for event coverage.
- **URLLC**: Ultra reliable low latency communications (URLLC) supports extremely low latency requirement applications and low- to moderate-bandwidth applications such as critical communications, control signals, industry 4.0 applications, network slicing, robotic controls, autonomous vehicles, remote medical surgery, emergency management, and remote network maintenance.
- **mMTC**: Massive machine type communications (mMTC) supports large devices, up to 1 million devices per km^2, and low bandwidth requirement applications. For example, IoT applications require mMTC capabilities, such as smart lighting, health monitoring devices, smart homes, and smart grids.

There is a wide range of applications being developed around the world to monetize the technological advancements that are solely dependent on the 5G use case capabilities. Figure 3.1 shows that 5G plays a key role in improving the telecommunication infrastructure and monetizing the services, which in turn would improve the feasibility of the network deployment. 5G supports several applications, such as personal communication, smart city, smart agriculture and farming, smart village, industry 4.0, autonomous vehicles, healthcare, education, remote work, and tourism. The real-time data is collected and transmitted using the 5G network for analyzing the data and improving the QoS in terms of resource management.

Smart village applications help accelerate infrastructure development to meet the SDGs outlined by ITU [14]. This is a high priority needed to improve the quality of living in rural areas and to slow down the urban areas from overcrowding. 5G services help drive the revenue of service providers directly and other dependent businesses indirectly. Local businesses can build their online presence via good Internet access, which helps them compete globally. Similarly, large businesses would get access to the local market, which would otherwise not be possible without Internet connectivity. 5G applications of local industries, such as industry, water boards, and electrical grids, would also boost connectivity. Some areas are quite popular as tourist spots and could monetize from tourism by enhancing the experience of tourists by having an online presence. Tourists could remotely work from rural areas and share experiences online with excellent Internet connectivity.

3.1.3 5G Timeline

The time it takes for a technology to progress from the research and development (R&D) to commercialization is striking. The *early R&D stage*, which involves initial concept development emerging from limitations of 4G, outlining the visions through IMT-2020, and development of newer technologies capabilities in terms of waveform, algorithms, frequencies, massive MIMO (mMIMO), and mmWave. The next stage involves *standardization* by ITU, 3GPP, and testing, which includes demonstrations of 5G globally for different use cases and pilot pre-commercial test trials by South Korea, Japan, and the USA. This stage also included the completion of SA architecture in Release 15 by 3GPP, followed by telecom companies and equipment manufacturers preparing for massive rollouts.

Since 2019, *commercial rollouts* of 5G began in countries such as South Korea, the USA, and China, with the main focus being eMBB. This was followed by deployments of 5G in over 60 countries, which was driven by the onset of COVID-19 pandemic for different applications; additionally, the deployment for SA begins to support low latency and other key use cases of 5G, such as URLLC and auctions of mmWave spectrum bands.

From 2022, *expansion and advanced use cases* have been driving the deployment of 5G private networks for industry use cases, broadcasting, and other applications requiring higher spectrum efficiency and low latency. There was a growth of handsets supporting 5G and related applications, which would require higher network capabilities. Finally, as part of *future developments*, there is ongoing work toward Release 18 laying the groundwork for 6G, energy-efficient 5G networks, bridging the digital divide using 5G, integration with AI, ML, and satellites, and other advanced capabilities offered by 5G integration with other technologies.

3.1.4 Key Technologies Powering 5G

Figure 3.5 illustrates a wide range of technologies that support 5G, rely on it for optimal performance, and are essential for extending coverage to rural areas.

- **Cloud Computing**: According to National Institute of Standards and Technology (NIST), the working definition of cloud computing is "*Cloud computing is a model for enabling ubiquitous, convenient, on-demand network access to a shared pool of configurable computing resources (e.g., networks, servers, storage, applications, and services) that can be rapidly provisioned and released with minimal management effort or service provider interaction*" [15]. Cloud technologies can potentially expand services in rural areas by leveraging Internet connectivity. For instance, in rural China, where approximately half a million people reside, providing accessible healthcare services remains a significant challenge. Cloud computing can be crucial in addressing this issue by empowering doctors and improving healthcare facilities. In underserved regions with low digital literacy, cloud computing helps bridge the digital divide by enabling the storage and retrieval of health records in areas with limited or no Internet access [16].
  Cloud computing can likewise transform regions with unreliable connectivity by granting on-demand access to software, services, and information at a fraction of the usual cost. This approach not only makes technology more affordable but also helps increase digital literacy, empowering individuals to use technology more effectively. Furthermore, it can stimulate demand for Internet connectivity and technological adoption, driving broader digital inclusion [17].
- **NFV**: The working definition of this technology is "The network function virtualisation (NFV) concept provides a new way to abstract and virtualizes the network functions, enabling them to be created, operated, distributed, and controlled by software running on standard servers" [18]. The key benefits offered by NFV are greater efficiency in the network capacity while minimizing the physical resources, reduced vendor lock-in, and flexibility for faster roll-out in the market. However, the key challenge lies in managing such a complex system. NFV greatly improves flexibility and scalability and is essential to the implementation and administration of 5G networks. NFV enables operators to run virtualized services on ordinary servers by separating network operations from proprietary hardware, which lowers costs and improves operational efficiency.
- **SDN**: Typically, the transfer of data packets on the network involves the transmission of the user plane or forwarding plane messages, which consist of operations such as discovering the entire network topology, shortest path, routing, etc., with the

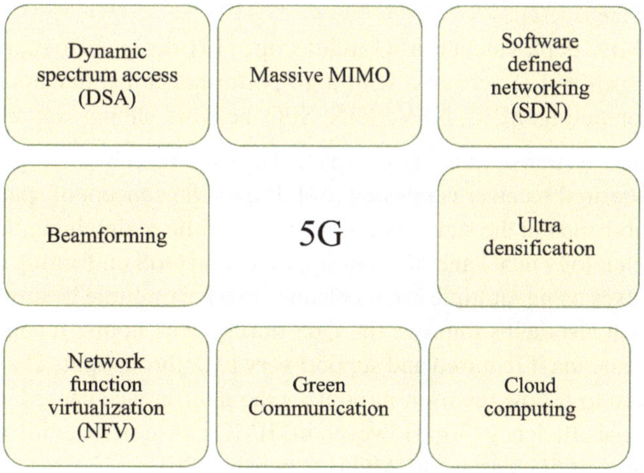

Fig. 3.5 Key technologies powering 5G

help of information exchange on the control plane [19]. This configuration of the network encouraged the development of software-defined networking (SDN) [20]. According to the Open Network Foundation, "SDN is an emerging architecture that is dynamic, manageable, cost-effective, and adaptable, making it ideal for the high-bandwidth, dynamic nature of today's applications. SDN architecture decouples the network control and forwarding functions, enabling the network control to become directly programmable and the underlying infrastructure to be abstracted for applications and network services" [21]. The key functionalities of SDN are managing the routing of network traffic, providing the components an interface to interact with the network, monitoring network performance and predictive maintenance, and automating the network management tasks [22].

- **SFC**: This helps to dynamically create network services and steer traffic among them, and unlike NFV, it involves end-to-end services across a chain of services in a particular order in the network [19]. The service function chaining (SFC) could be used to chain virtual network functions (VNFs) as well as physical network functions existing in the network service delivery architectures. SFC plays a crucial role in enabling flexible and efficient management of 5G network services, such as supporting dynamic service delivery and quick adaptation to changing user requirements and network conditions [23]. Furthermore, SFC optimizes the 5G network resource allocation by chaining with NFV and physical resources, minimizing latency, and improving network performance. SFC also provides support to other 5G capabilities such as network slicing, security, QoS, edge computing, and reliability [24, 25].
- **Network Slicing**: It is a technology that supports network virtualization and consists of independent logical networks, called slices, on the 5G network [26–28]. The key technology enablers for network slicing are SDN and NFV, which provide the lifecycle management of network slices by dynamically instantiating, modifying, and terminating the slices as per the end user requirements [19].
  Network slicing is integrated with MEC to combine the benefits offered by SDN, NFV, and SFC. This integration helps to address the static resource allocation issue by converting it to dynamic resource allocation while continuing to satisfy the network performance demands of the slice users [26]. Network slicing enables operators to configure network resources to meet the exact needs of each use case, effectively partitioning the network into customized slices that can evolve independently. This level of customization enhances service quality and allows operators to optimize resource utilization across multiple services.
- **MEC**: It is a key enabler for operators' end user applications that require ultra-low latency and is very critical for their operations. MEC has multiple architectures for different generations of mobile communication, such as 3G, 4G, and 5G. MEC is a universal access solution that supports reduced latency needs wherever they are required. MEC is the latest paradigm that uses cloud computing capabilities to process the data as close as possible to the network edge rather than the traditional way of data processing at the application server [29]. In the telecommunication industry, the MEC servers can be deployed within the RAN, and other network edge components such as BSs, access points, and network elements [30]. In areas that are sparsely populated, the cost consideration for MEC deployment should also be considered. A recent study showed that mobile networks are reaping benefits from technologies such as MEC, NFV, SDN, SFC, and network slicing [31]. The key to scalability in these solutions lies in effectively managing resource utilization and ensuring energy efficiency, particularly in remote and rural areas [32].
  Figure 3.6 shows the integrated MEC deployment in a 5G architecture [19]; designed to support edge computing, low-latency, and mission-critical use cases such as IoT services, with high-performance parameters and QoE [33]. This architecture highlights the usage of cloud computing, MEC, NFV, SDN, SFC, network slicing, and mMIMO.
- **mMIMO**: This technology focuses on transmitting the energy to the desired receiver using the concept of multiple reflections by the various objects until the desired receiver is reached [34]. It uses the concept of spatial multiplexing, which supports data to be transmitted on the sub-bands in the same frequency range while maintaining the integrity of the information.
  The mMIMO is a multi-user technology in 5G and 6G that supports good QoS uniformity for each user in regions with high mobility [35]. The concept involves using multiple array antennas to have multiple beamwidth, which can be electronically steered as the user density, which also helps improve the QoS during peak hours. It supports up to a 256×256 antenna array, which can act as a single antenna if required and support very high throughput. The mMIMO helps improve the QoS at the cell edge, transmits a beam to follow the user, minimizes the interference caused due to the signal attenuation, and increases the throughput and signal efficiency [36]. However, mMIMO consumes significant energy in performing complex computations, and the deployment of 5G massive MIMO is expensive [36].
  In [37], the research explores how Television (TV) station towers can be upgraded to mMIMO for bridging the digital divide. In [38], FWA fixed broadband technology at 3.5 GHz with massive MIMO is being explored to bridge the digital divide. MIMO systems operating at TVWS and ultra high frequency (UHF) are being explored for long-range solutions to

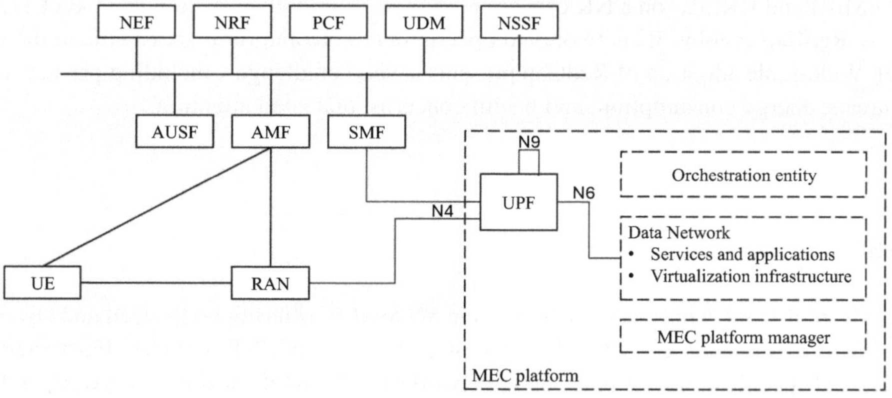

Fig. 3.6 5G architecture integrated with MEC and other enabling technologies

bridge the digital divide [39]. Several pilot studies have been conducted in various rural areas around the world, validating the solution's potential for addressing connectivity challenges in these regions [39].

- **5G RANGE**: Remote area Access Network for 5th GEneration (RANGE) is an extension of 5G that has been modified for rural and remote areas [39, 40]. 5G-RANGE will overcome the current limitations in terms of coverage and reliability by offering 50 km coverage and at least 100 Mbps data rate at the cell edge. This model uses cognitive radios on licensed and unlicensed bands along with UHF TVWS bands [41]. The key uses that would be supported by this model are voice and data, backhaul technology, smart agriculture, and e-healthcare. There are two deployment models for 5G-RANGE:

  - **Connectivity Model**: where 5G-RANGE would be used as the only RAN access technology.
  - **Backhaul Model**: where 5G-RANGE is used as a backhaul technology for 4G sites only.

  In this model, the MNO is expected to make decisions on settlement locations, technology selection, sales strategies for the end users, customer management, and its own licensed spectrum bands. While the radio mobile infrastructure operator (RMIO) would operate the network on the MNO's licensed spectrum bands. The role of RMIO would involve site selection, technology selection, network deployment (RAN, infrastructure, backhaul, and transport), operations and maintenance, and interference mitigation. The 5G-RANGE initiative is being explored as part of 5G and beyond standards to provide connectivity in underserved regions with poor viability. It encompasses various aspects, ranging from the design of physical, medium access control (MAC), and network layers to field deployments [42]. This solution was also explored as part of the Horizon 2020 project, and studies have shown that this technology has the potential to use access technology that can provide 100 Mbps in a cell radius of 50 km and also consists of small UAVs [43]. This solution used NFV to provide an overlay technology that could be integrated with wireless access points (APs) to offer connectivity to the end users located close to it.

- **FWA**: It has been an upcoming technology to support the need for high-speed, reliable Internet connectivity in areas where it is difficult to lay wired infrastructure to provide broadband services. This technology has been employed by several operators around the world [44]. FWA involves a single BS connected to a fixed network, has multiple PTMP end user units [45]. The location of these units is selected to provide maximum coverage and minimize blind spots.
  By the end of 2023, there were be more than 130 million FWA connections, and they are expected to reach 330 million by 2029 [46]. FWA operates on different technologies and different spectrum bands as well; for example, IEEE 802.11ay supports 60 GHz mmWave bands and is cost-efficient over conventional FWA and could potentially help in solving rural connectivity [47]. FWA has been studied as a solution to address the digital divide, and it has been shown that FWA can serve as an effective backhaul solution in both urban and rural areas [48].

- **RedCap**: 5G can be optimized to support a new device type with reduced capability (RedCap) NR aiming at lower cost and complexity, smaller physical size, and longer battery life compared to regular NR devices [49]. RedCap offers better coexistence with NR eMBB and URLLC on a NR carrier and provide a smooth transition from 4G LTE based solutions to 5G NR based solutions. RedCap enables 5G networks to operate in remote and rural areas without the need for additional spectrum license [50]. Wide-scale adoption of RedCap presents several challenges, including privacy and data protection, accessibility, cyber threats, energy consumption, and health concerns, that need attention.

3.2 Worldwide 5G Deployments

According to a recent study, mid-band deployment is driving the 5G market, offering both speed and coverage to consumers. The widely launched 5G services include eMBB, FWA, gaming, and some AR/VR services. Furthermore, with 3GPP the establishment of the technical specifications for non-public networks (NPN) to allow private network architecture, there has been a surge in the implementation of 5G private networks for industrial vertical applications in recent times [51, 52].

3.2.1 5G Deployment Across the Globe

The first country to deploy a large-scale 5G network was South Korea in April 2019. This was followed by 5G deployments in several countries around the world. Data released on penetration rate of 5G subscriptions worldwide by the end of 2023 [53] indicates that North America has the highest, at 61%, North East Asia at 41%, the Gulf Cooperation Council (GCC) at 34%, Western Europe at 25%, India at 11%, and other regions follow, with a combined 18% of all mobile subscriptions, or 1.6 billion people, being served. The study forecasts an increase in 5G subscriptions to reach 58% of overall subscriptions by the end of 2028. Several African countries commercially deployed 5G networks in the first half of 2023. 5G deployment in Africa is backed by strong government support for spectrum policies and telecommunication regulations.

Despite the setback due to the COVID-19 pandemic, Asian countries have successfully deployed large-scale commercial 5G networks, with South Korea, Japan, and China leading the race. The mid-band 5G deployments are very popular in these regions as they offer both coverage and capacity to the end user, and the government is formulating supporting policies to encourage wide-scale deployments in Asian countries. Southeast Asian countries are deploying 5G FWA to bridge the digital divide. Asian countries such as China, Japan, India, South Korea, the Philippines, Malaysia, Singapore, Saudi Arabia, Kuwait, and the United Arab Emirates have an average 5G download speed of over 200 Mbps per user. The influence of 5G and associated technologies, including network slicing, edge computing, AI, and cloud capabilities, on businesses drives revenue growth.

By 2020, 5G was deployed in at least 1 city in each European country. The target for 2025 is to provide 5G services in all urban areas in Europe, followed by 100% coverage by the end of 2030 to connect each house with a gigabit connection [4]. Though each country has its 5G network, the spectrum allocation process has been slower, which has caused a delay in the network uptake across Europe. However, European 5G networks have explored the potential revenue opportunities and applications for industrial verticals, especially connected automotive. Around 11% of the European population was using 5G services in 2022, and it is expected to reach 89% by the end of 2029.

5G was first deployed in the USA and Canada in 2018 and 2020, respectively. 5G services are widely used by end users in the USA compared to Canada. The dominant operational spectrum frequency for 5G in the USA is mid-band, offering average data rates over 300 Mbps per user, whereas Canada prefers to use frequencies below 3 GHz to provide 5G services with average data rates around 140 Mbps. The key reason for this difference is the availability of the bandwidth in the 5G mid-band frequency range. By 2027, the USA's 5G subscriber penetration is projected to reach 90%, thereby impacting the country's GDP significantly. By 2035, the Canadian government anticipates that 5G and related services will boost Western Canada's GDP by around $34 billion. In 2023, 31% of Canada's population had 5G coverage, compared to 62% in the USA.

The countries in Oceania are also actively participating in the race to adopt 5G on a wide scale [54]. Australia has 5G smartphone penetration of 80% with an average 5G download speed exceeding 200 Mbps per user and with 5G availability rate of around 80% in 2025. Similarly, New Zealand operators are planning to bridge the digital divide using FWA and provide coverage to 90% of their population.

5G was first launched in South America in 2019, but only in Uruguay and Puerto Rico. 5G was launched commercially in other parts of South America in 2021. Currently, around 15% of mobile subscribers are using 5G services in South America. The forecasts show that by 2030, 60% of subscribers will be using 5G. Several studies have pointed out the high cost of providing 5G services in South America, mainly because of challenging terrain and uninhabitable regions. Brazil has higher average data rates compared to other South American countries and uses 5G SA deployment in the mid-band frequencies.

Due to the economic slowdown, the 5G deployment in Africa was delayed. It is expected that until 2028, most users will be using 4G. The first deployment of 5G in mid-bands was in 2020, but it couldn't offer commercial services due to a lack of spectrum in the 3.5 GHz band. However, the second phase of 5G deployments in Africa began in 2022, with Botswana and Zimbabwe in February, Réunion in June, Nigeria and Tanzania in September, and Zambia in November. The 5G subscriber base in Africa is expected to reach 180 million in the year 2029 [4].

3.2.2 5G Rural Deployments

A combination of these technologies with 5G is deployed in several pilot field trials to address the digital divide. With the help of 5G systems' eMBB services, the study in [55] explores network slicing for broadband-related applications to improve rural broadband services. The study demonstrates the discussions about addressing the challenges of the digital divide and focusing on multi-tenancy infrastructure sharing to provide sustainable rural access. According to [56], one slice should be used to provide basic, free Internet access to those in the Democratic Republic of the Congo (DRC) who cannot afford it, with the remaining slices being used for commercial activities. The approach improves digital literacy and enlightens people on the benefits of having Internet connectivity. The concept of using 5G NHN for industry verticals is being tested in a pilot project in the Orkney Islands, United Kingdom. In the trial, one of the network slice was used for BBC's broadcasting applications and the other slices for various rural industrial applications [57]. Although 5G network slicing is an innovative approach to advancing rural connectivity, the trial's findings reveal its commercial viability is hindered by legal requirements.

Several 5G pilot trials and initiatives have been conducted across the globe to understand the suitability of technologies driven by 5G to bridge the digital divide. These studies have provided insights into solutions that are feasible in low-density, rural environments. Some of the key collaborations between different stakeholders include:

- **India**: Reliance Jio pilot study to test 5G capabilities in areas where deploying traditional fiber backhaul is expensive [58]. They are keen on exploring the FWA market solutions.
- **USA**: T-Mobile US and Ericsson were exploring the use of 5G NSA and SA for rural vertical use cases [59].
- **China**: China Mobile is conducting several 5G pilot trials in rural and remote areas to enhance broadband access, and reduce the digital divide [60].
- **UK**: The UK government has funded several small and SMEs, universities, and local authorities with a series of 5G testbed deployments and trials in the rural part of the UK to provide new opportunities in agriculture and rural business [61].
- **Europe**: There are several projects funded by the European Commission to explore 5G to bridge the digital divide, study the impact of private 5G networks in rural areas, and 5G for rural industry 4.0 applications such as IoT, agriculture, and healthcare [62].

3.3 5G: Accessibility, Affordability, and Climate

Given the broad scope of the network, economic, and environmental frameworks for 5G in rural areas, we will focus on providing a high-level overview of these important topics. While numerous detailed models and approaches contribute to these frameworks, it's not possible to cover all of them here. However, we'll highlight the key concepts and offer a glimpse into the exciting possibilities these models present for the future of rural connectivity.

Table 3.1 SWOT analysis of 5G rural networks

Strengths	Weaknesses
Higher data rates Low latency Reliable and resilient networks Advanced technological feature Energy efficient Improve spectrum efficiency and resource utilization	Backward compatibility with older technologies Need for a higher number of antennas and gNodeBs Increased energy consumption due to an increase in the number of gNodeBs Newer infrastructure
Opportunities	**Threats**
Improved QoS Job and revenue opportunities Newer business models Tourism and vertical applications such as farming and fisheries Enhance the quality of life for residents	Significant cost toward infrastructure investment Capacity-limited backhaul Vendor incompatibility Spectrum access challenges Costs for user equipment

3.3.1 Economic Frameworks

5G deployment would open new business models and boost the economic growth in the region of deployment. Some of the key socioeconomic benefits that 5G brings to rural areas are

- **Economic Growth and Job Creation**: 5G would provide faster connectivity and enhance the productivity of users by enabling faster and more reliable connectivity, which should be combined with an increase in digital literacy among the users to reap the benefits in terms of increased revenue and productivity, job creation, and innovation across different sectors. It would promote travel, regional business expansion, and global market expansion.
- **Investment and Infrastructure**: The deployment of 5G and providing higher QoS at low latency involves the deployment of the access, midhaul, backhaul, and transport networks in rural areas, which requires significant investment in terms of cost.
- **Quality of Life in Rural Areas**: By facilitating applications that improve quality of life, digital connectivity will benefit the population across areas from healthcare and personal communication to farming and agribusiness.

3.3.1.1 Business Models

The business models and the value for services related to 5G in rural scenarios are dependent on the role played by the stakeholders [63, 64]. The successful deployment and adoption of 5G involve the diverse stakeholders collaborating in rural deployments to realize the full potential of 5G and guarantee its sustainable impact on the global economy and society. Why is it important? Mainly because there is a significant challenge in terms of economic feasibility in these deployments and studies have shown that 5G is not economical feasible for addressing the digital divide [3, 65]. Hence, to improve feasibility and functionality of 5G deployment in underserved regions, researchers are exploring advanced capabilities in remote monitoring and maintenance of the network, neutral host network (NHN), spectrum sharing, and virtualization to optimize resource allocation for the end users [5, 43]. Newer business models also encourage community engagement to lower the deployment and increase the solution feasibility [66].

3.3.1.2 SWOT Analysis

Table 3.1 shows the SWOT analysis of generic 5G technology for rural areas, helping to understand its potential impact on various sectors, including business, government, and society.

3.3.1.3 Value Chains

The technical tasks and the roles of stakeholders in rural 5G deployments are highlighted by examining VNC. Figure 3.7 illustrates the variety of actions that stakeholders perform in the rural 5G NHN deployment [5].

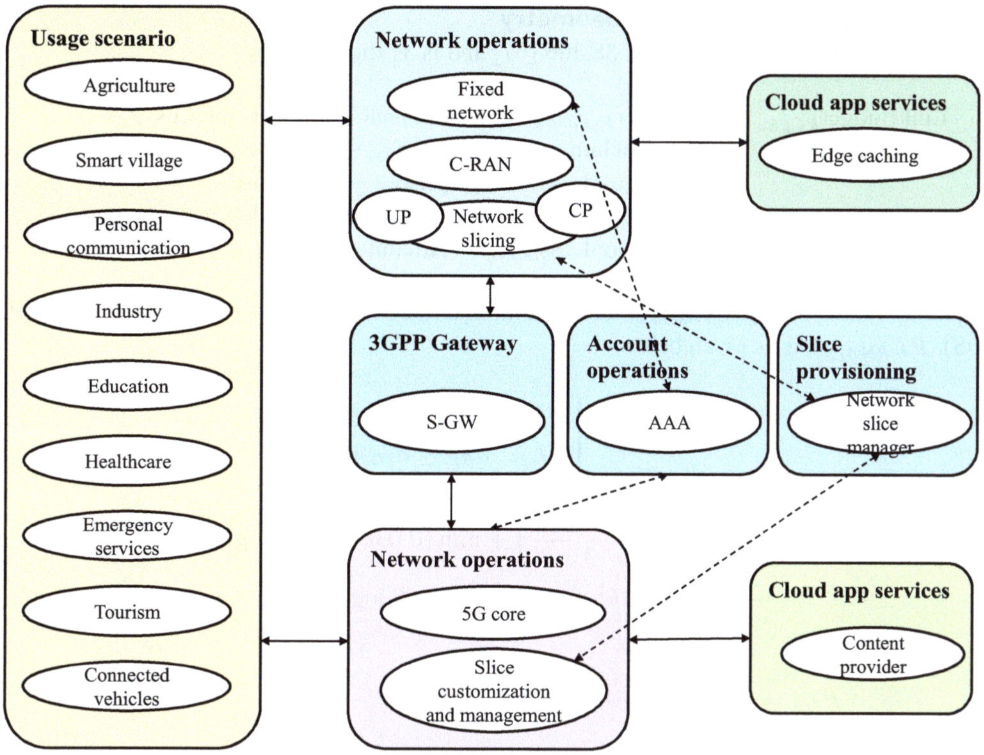

Fig. 3.7 5G VNC

Depending on which technical task the stakeholders perform, the 5G network would offer different values. The specific responsibilities assigned to each of these entities will influence how 5G is deployed and how it will help rural communities. These responsibilities might include everything from the technical setup and management of 5G infrastructure to implementing regulations and policies, encouraging regional adoption, and guaranteeing fair network access. Different sets of values and benefits might be offered by the 5G network in rural regions, depending on which technical task the stakeholders focus on. The impacts that the 5G network would have on rural areas depend on the collaborative efforts among the stakeholders and a tailored approach for each region.

The technical specification obtained from these analyses describes each of the components required for deployment, including spectrum allocation, hardware, software, network infrastructure, and support services. This guarantees that every facet of the network is covered and that the deployment can proceed without hiccups. By using a comprehensive approach, stakeholders can make sure that the 5G network is both technically sound and strategically in line with the objectives of improving rural communities' connection, economic growth, and social inclusion.

3.3.1.4 Channel Modeling Using Stochastic Geometry

5G NR link budget is given in the 3GPP document 38.306 [67] and is as shown below

$$\begin{aligned} \text{LinkBudget}_{Rx,dBm} = {} & \text{TxPower}_{TxBandwidth} + \text{AntennaGain}_{Tx} - \text{CableLoss}_{Tx} + \\ & \text{AntennaGain}_{Rx} - \text{CableLoss}_{Rx} - \text{PathLoss}_{propagationModel} - \\ & \text{PenetrationLoss} - \text{FoliageLoss} - \text{BodyLoss} - \\ & \text{InterferenceMargin} - \text{RainIceMargin} - \text{SlowFadeMarging} - \\ & \text{PenetrationLoss}_{indoor} - \text{AttenutationLoss}_{indoor} \end{aligned} \tag{3.1}$$

The 5G pathloss equation ($\text{PathLoss}_{\text{propagationModel}}$) for the rural macro cell scenario as defined in 3GPP 38.901 standards for a line of sight (LOS), $PL_{RMA_{LOS}}$, is as given below

$$PL_{RMA_{LOS}} = \begin{cases} PL_1, & 10m \le d_{2D} \le d_{BP}. \\ PL_2, & d_{BP} \le d_{2D} \le 10km. \end{cases} \tag{3.2}$$

$$\begin{aligned} PL_1 = {} & 20\log_{10}\left(\frac{40\pi d_{3D} f_c}{3}\right) + \min\left(0.03h^{1.72}, 19\right)\log_{10}(d_{3D}) \\ & - \min\left(0.044h^{1.72}, 14.77\right) + 0.002\log_{10}(h)d_{3D} \end{aligned}$$

$$PL_2 = PL_1(d_{BP}) + 40\log_{10}\left(\frac{d_{3D}}{d_{BP}}\right) \tag{3.3}$$

$$d_{BP} = \frac{2\pi h_{BS} h_{UT} f_c}{c}$$

$$d_{3D} = \sqrt{(h_{BS} - h_{UT})^2 + d_{2D}^2}$$

where c is the speed of light, d_{2D} is the ground distance between BS and UE, h_{BS} and h_{UT} are the height of the BS and UE, respectively, and f_c is the center frequency in Hz. For PL_1 has a shadow fading, $\sigma_{SF} = 4$, $h_{BS} = 35$m, $h_{UT} = 1.5$m, while for PL_1 has a shadow fading, $\sigma_{SF} = 6$. These formulas are valid for $10\text{m} \le h_{BS} \le 150\text{m}$ and $1\text{m} \le h_{UT} \le 10\text{m}$. The SINR to modulation code conversion is as given in Table 3.2. For the detailed mathematical description of non-line of sight (NLOS) mode, refer to [67].

Long-term capacity forecasting is a crucial component that needs to be properly taken into account while designing and establishing a 5G network, particularly in rural regions. Anticipating the network's capacity needs over the next five to ten years is crucial given the speed at which technology is developing and the anticipated increase in demand for high-speed Internet. By forecasting the network demand, one can make sure that the infrastructure is scalable, future-proof, and able to meet both present and expanding demands at a reasonable cost.

Table 3.2 Mapping SINR ranges to modulation codes

Modulation code	SINR range
QPSK	4–14
4 QAM	14–20
16 QAM	20–26
64 QAM	26–32
256 QAM	32–34

3.3.1.5 Techno-economic Assessment

Once the technical specification of the 5G network is thoroughly assessed, the next crucial step is to explore the economic assessment for the proposed network architecture. Understanding the economic feasibility of deploying a 5G network is essential to ensure that the solution is not only technically sound but also economically viable and sustainable. TEA evaluates both the technical requirements and the costs associated with deploying and maintaining the network, balancing the need for advanced technology with the financial realities of rural deployments.

This section presents the mathematical modeling TEA for a generic 5G network, which could be extended to the rural scenario as well [5]. First, let's look at the investment and cost analysis:

- **CAPEX**: This accounts for the initial investment required for 5G infrastructure deployment, which is a function of several factors, such as the number of BSs, antennas, backhaul, debt, core, and network components. The CAPEX C_c, includes installation of a new BS with a tower (C_{bs}), multi-carrier **BS** (C_{mbs}), fiber (C_f), backhaul (C_{bh}), 5G core (C_{co}), and other equipment such as batteries (C_{bt}) and backup generator support for power supply (C_{gt}), and technology licensing (C_{tech}), and is given as

$$\begin{aligned} C_c = C_{bs} + C_{mbs} + C_f + C_{bh} + C_{co} + C_{bt} + \\ C_{gt} + C_{tech} \end{aligned} \tag{3.4}$$

- **OPEX**: This includes the ongoing costs toward the operations and maintenance of the 5G networks, such as energy consumption, spectrum, rentals, labor, predictive maintenance, routine checks, and marketing. The OPEX $C_{o,i}$ per year i, which includes day-to-day maintenance such as BS OPEX (C_{bso}), site lease (C_l), network maintenance (C_m), backhaul (C_{bho}), spectrum (C_s), and rentals (C_{rent}), is given:

$$C_{o,i} = C_m + C_{bho} + C_{rent} + C_{bso} + C_l + C_s \tag{3.5}$$

The TCO accounts for the entire cost spent toward the 5G deployment during the life of the network in T years, (C_T), are given as follows:

$$C_T = C_c + \sum_{i=1}^{T} C_{o,i} \tag{3.6}$$

Next, we need to explore the ***revenue modeling*** for 5G services that would be provided after the deployment, which would help us understand the demand market for the solution. The mathematical modeling for the revenue generated for greenfield deployment based on a subscription model, R, from different 5G services depends on the user subscription, data usage, service pricing, market penetration, and service adoption, and is expressed as

$$R = \sum_{j=1}^{T} 12 Z_n^j S_n^j T \tag{3.7}$$

where S_n^j is the number of subscribers, and Z_n^j is the ARPU per month for n^{th} slice tenant for j^{th} year. Similarly, for a brownfield deployment, the total revenue, R, over the period T for ARPU Z_{5G} for 5G services is calculated as

$$R^n = \sum_{i=1}^{T} 12\rho_{up_{sub},i}(Z_{5G,i} - Z_{old,i}) + 12\rho_{new5G,i}(Z_{5G,i}) \tag{3.8}$$

where i is the year of study, $Z_{old,i}$ shows the ARPU for existing infrastructure, $Z_{new5G,i}$ are the additional new subscribers joining the network who require 5G KPIs for their applications, and $\rho_{up_{sub},i}$ be the existing subscribers who would upgrade their services [68].

Once the necessary data regarding the capacity, demand, and cost structure of the network is obtained, the next step is to assess the ROI, denoted as μ. It allows stakeholders to evaluate whether the benefits derived from the 5G network outweigh the initial and ongoing investment costs.

$$\mu = \sum_{i}^{T} \frac{R - C_{o,i}}{C_c} \tag{3.9}$$

These models could certainly be enhanced and explored in greater depth by mathematically modeling various strategies and analyzing their impact on network performance. However, such detailed mathematical modeling and strategy exploration are beyond the scope of this discussion and will be reserved for future research or more in-depth studies. The aim here is to provide a high-level overview of the key considerations without delving into the complexities of advanced modeling techniques, which can significantly extend the analysis.

3.3.2 Environmental Frameworks

The environmental framework focuses on the environmental impact of 5G networks. The framework promotes green technology, which is necessary in underserved regions where the electricity supply is unreliable, and guarantees that 5G networks meet sustainable targets in terms of resource usage, energy consumption, and carbon footprints. Some of the key components of the environmental framework for 5G include

- **Energy Efficiency and Sustainability**: One of the goals of 5G networks is to reduce energy consumption and improve energy efficiency compared to the previous generations. Technologies such as massive MIMO, beamforming, and NFV have to become more energy efficient while supporting a higher number of users. Some of the alternatives to reduce energy consumption and carbon emissions for 5G networks include switching to green RAN, low-power base stations (BSs), and energy-efficient components and integration with renewable energy sources.
- **Carbon Footprint and Emissions**: The entire lifecycle of 5G technologies, ranging from manufacturing to operations and finally decommissioning, has an environmental impact and could be estimated using the carbon footprint studies. There is a need to reduce the carbon footprint by ensuring the network is functioning optimally, maximizing resource utilization, efficient traffic management, dynamic load balancing, and minimizing unnecessary energy consumption.
- **E-waste and Recycling**: The rollout of newer technologies generally leads to the retirement of older technologies, contributing to e-waste. Therefore, strategies need to be developed to minimize e-waste. Furthermore, 5G networks should be designed to support the circular economy, which involves reuse, refurbishment, and recycling, as well as upgrades to advanced technologies with software upgrades.

Quantifying these observations on a uniform scale is the next step to facilitate accurate data understanding. In the following topics, let's explore the mathematical modeling for some of these concepts.

3.3.2.1 Energy Consumption

The power consumption of a 5G network can be modeled as a power-consuming coefficient for each type of equipment [69, 70]. The total power consumption of 5G networks, P_{5G} is given as

$$P_{5G} = P_{BS} \cdot N_{BS} + P_{SC} \cdot N_{SC} + P_C \cdot N_C \tag{3.10}$$

where P_{BS}, P_{SC}, and P_C represent the power consumption per BS, small cells, and core network elements, respectively. Similarly, N_{BS}, N_{SC}, and N_C represent the number of BSs, small cells, and core network elements, respectively.

BSs are the primary source of energy consumption in 5G networks, and power consumption depends on the traffic handled and can be expressed as

$$P_{BS} = P_{Tx} + P_{Rx} + R_{Idle} + P_{cooling} \tag{3.11}$$

where P_{Tx} is for power consumption for transmission, P_{Rx} is for power consumption for reception, P_{Idle} is for power consumption when the BS is idle, and $P_{cooling}$ is for power consumption toward cooling systems. This concept can be extended to the small cell as well.

This can be further broken down to power consumption from the active antenna unit (AAU) and the baseband unit (BBU), which constitutes dynamic and static power consumption. Typically, the AAU power consumption varies with traffic, while the BBU power consumption remains constant. This behavior can be modeled as [71]:

$$P_i^t = a_i + b_i.T_i^t \tag{3.12}$$

where P_i^t is the power consumption of BS_i at time t, a_i and b_i are the static and maximum dynamic power of the BS_i, and T_i^t is the communication traffic of BS_i at time t.

$$T_i^t = \frac{L_i^t}{U_i^{max}} \tag{3.13}$$

where L_i is the number of UEs connected with the BS_i at time t and U_i^{max} is the maximum number of users served by the BS_i at time t.

The core network handles data routing, switching, authentication, and management functions, and its power consumption model is given as

$$P_{core} = P_{processing} + P_{storage} + P_{networking} \tag{3.14}$$

where $P_{processing}$ is the power consumption toward data processing, $P_{storage}$ is the power consumption for data storage and authentication, and $P_{networking}$ is the power consumption for networking equipment.

The energy consumption E can be modeled as

$$E = \int_{t_0}^{t_1} P_{5G}(t)\, dt \tag{3.15}$$

where the time interval for study lies between t_0 and t_1, and $P_{5G}(t)$ is the power consumption of the 5G network at time t.

3.3.2.2 Carbon Emissions

The next critical factor is the carbon emissions of the network, which can be associated with the power consumption of the network. The total carbon footprint, CE, for a 5G network can be expressed as [72]:

$$CE = CE_p + CE_t + CE_c + CE_o \tag{3.16}$$

where CE_p is the carbon emissions of the production phase, CE_t is the carbon emissions during the transportation phase, CE_c is the carbon emissions during the construction phase, and CE_o is the carbon emissions during the operations phase. The detailed mathematical modeling for each of these parameters is present [72], and is briefly described below

$$\begin{aligned} CE_p &= \textstyle\sum_{i=1}^{n} q_i \times c_i \\ CE_t &= \textstyle\sum_{i=1}^{n} q_i \times d_i \times f_i \\ CE_c &= \textstyle\sum_{i=1}^{n} \sum_{j=1}^{n} (q_i \times c_i + m_j \times g_j) \\ CE_o &= \textstyle\sum_{j=1} m_j \times g_j \times \alpha \end{aligned} \tag{3.17}$$

where i refers to the type of materials, q_i is the quantity of i, c_i is the carbon emission factor of i, d_i is the distance from the place of production of i to the BS site, f_i refers to the carbon emissions factor of transportation vehicles of i, j refers to the type of equipment, m_j is the energy consumption of equipment of j, g_j is the energy emission factor of j, and α is the equipment utilization.

3.4 Summary

This chapter focused on comprehending 5G as a technology. It presents a monumental leap in network performance of wireless technology, with the chapter focusing on global deployments, advanced architecture, key enabling technologies, and industry-specific applications, all with an emphasis on bridging the digital divide. Its role in bridging the digital divide hinges on how effectively it employs the strategies to lower the TCO, while leveraging the economic and environmental modeling to build a sustainable and scalable 5G network that is community-focused and inclusive. By addressing not only coverage and affordability but also meaningful usage of the network in underserved regions, 5G holds the potential to address the digital divide. However, without deliberate efforts, it could become a technology that would widen the disparity further.

Exercises

3.1 A 5G network operates on three spectrum bands: low-band, mid-band, and high-band, with respective bandwidths of 20 MHz, 100 MHz, and 500 MHz. Assume the spectral efficiency for these bands is 1.5 bps/Hz, 4 bps/Hz, and 8 bps/Hz, respectively. Calculate:

(a) The total data rate achievable by the 5G network.

(b) The percentage contribution of each spectrum band to the total data rate.

Use the formula: Data Rate = Bandwidth × Spectral Efficiency

3.2 A telecom company is deploying 5G in a rural area with the following cost components:

- Tower installation: $50,000 per tower.
- Fiber backhaul setup: $20,000 per km.
- Operational cost per year: $10,000 per tower.

For a network covering 5 towers connected by 10 km of backhaul, calculate:

(a) The initial deployment cost.
(b) The total cost over a 5-year period, including OPEX.

3.3 A 5G base station consumes 1.5 kW of power and serves an average data rate of 10 Gbps. Calculate:

(a) The energy efficiency of the base station in bits per joule.
Use the formula: Energy efficiency = $\frac{\text{Data Rate (bits/sec)}}{\text{Power Consumption (watts)}}$.
(b) How the energy efficiency changes if the base station's power consumption increases to 2 kW while maintaining the same data rate.

3.4 Evaluate the suitability of FWA for providing connectivity in rural areas. Discuss:

(a) How spectrum band selection impacts FWA performance.
(b) The advantages and limitations of using FWA compared to traditional fiber deployment.

Provide real-world data or hypothetical scenarios to justify your analysis.

3.5 Write pseudo-code for modeling the cost estimation of 5G networks that optimize resource utilization in rural areas by maximizing the network coverage. Consider the network operates only using 700 MHz and the region of interest is surrounded by farms and hills with an area of 10 km^2.

3.6 A telecom operator is evaluating two options for deploying a rural network:

- Option A: Installing fiber optics, with a fixed cost of $200,000 and a maintenance cost of $20,000 per year.
- Option B: Deploying a wireless solution, with an initial cost of $120,000, annual maintenance costs of $35,000, and a replacement cost of $60,000 after 5 years.

Calculate:

(a) The total cost of each option over a 10-year period.
(b) The cost efficiency of each option, defined as Cost Efficiency = $\frac{\text{Total Cost}}{\text{Coverage Area (km}^2\text{)}}$
Assume that Option A covers 100 km^2, while Option B covers 60 km^2.

3.7 A company invests $1,000,000 in a 5G deployment project. Over 5 years, the expected revenue is $250,000 per year, with operational expenses of $50,000 per year. Calculate:

(a) The net profit over 5 years.
(b) The ROI using the formula:$\text{ROI} = \frac{\text{Net Profit}}{\text{Investment Cost}} \times 100$

3.8 A mobile operator plans to scale its network to support 10,000 additional users. The cost model is: Fixed infrastructure cost: $500,000, variable cost per user: $50. Calculate:

(a) The total cost for supporting the additional users.
(b) The average cost per user.

3.9 A base station consumes 3 kW of power and supports a data rate of 6 Gbps. Calculate:

(a) The energy efficiency in bits per joule.
(b) How the energy efficiency changes if the power consumption is reduced to 2.5 kW while maintaining the same data rate.

3.10 A telecom network emits 10 metric tons of CO_2 annually for every 100 kW of power consumed. If the total power consumption of the network is 5 MW:

(a) Calculate the total annual carbon emissions of the network.
(b) Determine the reduction in emissions if energy-efficient upgrades reduce power consumption by 20%.

3.11 A project involves an investment of $500,000 and is expected to generate cash flows of $120,000 per year for 5 years. If the discount rate is 8%, calculate:

(a) NPV
(b) Determine if the project is financially viable ($\text{NPV} \geq 0$).

3.12 Discuss the unique challenges of deploying 5G networks in rural areas. What innovative solutions can be implemented to overcome these challenges?

3.13 Develop a scenario plan for a rural 5G deployment project. Identify key uncertainties and potential outcomes.

3.14 Create a business model canvas for a 5G service provider focusing on rural areas. Highlight the key partnerships, value propositions, and revenue streams.

3.15 Discuss the importance of value chains and value network configuration in the context of 5G with active sharing in a rural scenario. How can these concepts help bridge the digital divide?

3.16 Evaluate the potential of 5G technology with spectrum sharing and infrastructure sharing to bridge the digital divide in rural areas. What are the main challenges and solutions?

3.17 Conduct a techno-economic assessment for a proposed 5G network in a rural area for a particular village of interest. What factors should be considered?

3.18 Calculate the total energy consumption of a 5G network using the expanded formula provided for a rural network in Malawi. Discuss the factors that influence energy consumption.

3.19 Write a case study on how to integrate renewable energy sources in rural 5G networks.

References

1. A. Narayanan, M.I. Rochman, A. Hassan, B.S. Firmansyah, V. Sathya, M. Ghosh, F. Qian, Z.L. Zhang, A comparative measurement study of commercial 5G mmWave deployments, in *IEEE INFOCOM 2022-IEEE Conference on Computer Communications* (IEEE, 2022), pp. 800–809
2. R. Bajracharya, R. Shrestha, H. Jung, H. Shin, Neutral host technology: the future of mobile network operators. IEEE Access **10**, 99221–99234 (2022)
3. W. Webb, *The 5G Myth: When Vision Decoupled from Reality* (De-G Press, 2018)
4. Ericsson, Network coverage data and forecasts—ericsson mobility report (2025). Accessed 27 Mar 2025
5. S.K.A. Kumar, 5G network slicing for rural connectivity: multi-tenancy in wireless networks (University of Strathclyde, 2023)
6. M. Shafi, A.F. Molisch, P.J. Smith, T. Haustein, P. Zhu, P. De Silva, F. Tufvesson, A. Benjebbour, G. Wunder, 5G: a tutorial overview of standards, trials, challenges, deployment, and practice. IEEE J. Sel. Areas Commun. **35**(6), 1201–1221 (2017)
7. N. Haddaji, A. Bayati, K.-K. Nguyen, M. Cheriet, Backhauling-as-a-service (BHaaS) for 5G optical sliced networks: an optimized TCO approach. J. Lightwave Technol. **36**(18), 4006–4017 (2018)
8. D. Xu, A. Zhou, X. Zhang, G. Wang, X. Liu, C. An, Y. Shi, L. Liu, H. Ma, Understanding operational 5G: a first measurement study on its coverage, performance and energy consumption, in *Proceedings of the Annual Conference of the ACM Special Interest Group on Data Communication on the Applications, Technologies, Architectures, and Protocols for Computer Communication* (2020), pp. 479–494
9. G. Varrall, *5G Spectrum and Standards* (Artech House, 2016)
10. S.M. Kim, J. Kim, H. Cha, M.S. Sim, J. Choi, S.W. Ko, C.B. Chae, S.L. Kim, Opportunism in spectrum sharing for beyond 5G with sub-6 GHz: a concept and its application to duplexing. IEEE Access **8**, 148877–148891 (2020)
11. R.A.K. Fezeu, J. Carpenter, C. Fiandrino, E. Ramadan, W. Ye, J. Widmer, F. Qian, Z.L. Zhang, Mid-band 5G: a measurement study in Europe and US (2023). arXiv preprint arXiv:2310.11000
12. Y. Wang, J. Li, L. Huang, Y. Jing, A. Georgakopoulos, P. Demestichas, 5G mobile: spectrum broadening to higher-frequency bands to support high data rates. IEEE Veh. Technol. Mag. **9**(3), 39–46 (2014)
13. C. Burrow, FCC announces 5G spectrum frontiers auctions. Mondaq Bus. Brief. (2018)
14. International Telecommunication Union, Smart village initiative (2025). Accessed 27 Mar 2025
15. H. Cloud, The NIST definition of cloud computing. Natl. Inst. Sci. Technol. Spec. Publ. **800**(2011), 145 (2011)
16. C.W. Lin, S.S. Abdul, D.L. Clinciu, J. Scholl, X. Jin, H. Lu, S.S. Chen, U. Iqbal, M.J. Heineck, Y.C. Li, Empowering village doctors and enhancing rural healthcare using cloud computing in a rural area of mainland China. Comput. Methods Programs Biomed. **113**(2), 585–592 (2014)
17. R. Sukanesh, A. Kanmani, Cloud computing in the field of rural education. Appl. Mech. Mater. **573**, 593–599 (2014)
18. A.M. Alwakeel, A.K. Alnaim, E.B. Fernandez, A survey of network function virtualization security, in *SoutheastCon 2018* (IEEE, 2018), pp. 1–8
19. A. Filali, A. Abouaomar, S. Cherkaoui, A. Kobbane, M. Guizani, Multi-access edge computing: a survey. IEEE Access **8**, 197017–197046 (2020)
20. M.K. Shin, K.H. Nam, H.J. Kim, Software-defined networking (SDN): a reference architecture and open APIs, in *International Conference on ICT Convergence (ICTC)* (IEEE, 2012), pp. 360–361
21. Open Networking Foundation, Software-defined networking: the new norm for networks. ONF White Paper **2**(2–6), 11 (2012)
22. D. Kreutz, F.M.V. Ramos, P.E. Verissimo, C.E. Rothenberg, S. Azodolmolky, S. Uhlig, Software-defined networking: a comprehensive survey. Proc. IEEE **103**(1), 14–76 (2014)
23. A. Farrel, Recent developments in service function chaining (SFC) and network slicing in backhaul and metro networks in support of 5G, in *20th International Conference on Transparent Optical Networks (ICTON)* (IEEE, 2018), pp. 1–4
24. J.L. Vieira, E.L.C. Macedo, A.L.E. Battisti, J. Noce, P.F. Pires, D.C. Muchaluat-Saade, A.C.B. Oliveira, F.C. Delicato, Mobility-aware SFC migration in dynamic 5G-edge networks. Comput. Netw., 110571 (2024)
25. D. Zhao, J. Ren, R. Lin, X. Shizhong, V. Chang, On orchestrating service function chains in 5G mobile network. IEEE Access **7**, 39402–39416 (2019)
26. S. Zhang, An overview of network slicing for 5G. IEEE Wirel. Commun. **26**(3), 111–117 (2019)
27. I. Afolabi, T. Taleb, K. Samdanis, A. Ksentini, H. Flinck, Network slicing and softwarization: a survey on principles, enabling technologies, and solutions. IEEE Commun. Surv. Tutor. **20**(3), 2429–2453 (2018)
28. NGMN Alliance, Description of network slicing concept. NGMN 5G P **1**(1), 1–11 (2016)
29. X. Chen, L. Jiao, W. Li, F. Xiaoming, Efficient multi-user computation offloading for mobile-edge cloud computing. IEEE/ACM Trans. Netw. **24**(5), 2795–2808 (2016)
30. Q.V. Pham, F. Fang, V.N. Ha, M.J. Piran, M. Le, L.B. Le, W.J. Hwang, Z. Ding, A survey of multi-access edge computing in 5G and beyond: fundamentals, technology integration, and state-of-the-art. IEEE Access **8**, 116974–117017 (2020)
31. T. Dlamini, S. Vilakati, Remote and rural connectivity: infrastructure and resource sharing principles. Wirel. Commun. Mob. Comput. **2021**(1), 6065119 (2021)
32. F. Spinelli, V. Mancuso, Toward enabled industrial verticals in 5G: a survey on MEC-based approaches to provisioning and flexibility. IEEE Commun. Surv. Tutor. **23**(1), 596–630 (2021)
33. F. Giust, G. Verin, K. Antevski, J. Chou, Y. Fang, W. Featherstone, F. Fontes, D. Frydman, A. Li, A. Manzalini et al., MEC deployments in 4G and evolution towards 5G. ETSI White paper **24**(2018), 1–24 (2018)
34. E. Biglieri, R. Calderbank, A. Constantinides, A. Goldsmith, A. Paulraj, H.V. Poor, *MIMO Wireless Communications* (Cambridge University Press, 2007)
35. E.G. Larsson, O. Edfors, F. Tufvesson, T.L. Marzetta, Massive MIMO for next generation wireless systems. IEEE Commun. Mag. **52**(2), 186–195 (2014)
36. L. Lu, G.Y. Li, A.L. Swindlehurst, A. Ashikhmin, R. Zhang, An overview of massive MIMO: benefits and challenges. IEEE J. Sel. Top. Signal Processing **8**(5), 742–758 (2014)

37. A. El Falou, M.-S. Alouini, Enhancement of rural connectivity by recycling TV towers with massive MIMO techniques. IEEE Commun. Mag. **61**(4), 78–83 (2022)
38. A. Lappalainen, Y. Zhang, C. Rosenberg, Planning 5G networks for rural fixed wireless access. IEEE Trans. Netw. Serv. Manage. **20**(1), 441–455 (2022)
39. A.M. Pessoa, B. Sokal, C.F.M.E. Silva, T.F. Maciel, A.L.F. De Almeida, F.R.P. Cavalcanti, A CDL-based channel model with dual-polarized antennas for 5G MIMO systems in rural remote areas. IEEE Access **8**, 163366–163379 (2020)
40. A. Ferreira, L. Mendes, W. Dias, T. Marins, D. Gaspar, A. Matos, C. Silva, B. Sokal, 5G-RANGE project field trial, in *2019 European Conference on Networks and Communications (EuCNC)* (IEEE, 2019), pp. 490–494
41. W. Dias, A. Ferreira, R. Kagami, J.S. Ferreira, D. Silva, L. Mendes, 5G-RANGE: a transceiver for remote areas based on software-defined radio, in *2020 European Conference on Networks and Communications (EuCNC)* (IEEE, 2020), pp. 100–104
42. L.L. Mendes, C.S. Moreno, M.V. Marquezini, A.M. Cavalcante, P. Neuhaus, J. Seki, N.F.T. Aniceto, H. Karvonen, I. Vidal, F. Valera et al., Enhanced remote areas communications: the missing scenario for 5G and beyond 5G networks. IEEE Access **8**, 219859–219880 (2020)
43. V. Sanchez-Aguero, I. Vidal, F. Valera, B. Nogales, L.L. Mendes, W. Damascena Dias, A.C. Ferreira, Deploying an NFV-based experimentation scenario for 5G solutions in underserved areas. Sensors **21**(5), 1897 (2021)
44. K. Topyan, M. Ulema, Architectural and financial considerations for deploying 5G based fixed wireless access, in *International Black Sea Conference on Communications and Networking (BlackSeaCom)* (IEEE, 2020), pp. 1–6
45. W.H. Lehr, J.M. Chapin, On the convergence of wired and wireless access network architectures. Inf. Econ. Policy **22**(1), 33–41 (2010)
46. P. Taylor, Fixed wireless access (FWA) connections worldwide from 2020 to 2029 (in millions). Accessed 29 Oct 2024
47. C. Chen, O. Kedem, C.R.C.M. da Silva, C. Cordeiro, Millimeter-wave fixed wireless access using IEEE 802.11 ay. IEEE Commun. Mag. **57**(12), 98–104 (2019)
48. M. Bhuiyan, Solutions for wireless internet connectivity in remote and rural areas. Master's thesis, M. Bhuiyan, 2020
49. S.N.K. Veedu, M. Mozaffari, A. Höglund, E.A. Yavuz, T. Tirronen, J. Bergman, Y.P.E. Wang, Toward smaller and lower-cost 5G devices with longer battery life: an overview of 3GPP release 17 RedCap. IEEE Commun. Stand. Mag. **6**(3), 84–90 (2022)
50. Alotcer IoT, Unleashing the power of 5G RedCap: evolution in connectivity (2025). Accessed 29 July 2025
51. L.M. Bartolín-Arnau, J. Vera-Pérez, V. Sempere-Payá, J. Silvestre-Blanes, Are European initiatives related to local spectrum allocation for private 5G networks ready for use in industrial cases? IEEE Access (2024)
52. J. Collin, J. Pellikka, J.T.J. Penttinen, Private networks, in *5G Innovations for Industry Transformation: Data-driven Use Cases* (2024)
53. Ericsson, Mobile subscriptions outlook. Accessed 29 Oct 2024
54. Ookla, 5G in Asia pacific: deployment momentum continues. Accessed 29 Oct 2024
55. P.P. Latapu, R. Lutui, S. Vaipuna, K. Moala, S. Fonua, W. Liu, J.A. Gutierrez, J. Song, L. Chiaraviglio, J. Wu et al., Bridging the digital divide in Tonga through a sustainable multi-tenancy broadband infrastructure: are we ready? in *International Conference on Environmental Engineering (EE)* (IEEE, 2018), pp. 1–6
56. J. Noll, S. Dixit, D. Radovanovic, M. Morshedi, C. Holst, A.S. Winkler, 5G network slicing for digital inclusion, in *2018 10th International Conference on Communication Systems & Networks (COMSNETS)* (IEEE, 2018), pp. 191–197
57. 5G RuralFirst, Project conclusion report (2019)
58. J.P. Tomas, Jio sees demand for its 5G FWA service from rural markets. Accessed 29 Oct 2024
59. P. Lipscombe, US cellular taps ericsson routers for rural 5G network expansion. Accessed 29 Oct 2024
60. E. Baark, China's new digital infrastructure: expanding 5G mobile communications. East Asian Policy **14**, 124–136 (2022)
61. DSIT and DCMS, 5G testbeds and trials programme: complete list of 5g projects. Accessed 29 Oct 2024
62. European Commission, 5G for rural smart communities of tomorrow. Accessed 29 Oct 2024
63. S. Kukliński, L. Tomaszewski, K. Kozłowski, S. Pietrzyk, Business models of network slicing, in *9th International Conference on the Network of the Future (NOF)* (IEEE, 2018), pp. 39–43
64. A.M. Cavalcante, M.V. Marquezini, L. Mendes, C.S. Moreno, 5G for remote areas: challenges, opportunities and business modeling for Brazil. IEEE Access **9**, 10829–10843 (2021)
65. L. Chiaraviglio, N. Blefari-Melazzi, W. Liu, J.A. Gutiérrez, J. Van De Beek, R. Birke, L. Chen, F. Idzikowski, D. Kilper, P. Monti et al., Bringing 5G into rural and low-income areas: Is it feasible? IEEE Commun. Stand. Mag. **1**(3), 50–57 (2017)
66. K. Mason, S. Wagg, D. Perez Ojeda, MANY project toolkit for community engagement in complex projects: lessons from a 5G digital infrastructure project. Mobile Access North Yorkshire (2022)
67. 3GPP, NR; User Equipment (UE) radio access capabilities—specification 38.306. Accessed 15 Oct 2024
68. S. Koratagere Anantha Kumar, E.J. Oughton, Techno-economic assessment of 5g infrastructure sharing business models in rural areas. Front. Comput. Sci. **5**, 1191853 (2023)
69. A.M. Busch, K. Eger, B. Richerzhagen, Comparison of power consumption models for 5G cellular network base stations, in *International Symposium on Intelligent and Distributed Computing* (Springer, 2023), pp. 199–212
70. D.M. CK, D. Selvamuthu, P. Kalita, Energy efficiency in a base station of 5G cellular networks using m/g/1 queue with multiple sleeps and n-policy. Methodol. Comput. Appl. Probab. **25**(2), 48 (2023)
71. J. Huttunen, M. Pärssinen, T. Heikkilä, O. Salmela, J. Manner, E. Pongracz, Base station energy use in dense urban and suburban areas. IEEE Access **11**, 2863–2874 (2023)
72. Y. Ding, H. Duan, M. Xie, R. Mao, J.J. Wang, W. Zhang, Carbon emissions and mitigation potentials of 5G base station in china. Resour. Conserv. Recycl. **182**, 106339 (2022)

Space Networks: Satellite

4

Space networks involve communication systems that operate beyond the Earth's atmosphere, which orbit the Earth at different altitudes and are mainly used for providing global coverage [1]. Space networks through satellites support global applications such as broadcasting, satellite TV, navigation, and Internet services; to offer robust and ubiquitous communication services that help provide connectivity in remote areas, oceans, and space. As introduced in Sect. 1.3.2.1, satellite communication is an emerging solution to connect regions where the deployment of a terrestrial network is challenging or impossible. Today, satellites are capable of providing KPIs similar to 4G in places where the deployment of a terrestrial network is challenging or impossible, such as oceans. The cost of service provisioning has been decreasing over the years, yet satellite connectivity remains relatively expensive for wider adoption among individuals who are yet to be connected [2]. The advent of satellite integration with terrestrial networks, as well as the deployment of Starlink and similar projects along with subsidized pricing, has enabled satellites to offer affordable and reliable services to bridge the digital divide [2, 3]. Therefore, this chapter focuses on exploring satellites for bridging the digital divide, highlighting the need for innovative business models, techno-economic assessments, and environmental modeling to assess their accessibility, availability, and affordability.

4.1 State-of-the-Art for Satellites

Satellite transmission involves using artificial satellites to provide communication links across different locations on the Earth and works on the LOS propagation between the ground station to satellite (uplink), and back to the earth station (downlink) [4]. Figure 4.1 shows the architecture for the satellite networks, where gateway link (GWL) represents the communication links with the ground stations from or to the satellite. The main architecture of the satellite networks consists of three parts [5]:

- **Space segment**: It consists of satellites and the onboard systems that operate in space, and consists of satellite payload, satellite bus, and frequency bands. Satellite payloads are mission-specific and depend on the application of interest, such as communication, sensing, and scientific instrumentation. A satellite bus is a platform that supports payload and ensures proper orientation in space and consists of power systems, amplifiers, attitutude and orientation control systems (AOCS), thermal systems, and propulsion systems. For enhancing the performance of a satellite constellation, intersatellite link (ISL) can be created for a group of satellites to work toward complicated requirements. The satellites communicate with each other via radio frequency (RF) or optical wireless technology [5].
- **Ground segment**: The ground segment, also known as the earth station or ground control station (GCS), includes various critical components such as gateway (GW) stations for network access and backhauling, maintained by network operators, and large ground control facilities that manage network operations. It also features high-gain antennas for transmission and reception, ensuring strong and clear signals, along with controllers and orbit tracking systems that handle telemetry, payload configuration, and orbit tracking. Additionally, data collection and distribution systems gather and distribute data efficiently. The ground segment also comprises telemetry, tracking and control (TT&C) stations, which are responsible for running updates and configurations, are maintained by satellite operators. The network topology in the ground segment would be either a star, where all ground stations communicate through a central hub, or a mesh, where all ground stations can communicate with each other without a central hub. Star topology is preferred for sensor data collections, whereas mesh topology is preferred for video and latency-sensitive communication, which would require intelligent routing [5].

S. Koratagere Anantha Kumar, *Wireless Broadband Access to Bridge the Digital Divide*,
Textbooks in Telecommunication Engineering, https://doi.org/10.1007/978-3-032-09261-8_4

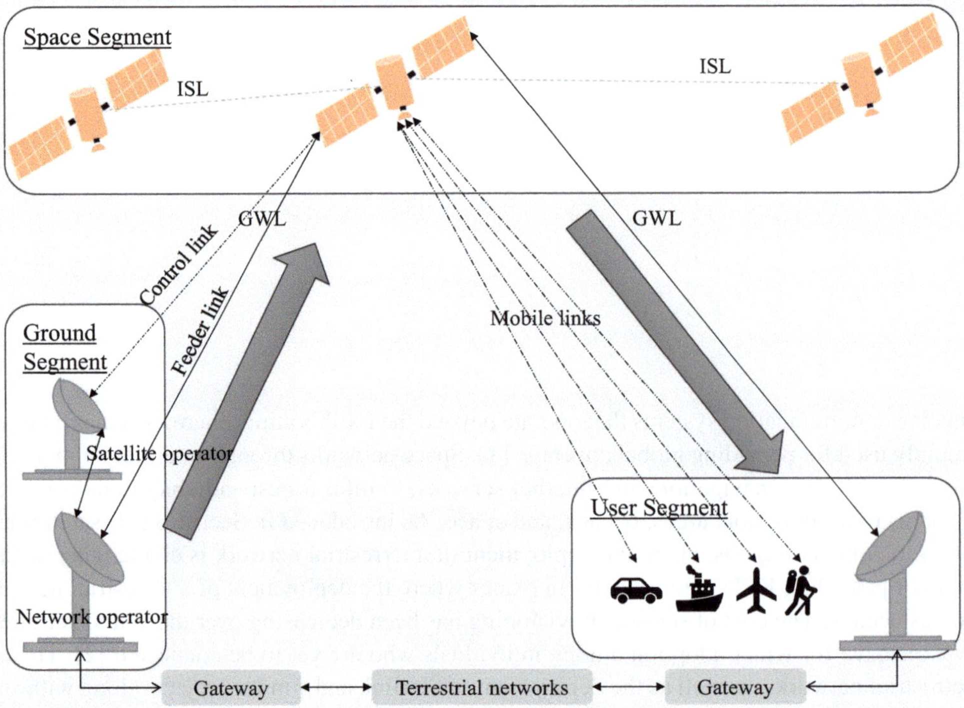

Fig. 4.1 Satellite architecture

- **User segment**: This refers to the end user devices that would communicate directly with the satellite systems, such as satellite phones, very small aperture terminals (VSAT), IoT devices, satellite TV receivers, ground-based, airborne, and maritime receivers, and transmitters. These are mainly used for applications such as communication, weather monitoring, navigation, disaster management, equipment health, defense, and precise location tracking. Lately, there has been a significant advancement in enhancing the smartphone's capabilities to support direct communication with satellites.

Current satellite technologies focus on high throughput for better data rates and lower costs, and mega-constellations that consist of hundreds of small satellites in LEO. Mega-constellations are capable of providing global coverage, ultra-low latency, are scalable in nature, and support upgrades and expansions. Mega-constellations include, for instance, OneWeb, which intends to reach over 1,000 satellites, and Starlink, whose goal is to install over 30,000 satellites.

4.1.1 Spectrum Bands

Satellite operates in frequency bands ranging from 1 GHz to 90 GHz [6]. The key advantages and disadvantages of different satellite spectrum bands are summarized in Table 4.1 [7].

The following list provides the details of the frequency range and the applications supported on those bands [7]:

- **L-band (1–2 GHz)**: Mobile satellite services (MSS) operate in this band (1.518–1.675 GHz), supporting civilian mobile communication systems using satellite phones and GPS by Iridium; and Inmarsat provides communications and surveillance at sea, land, and air [10]. This band is also used by WorldSpace satellite radio.
- **S-band (2–4 GHz)**: In this band also MSS operates but at 1.97–2.69 GHz for applications related to weather radar, surface ship radar, airport surveillance radar for air traffic control, space communication with International Space Station (ISS) and Space Shuttle, satellite television, mobile broadband services, radio broadcasting, and in-flight connectivity [11]. This band is also used by terrestrial networks using spectrum sharing technologies [12].
- **C-band (4–8 GHz)**: This band is used by fixed satellite service (FSS) ranging from 3.4 to 7.025 GHz for applications such as full-time satellite TV networks, raw satellite feeds, data services, critical applications such as distance learning,

Table 4.1 Characteristics of satellite spectrum bands properties

Spectrum band	Advantages	Disadvantages
L	Excellent penetration capabilities, less susceptible to climatic conditions	Low data rates, and large antennas
S	Significant resistance to rain, ice, and snow, low attenuation, high coverage	Low throughput, narrowband spectrum, and high antenna size
C	Support high-speed and low-latency applications of 5G, higher capacity, 5G traffic offloading, higher reliability	Not suitable for IMT systems and long-distance transmission, and cannot penetrate through buildings, less throughput
X	Suitable for locations with extreme weather conditions due to immunity to weather conditions, narrow bandwidth, and high data transfer rates, enabling it to deliver precise, low-cost, and high-resolution data for specialized purposes	spectrum band above 10 GHz has attenuation due to rain, snow, and other weather conditions, support limited air measurements
Ku	Higher throughput, smaller antenna size, wider coverage	Severe water droplet fading
Ka	Very high throughput, smaller antenna, lesser susceptibility to rain fading, high power transmission	Severe water droplet fading, higher propagation delay
Q/V	Higher data rates	Loud attenuation, some fraction of rain attenuation, scintillation, and depolarization events [8]
E	High-speed data rates, long-distance space communication, low attenuation in space [9]	Attenuation due to gases and water droplets

telemedicine, and universal access services; backhaul services [13]; VSAT data links such as bank transactions or corporate networks; mobile-satellite service feeder links, and emergency links, including disaster recovery services and meteorological tracking. These services need great reliability and extensive geographic coverage, which can only be provided in the C-band because the signal attenuation is low in this band [14].

- **X-band (8–12 GHz)**: This band is used by FSS between 7.25 and 8.44 GHz for applications primarily for military use, and includes military services, pulsed radar, synthetic operative radars, continuous wave radar, weather monitoring, air traffic control, maritime traffic control, defense tracking, and vehicle speed detection for monitoring over-speeding [15].
- **Ku-band (12–18 GHz)**: Ku-band is in the frequency range of 12–18 GHz. It is one of the major technologies used today for high-speed satellite Internet [15]. The band is split into multiple segments to be used by different geographical regions (the Americas, Europe, Africa, Australia, and Indonesia). Other uses co-existing within Ku-band include fixed service on microwave bands, radio astronomy, mobile service, mobile-satellite service, radio-location service, amateur radio service, and radio navigation.
- **Ka-band (26–40 GHz)**: Ka-band supports increased data transmission rates, improved signal resolution, and smaller antenna sizes, which operate at 17.3–30 GHz [16]. The applications supported in this band include mobile and fixed two-way broadband services, fixed satellite television, close-range targeting radars in military systems, earth observation and remote sensing, space exploration, autonomous vehicles, telemedicine, and in-flight connectivity.
- **Q/V band (30–80 GHz)**: Q band (33–50 GHz) and V band (40–75 GHz) support MSS and BSS for high-speed broadband services [17].
- **E-band (60–90 GHz)**: In the E-band, 71–76 GHz has been assigned for FSS (downlink) and 81–86 GHz has been assigned for MSS (uplink). The applications for this band include high-speed communications in space, high-resolution radar for planet surface exploration, and docking operations, as well as space debris detection. Additionally, it supports high-speed 5G and 6G communications [9].

4.1.2 Satellite Timelines

When Arthur C. Clarke first proposed the concept of satellite communication in a science fiction book in 1945, he perceived three GEO satellites being used to provide relay communication via satellite in locations where establishing a terrestrial network would be impractical. Roughly 12 years later, the Union of Soviet Socialist Republics (USSR) first launched Sputnik 1 into space to measure atmospheric density, radio signal propagation in the atmosphere, and performance of radio as well as optical orbit tracking techniques [18]. Figure 4.2 shows the evolution of communication satellites over the years [14, 18, 19].

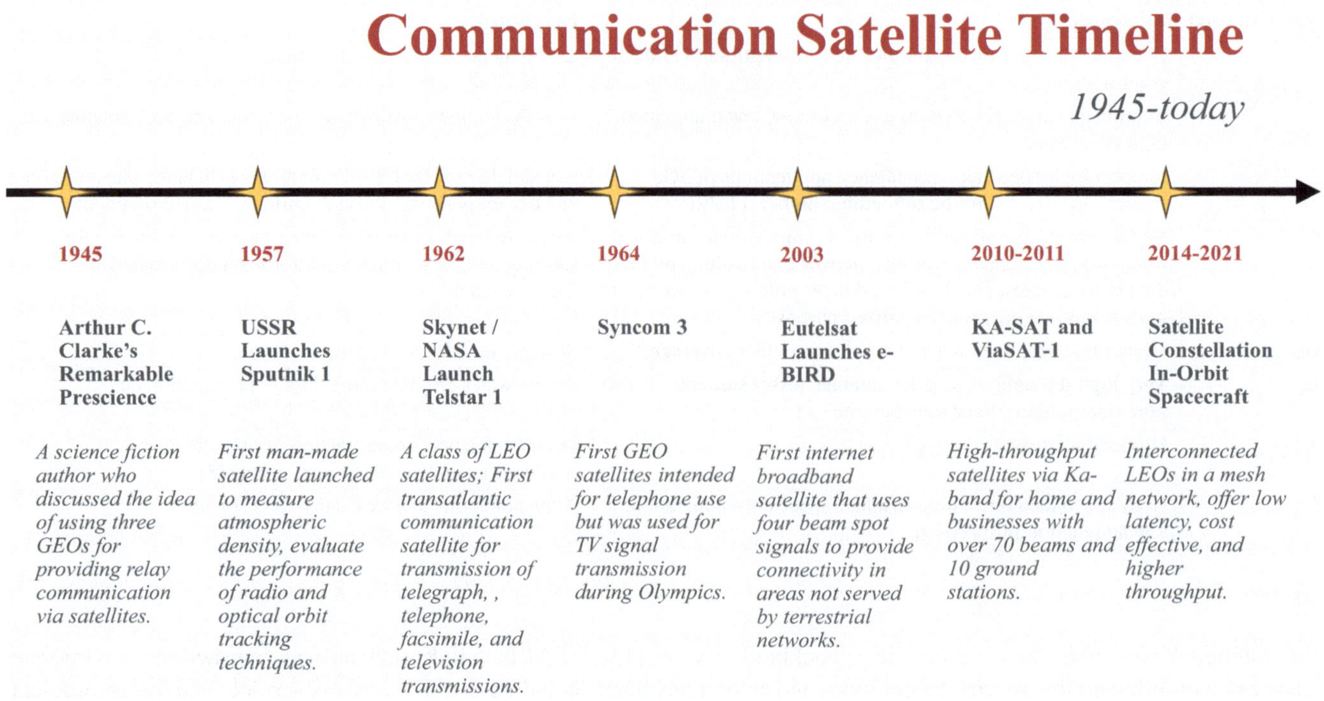

Fig. 4.2 Communication satellite timeline

The current generation of satellites can provide ultra-high-speed broadband connectivity to several sectors. There is ongoing research to explore the use of LEO, very low earth orbit (VLEO), and nanosatellites as well as increase their data rates for smooth integration with terrestrial network technologies such as 5G and 6G, and support use cases such as eMBB, URLLC, and mMTC. Since nanosatellites are small, they are cost-effective and easy to launch [20].

Each satellite firm would want to launch as many satellites as it could to increase capacity, availability, and competition, which would drive down prices. As of today, technology exists to bridge the digital divide, but the challenge lies in terms of digital literacy and affordability. The key stakeholders believe that the government plays an important role in bridging this gap; however, the question lies in who they should fund - consumers or firms. The policy and funding for satellite communication require real solutions to bridge this gap.

4.1.2.1 Communication Satellites Use Cases

Communication satellites are employed for an assortment of applications ranging from personal communications to industrial vertical use cases, as well as being widely suitable for live broadcasting of events and media streaming content [21]. Communication satellites play a vital role in bridging the digital divide, particularly in regions where terrestrial networks are absent or impractical. While terrestrial networks deliver a broad array of services, their deployment remains challenging in rural, densely forested, isolated regions, rugged terrains, oceanic expanses, and even space environments. In these regions, NTN plays a crucial role by seamlessly integrating with terrestrial and aerial networks to extend coverage and deliver essential services to underserved and remote areas. NTN also acts as a backup when there is a network failure for the terrestrial networks. The three major reasons for 5G integrations with NTN focus on network reliability, service ubiquity, and service scalability [22].

Other use cases include network backhaul technologies in challenging terrain, offloading terrestrial IoT traffic, wide area IoT services for surveillance, fleet management, energy companies site monitoring, livestock management, agriculture, industrial equipment health monitoring, industry vertical data collection such as transmitting real-time health of sensors on site, autonomous vehicles, remote health monitoring and surgery [20, 23].

4.1.2.2 Why Satellites are Booming as an Alternative to Terrestrial Networks?

Satellites are becoming a crucial component of many use cases, primarily because of their capacity to provide KPIs similar to 4G in places with very low or no population density. Satellite research is underway and will contribute to bringing satellite capabilities up to par with 5G [24]. Every satellite has the ability to cover a large area and has the capability of altering the capacity and coverage according to the needs of the end users in each location [25]. The cost of service provisioning for direct end user services has been decreasing over the years [26]. Research has indicated that although the price of satellites has dropped, the cost of launching the satellites has reduced significantly due to the advancements in technology and reusable rockets. However, as space becomes more crowded, maintenance issues, spectrum charges, space debris, and collision danger increase the expense of retaining the satellite in the desired position.

There are several other reasons why satellites are becoming an alternative to terrestrial networks [27]. Satellite communication offers several advantages that make it a reliable and versatile technology for wide area communication. It is independent of the distance bridged within the satellite's coverage area, ensuring consistent connectivity across vast regions. Supporting both broadcast and multicast services, satellites enable direct communication links for end users, whether stationary or mobile. Additionally, satellite connectivity can be established quickly and with minimal infrastructure requirements. This makes it a cost-efficient and economical solution, particularly during disaster management and for providing connectivity in remote and rural areas, making it a critical factor in bridging the global digital divide [28]. Furthermore, satellites can accommodate a large number of users with diverse demands, functioning effectively as broadband technology with reduced latency using LEOs. Satellite networks are being integrated with the terrestrial networks, such as 5G and 6G, to complement the coverage and improve the overall QoS.

4.1.3 Key Technologies Enabling Satellite Communication Systems

There are several key technologies empowering satellite communication systems to advance in the direction of high data throughput, low latency, and reliable networks. Some of the key technological innovations that are driving satellite deployments include small satellites, newer constellations, satellite IoT, in-orbit services, AI, advanced payload systems, spacecraft propulsion, VHTS, and digital twins. Some of the cutting-edge technologies are discussed below:

- **Advanced satellite constellation**: The deployment of a next-generation constellation of satellites consisting of thousands of LEO satellites is poised to revolutionize the communication systems in space. These satellites offer low-latency and high-speed communication systems. These advanced satellite constellations are being deployed to provide global coverage and seamless connectivity. These large networks of satellites consist of satellites in different orbits and drive technologies such as NTN integration with terrestrial networks, military communication, earth observation, remote region connectivity, environmental studies, and advanced remote sensing.
- **AI and autonomous operations**: Integration of AI and ML with satellite systems has improved their performance [29]. AI-powered satellites can handle a vast volume of data independently and boost their performance in real-time to maximize the QoS and optimize mission planning, routing, resource allocation, and anomaly detection. AI also assists the operator in earning better profits by evaluating network usage data and proposing appropriate modifications to improve the dependability, performance, reliability, responsiveness, and utilization [30]. Another interesting feature supported by AI and ML is the traffic management and network efficiency.
- **THz communications and optical links**: Satellites are expected to employ THz communication to provide up to 1 Tbps per user, meeting the 6G minimum data standard [30]. This will fuel the integration of 6G and satellites to support a truly ubiquitous network. Another interesting technology is the optical links using lasers to offer interference-free transmissions. This is mainly because lasers are highly directional rays. These technologies help minimize the latency by enabling high-speed ISL and faster satellite-to-ground communications.
- **mMIMO and beamforming**: The integration of mMIMO and beamforming will also help manage interference for a mega-constellation of satellites and boost spectrum and energy efficiency [31].
- **Software-defined satellites (SDSs)**: Another major technology driving the satellite industry is SDS, which is added to the space platform, and would help the system to perform channel modulation, demodulation, and complete signal processing before further transmission to the earth station [32]. SDS would also boost the intelligence of satellites to support critical communications while reducing the link budget to improve the QoS. SDS allows satellite operators to reallocate and reconfigure satellites for different assignments, regions, and users, providing dynamic services by adjusting the satellite's

bandwidth and beamwidth. SDS utilizes the advancements offered by SDR using FPGAs and graphic processing units (GPUs) to the entire payload while meeting the demand for additional processing power requirements [33].

- **Satellite IoT integration**: In the future, satellite communication systems are expected to support massive IoT deployments and offer connectivity for applications across urban and rural areas [34]. They are expected to offer low latency for IoT applications. Satellite IoT is projected to facilitate several agriculture use cases in rural settings, such as smart agriculture, energy farm sensing, logistics, environmental monitoring, real-time environment monitoring, enhanced agricultural decision-making, and animal management.
- **Reconfigurable selective surfaces (RSS) and reconfigurable intelligent surfaces (RIS)**: The deployment of mega-constellations will require minimizing satellite power consumption and maximizing power efficiency. RSS is a thin, lightweight metasurface integrated with passive electronics components to provide unique and controlled manipulation of the wireless signals [35]. It supports energy-efficient enabling technology by passively manipulating the amplitude and phase of the signal to target its desired receiver. RSS is a category of RIS when applied on a satellite platform to perform signal processing on channel signals in space. RSS improve the QoS for satellites by enhancing the signals toward the desired targets.
  RIS is also a smart surface technology that consists of a surface made of small, individually controlled elements that can absorb, reflect, and transmit electromagnetic waves to support different requirements of operators to boost their signal strength [36]. RIS enables real-time control of the metasurface properties when applied on antennas, filters, polarizers, and beamformers, to dynamically adapt to end users' demands.
- **Quantum communication advancements**: Communication satellites are leveraging the principle of quantum physics for data transmission in the space layer. Quantum communication ensures that the communication is secure by employing quantum cryptography, mainly quantum key distribution (QKD) [37]. The key benefits offered by QKD for satellite communication are in the detection of eavesdropping and immediate notification of data breaches. Several countries are launching their quantum satellite to ensure secure data and key transmission, to ensure that data interception is virtually impossible. As per the study, the working principle is explained as, "Quantum communication over a satellite or UAV is possible because of an incident laser beam, sourced from the earth, over a crystal and thus generates entangled photons with opposite polarization states. The major stage is to further split the entangled photon and focus back on Earth at two distinct base stations" [37]. The challenges that pose a threat to quantum communication in satellites are authentication, cost, denial-of-service for legitimate users, atmospheric conditions such as turbulence, weather, noise, radiation, thermal, vacuum, and shock [37].
- **Smart antennas and gateways**: Smart antennas, such as advanced phased array and electronically steerable antennas, enable dynamic steering of the satellite beamwidth to target the signal on the specific region or user(s) by providing advanced and precision control of the beamforming capabilities [38, 39]. This helps in building a secure network and reduces the interference as well as the interception of the signal. This technology also supports isolated communication for multiple users and enables efficient spectrum and resource utilization, as well as offers a secure communication channel. These features support rural connectivity and higher data rates.
- **Cybersecurity enhancements**: Advanced cybersecurity features are necessary to prevent sophisticated threats. Satellite communications are extremely vulnerable to attacks from adversaries who target user terminals, which lack even the basic protection that is present in modems and routers. Building cybersecure space links is critical in satellite communication as it relies upon IP-based communication. There is a need to build resilient satellite networks to protect them from attacks such as denial-of-service (DoS), jamming, and other sophisticated root access exploits [40]. In many scenarios, the attack originates from the end user devices compromised through the satellite network, which can provide an opening for malicious activities. Therefore, robust security measures must be integrated into the satellite systems to boost cybersecurity, such as end-to-end encryption, device authentication, and network monitoring. With the expansion of the satellite industry, there is a need to increase the knowledge of digital protection and adhere to strict security policies on both the operator side and end user devices in protecting the satellite network and critical infrastructure.
- **Modern integrated circuit (IC)**: With the advancements in the IC, the satellites are now able to handle the harsh operational conditions in space, such as solar flares, extreme temperatures, radiation, and limited power availability [41]. They are also designed to maintain their reliability, QoS, and performance metrics. IC are key components of different parts of satellites, such as filters, signal processing, data processing, control systems, power consumption, security, power efficiency, spectrum utilization, and system integration. Modern IC are leading the development of miniaturization of satellites to support emerging applications by integrating with the terrestrial networks and IoTs to offer global ubiquitous connectivity [42].

- **SmallSats and CubeSats**: Over recent years, small-size satellites have gained popularity to support different applications [43]. SmallSats are spacecraft that weigh less than 180 kg and are the size of a large kitchen fridge [44]. These are classified as follows:

  – Minisatellite: 100–180 kg
  – Microsatellite: 10–100 kg
  – Nanosatellite: 1–10 kg
  – Picosatellite: 0.01–1 kg
  – Femtosatellite: 0.001–0.01 kg

  Cube satellite (CubeSats) are nanosatellites that use a standard size and factor, and are different compared to SmallSats; [45]. For example, 1U or one unit of CubeSats measures 10x10x10 cm, weighing a maximum of 1 kg, and is available in different sizes such as 1.5U, 2U, 3U, 6U, and 12U. Both these satellites require a much smaller launch vehicle dedicated to launching small satellites. SmallSats and CubeSats are miniaturized, low-cost satellites that are gaining research interest for rural connectivity and industry applications. These are flexible, scalable constellation solutions accessible to underserved geography areas while allowing broadband services for remote regions, which can help to bridge the digital divide in underserved areas [46, 47].

4.1.4 Integration with Terrestrial Networks

Terrestrial networks and NTN have been developed independently, with different use cases to fulfill. However, over time, industry and academia recognized the importance of integrating both to provide universal coverage for use cases such as autonomous vehicles, direct-to-satellite connectivity, remote area coverage, and connectivity in areas where terrestrial networks cannot be deployed, such as oceans, mountain ranges, and desolate landscapes. Integration of satellite with the terrestrial networks is one of the key pillars for 6G networks [48]. This has also been explored as part of 4G and 5G networks as well [49]. With the rise of LEO satellites providing high-speed Internet with low latency, terrestrial networks and NTN are working together to complement each other and expand global coverage.

Therefore, these integrations would help offer a truly ubiquitous network across the globe.

- **Integration with 4G**: To integrate satellites with 4G networks, the approach focuses on deploying a 4G access network as a fronthaul network and connecting with the satellite as a backhaul network [2]. The integration framework and architecture vary depending on the role and capabilities required; the three main frameworks are satellite as a backhaul for 4G network to connect remote gNodeB to the core network, complementary coverage in areas with no terrestrial networks, and dual-mode devices for continuous connectivity services when the user is switching between terrestrial and NTN depending on the QoS. For example, Fig. 4.3 shows the complete network integration for complementary coverage in remote areas [2].
  The integration of satellites with 4G has a different set of challenges, such as satellite latency being higher compared to the terrestrial networks, TCO for satellite being high, and a lack of efficient spectrum management tools to minimize the interference between 4G and NTN systems.
- **Integration with 5G**: With the advent of 5G, the digital divide gap still existed, as the cost of deploying terrestrial 5G networks using the traditional deployment strategy became expensive [49]. Therefore, academia and industry explored the role of satellite communication in 5G as a complementary coverage solution for ubiquitous connectivity, broadcast and multicast, emergency and disaster recovery, machine-type communication, aeronautical and maritime communication, remote sensing, and a potential solution to bridge the digital divide at affordable prices.
  The advancements in satellite technologies enabled the development of high throughput satellite (HTS) systems using GEO and LEO orbits, driving the launch of mega-constellations of LEO satellites that offered terabit per second (Tbps) capacity across the globe. The evolved satellite systems will provide support of 5G RAN along with other wireless technologies, offering seamless handover, interoperability, reliability, secure communication, and extending support to different access technologies as shown in Fig. 4.4 [49]. There are different standards and interface protocols developed to drive the integrations of satellites with 5G, such as 3GPP Release 15, Release 16, ITU-R Working Group 4B, IEEE Future Directions Initiative, 5G India, EU H2020 and European Space Agency (ESA) (shared access terrestrial-satellite backhaul network

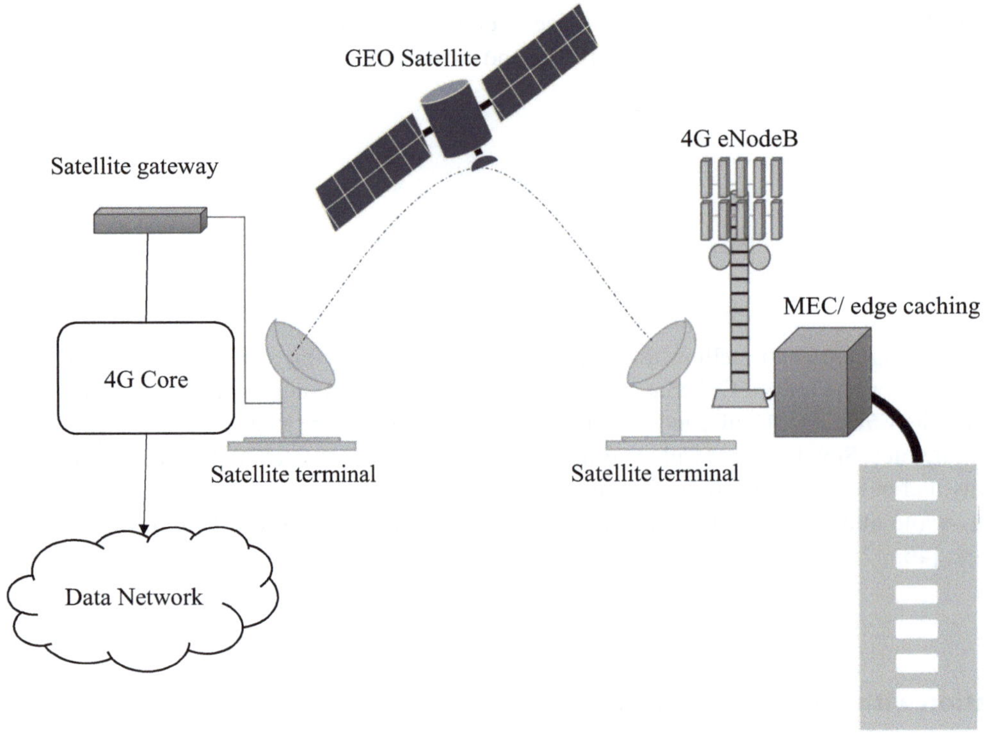

Fig. 4.3 Satellite integration with 4G networks

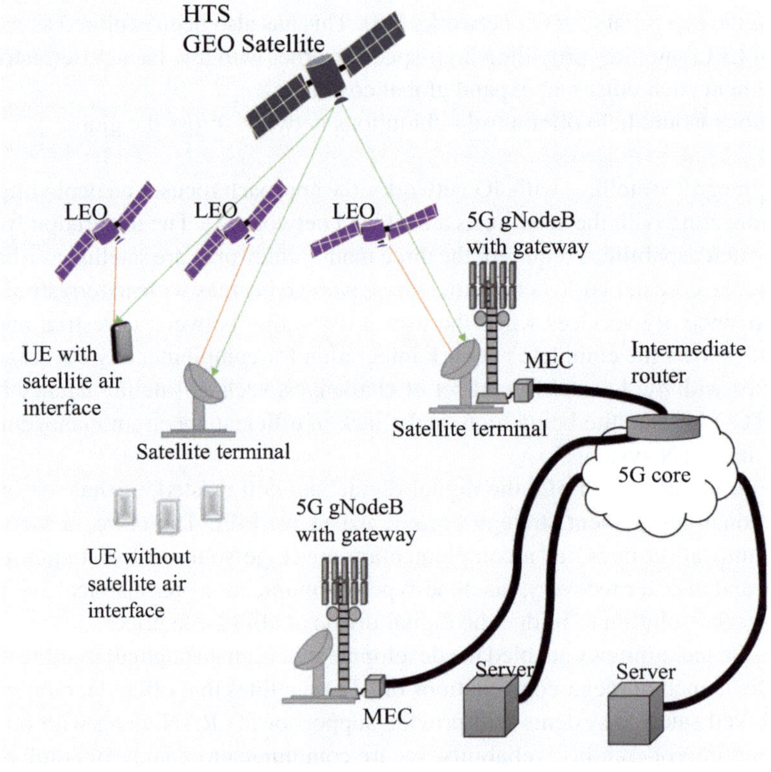

Fig. 4.4 Integration of satellites with 5G networks

enabled by smart antennas (SANSA), virtualized hybrid satellite-terrestrial systems for resilient and flexible future networks (VITAL), Sat5G, CloudSat, and satellite-terrestrial integration in the 5G (SATis5)), and US funded projected such as 5G-enabling mobile and enhanced resilient ground-to-space ecosystem (EMERGE), 5G hybrid open satellite-terrestrial system (5G-HOSTS-SAT), and 5G military applications (5G.MIL).

Satellite integration with 5G comprises a different framework compared to 5G [49]:

- HTS GEO in Ka-band offering over 300 Gbps of total network capacity with dynamic auto shifting of traffic over multiple GWs, and these are capable of providing 100 Mbps broadband services for residential Internet services and 1 Gbps for maritime services.
- Similarly, mega-constellations of MEO satellites on Ka-band offer broadband access in African and Latin American regions to bridge digital connectivity in remote areas and maritime use cases. It has a latency of 179 ms for voice and 140 ms for data services.
- LEO satellites operating in Ka-band and supporting 10 steerable antennas may deliver data rates up to 1.6 Gbps each; two high-performance steerable antennas could offer up to 5.2 Gbps capacity each; and 4 ISL to decrease latency similar to terrestrial networks. Using this constellation, the end user can use up to 500 MHz of bandwidth for both uplink and downlink.

  Though integrating satellite with 5G provided greater functionality, it also presented certain challenges that must be overcome before these functionalities can be widely adopted. Some of the primary problems include finding the correct balance between harmonization and specialization of network services for the various terrestrial and satellite domains. The problem is enabling continuous communication through synchronization, complicated antenna tracking, and double antennas on Earth. End-user devices should be capable of constantly evaluating signal quality to ensure a seamless transition from one technology to another. These shortcomings add to the overall cost for operators and end users, rendering it less economical in the short term.

- **Integration with 6G**: To address the integration challenges with the older technologies, satellite integration is one of the pillars of 6G connectivity, which is being explored by stakeholders. It is a key enabler to achieve the vision of

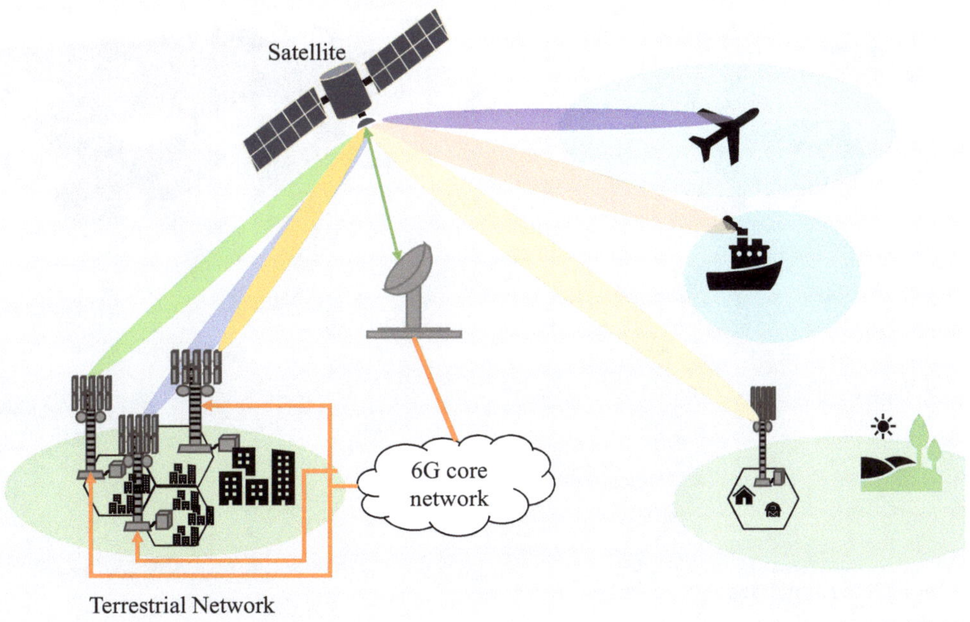

Fig. 4.5 Satellite integration with 6G networks

global, seamless, ubiquitous, and intelligent connectivity. 6G is the seamless integration of terrestrial networks, space networks, and short-distance wireless communications [50, 51]. Figure 4.5 shows the hybrid architecture for satellite and 6G integration [52].

6G networks are expected to offer intelligent mobility management across the globe and provide high-speed broadband connectivity across land, water, and air. As 6G is still an ongoing research work, there are six key areas where the integration is a topic of interest, such as standard, terminal, platform, network, frequency, and resource management [51].

4.2 Worldwide Satellite Deployments

There has been a surge in satellite deployment across the globe in recent years, driven by technological advancements, especially in LEOs, to cater to the increased demand in terms of latency and high data throughput for connectivity. This has been a revolution in the space communication segment, reshaping communication and navigation worldwide.

Figure 4.6 shows the number of active satellites per country as of November 2024. The United States leads the world with approximately 8,668 operational satellites, followed by the former USSR with 1,542, the People's Republic of China with 739, the United Kingdom with 658, Japan with 206, and France and India with 104 satellites each. The European Space Agency maintains a fleet of 87 satellites [53]. The leading satellite operators in the USA are SpaceX (as part of Starlink mega-constellation projects), national aeronautics and space administration (NASA), and the department of defense (DoD); and they contribute significantly toward deploying satellites for communication, research, navigation, and surveillance purposes.

Figure 4.7 shows the number of active satellites per category. The increasing diversity of satellite operators worldwide highlights how important space is for fostering innovation, facilitating global connection, and tackling issues like disaster relief and climate change. As nations and commercial organizations continue to invest in space technology to increase coverage, enhance capabilities, and satisfy the growing demand for connection and data services, deployment patterns are anticipated to increase. Several satellite operators are aiming to make personal communication via satellites available to the end user

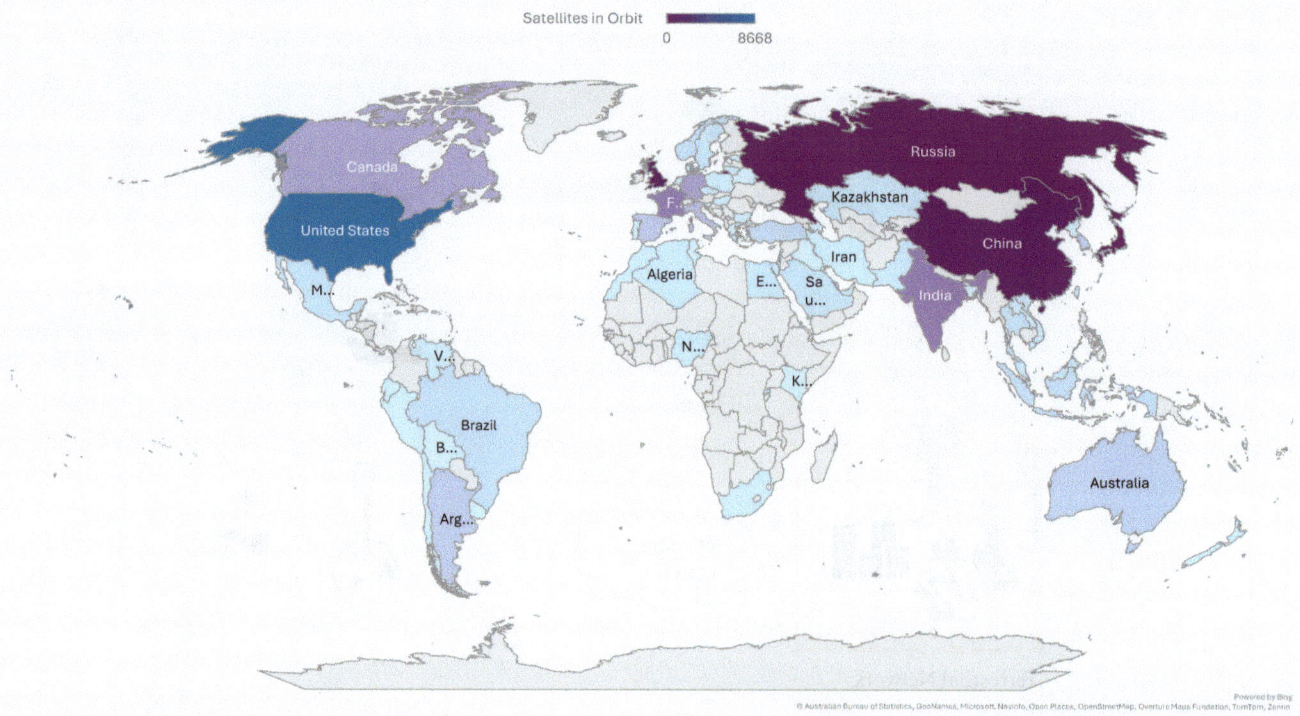

Fig. 4.6 Number of satellites per country

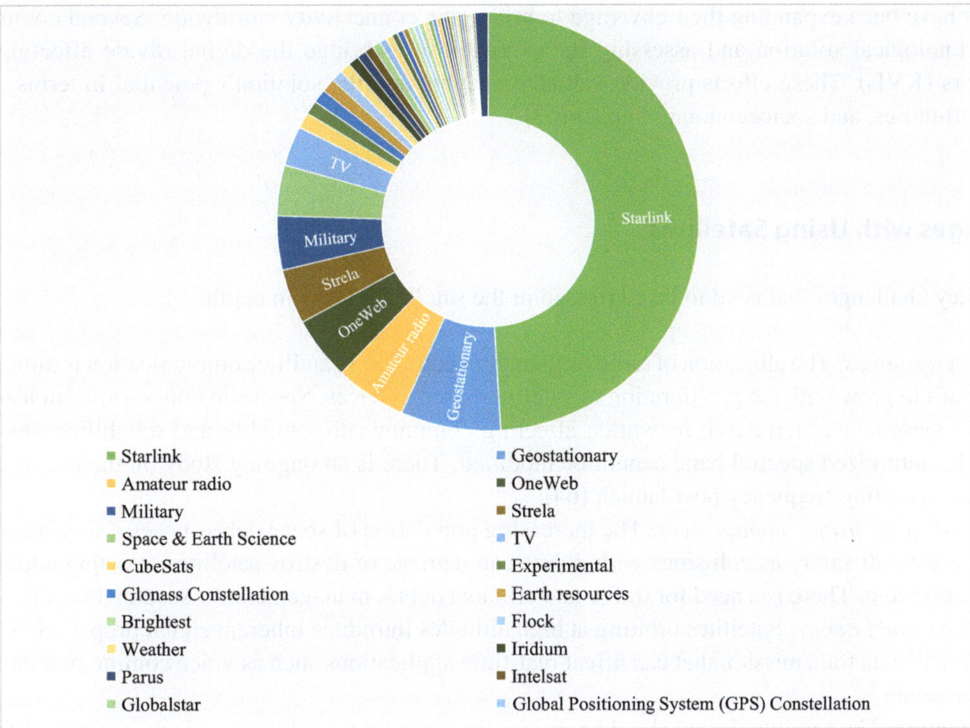

Fig. 4.7 Number of satellites per category

handsets directly and at affordable rates for pricing services as well as handsets. As of 2024, satellite handsets are expensive, and the pricing plans are significantly higher to offer affordable solutions for bridging the digital divide.

The operators are deploying constellations of satellites to support myriad use cases. Almost 50% of the active satellites are deployed by Starlink, followed by Geostationary, Amateur Radio, OneWeb, Strela, Military, Space & Earth, and TV [54, 55]. Each operator deploys satellites for different use cases. For example:

- *Starlink (SpaceX)*: These mega-constellation projects are deployed to provide worldwide coverage with high-speed data rates for residential areas, remote businesses, connectivity during disasters, and personal communications, especially in rural and remote regions [56]. They are considered a pioneer in scalable and low-latency satellite deployments.
- *Amateur radio*: These satellites provide intercontinental communication between low-power amateur radio stations worldwide and transmit voice, data, and TV signals [57, 58].
- *OneWeb*: They specialize in space communication technology, deploying LEO satellites and focus on GEO-LEO integration to deliver low latency and high throughput services for applications such as maritime, land only, aviation, governments, and enterprise [59, 60].
- *Strela*: These consist of LEO satellites that act as mailboxes for Russian military communications [55].
- *Military*: These satellites are mainly used for military purposes and have variable payloads depending on the purpose they are built for, such as imaging, intelligence gathering, navigation, and communication [61].
- *Space and earth science*: These satellites are launched into space to observe, study, and measure space phenomena, Earth's atmosphere, oceans, lands, forests, and rivers. These results are important for understanding planetary systems, monitoring weather, accidents, and the environment, performing scientific research, and aiding in disaster management scenarios.
- *TV*: Broadcasting continues to grow rapidly and plays a vital role in delivering content to millions of viewers [62]. Live content is broadcast in different modes of transportation and is earning newer streams of revenue for the operators.

Many satellites have been expanding their coverage to bridge the connectivity worldwide. Several countries are actively exploring this technological solution and assessing its affordability to bridge the digital divide affecting populations in key value indicators (KVIs). These efforts provide valuable insights into the solution's potential in terms of sustainability, development opportunities, and socioeconomic impact [63].

4.2.1 Challenges with Using Satellites

Let's explore the key challenges that need to be addressed in the satellite sectors in depth:

- *Limited spectrum resources*: The allocation of radio frequency spectrum for satellite communication is finite, and demand for spectrum continues to grow with the proliferation of satellite-based services. Spectrum congestion can lead to interference between satellite systems and terrestrial networks, affecting communication quality and reliability. Once a satellite has been launched, its authorized spectral band cannot be modified. There is an ongoing study on the use of software-defined radios to alter the operating frequency post-launch [64].
- *Orbital debris and space traffic management*: The increasing population of space debris poses a significant risk to satellite operations and spacecraft safety as collisions with debris can damage or destroy satellites, creating additional debris and exacerbating the problem. There is a need for traffic and efficient debris management for sustainable NTN communication.
- *Propagation delays and latency*: Satellites orbiting at high altitudes introduce inherent signal propagation delays, resulting in latency or delay in data transmission that can affect real-time applications such as voice communication, online gaming, and video conferencing.
- *Environmental factors*: The satellite design should consider the operating conditions and environmental factors impacting it, such as radiation, thermal fluctuations, micro-meteoroids, and solar activity. The satellite design should help withstand harsh operating conditions and provide reliable services. These factors are critical for the lifespan of the satellite and are one of the major challenges for equipment manufacturers and operators.
- *Launch costs and access to space*: As discussed earlier, launch cost is a critical factor and acts as an entry barrier for SME. There is also a need to ease the entry requirements for space communication firms by increasing access to space through reusable launch vehicles, ride-sharing opportunities, and public-private partnerships, and stimulate innovation in satellite communication.
- *Regulatory and licensing requirements*: Satellite communication systems are subject to regulatory norms by national and international agencies. Navigating the regulatory landscape, obtaining necessary approvals, and ensuring compliance with applicable regulations can be complex and time-consuming for satellite operators, particularly in multinational operations.
- *Interference and spectrum management*: Both intentional and unintentional interference from neighboring satellite systems, terrestrial transmitters, and atmospheric conditions can interfere with satellite communication systems. Therefore, there is a need for effective spectrum management to ensure reliable communication.
- *Economic viability and market dynamics*: Rapid technical innovation, changing customer demands, and dynamic business models are characteristics of the competitive market in which the satellite communication sector works. For satellite operators, investors, and stakeholders, achieving economic viability and profitability while navigating price pressures, shifting industry dynamics, and market uncertainty presents problems.
- *Affordable handsets*: As of today, satellite services-enabled handsets are not affordable to the wider population [45]. There is a need for low-cost smartphones or handsets with satellite connectivity to use this technology to bridge the digital divide. This would also improve the scalability of satellite technologies for other use cases and in urban areas as well.
- *Position, navigation, and timing (PNT)*: PNT are critical functions provided by satellite-based systems such as GPS, global navigation satellite system (GLONASS), Galileo, and BeiDou. While these systems offer invaluable services for various applications, they also face several challenges that can affect their accuracy, reliability, and availability. Some common issues related to satellite PNT that affect their QoS are signal interference, satellite constellation geometry, signal attenuation and obstruction, multi-path effects, clocking synchronization between transmitter and receiver, ionospheric and tropospheric effects, and cybersecurity issues.

Addressing these challenges requires ongoing research, development, and innovation in satellite technology, signal processing algorithms, receiver design, and system architecture. Collaborative efforts among government agencies, industry stakeholders, academia, and international organizations are essential to enhance the resilience, accuracy, and reliability of satellite systems for catering to diverse applications, user communities, and bridging the digital divide.

4.3 Satellite: Accessibility, Affordability, and Climate

This analysis focuses on the network, economic, and environmental underpinnings of satellite in rural regions. We'll outline the major principles and provide an overview of the intriguing possibilities these models provide for the future of rural connection. While several detailed models and methodologies contribute to these frameworks, we may not address them here.

4.3.1 Economic Frameworks

Satellite deployments attract stakeholders' attention to potentially open new business models and boost the economic growth of several industry verticals. In recent years, several research works and pilot trials have been exploring the idea of bridging the digital divide using satellite connectivity. According to the research on using satellites, developing an innovative business model for tackling the digital divide entails developing a system that is economical, trustworthy, resilient, and reliable, as well as providing high-speed Internet access to rural and underserved locations. Some of the key potential business models are covered and offer a glimpse into the future possibilities these models present in bridging the digital divide, which offers financial and social sustainability.

4.3.1.1 Business Models

Recent advancements in satellite communication technologies have empowered operators to penetrate previously inaccessible markets through innovative business models, thereby boosting service providers' revenues and stimulating economic growth in digitally underserved regions [65]. Satellites provide ubiquitous coverage across the globe for communication services, satellite TV, and community connectivity; and employ different business models to enhance accessibility, reduce user costs, and support deployment and affordability in underserved or remote areas. Revenue streams for the satellite technology include tiered pricing, pay-as-you-go, and freemium approaches, business-to-business (B2B) model, and subsidy model.

Satellite business models for rural connectivity must help in finding a sustainable business model that would bridge the digital divide. There is a need for tailored strategies and initiatives that allow service providers to access the underserved market and empower individuals. Furthermore, these solutions must be supplemented by low-cost end user devices to boost digital adoption. Although these business models are currently employed, several challenges remain unaddressed: operational requirements, technological advancements, QoS, interference management, space debris mitigation, cost modeling, revenue potential, digital adoption, affordability, spectrum pricing, and policy frameworks.

Innovative business models have to integrate financial rewards with decreasing carbon footprints and environmental effects [65]. Newer business models for addressing the digital gap should prioritize fair access to ICT and minimize electronic waste. The newer business models should also focus on building a resilient network that comprises a constellation of satellites of different sizes to achieve the aim of market expansion in underserved regions at an affordable price.

4.3.1.2 SWOT Analysis

Table 4.2 shows the SWOT analysis for the satellite business models in bridging the digital divide across the globe, and provides insight into its potential influence on business, government, and society.

4.3.1.3 Value Chains

The study of satellites VNC helps to understand and design a system that would focus on key roles to be played by different stakeholders to maximize the value derived from satellite connectivity. Figure 4.8 shows the VNC for satellite networks for bridging the digital divide. Each stakeholder's responsibilities play a key role in determining how satellites would help in bridging the gap. The responsibilities range from technological development to implementing policies and interfaces. This value chain helps the stakeholders understand the collaborative efforts required to enhance the economic growth for all the players.

Table 4.2 SWOT analysis for satellite networks in bridging the digital divide

Strengths	*Weaknesses*
Global coverage Scalable Reliable and resilient networks Advanced technological feature Energy efficient Versatile and supports a range of applications and industries	Expensive Market entry barriers Latency is higher than terrestrial networks Vulnerability to space debris Regulatory constraints and limitations due to frameworks Regulatory and compliance challenges
Opportunities	*Threats*
Improved QoS Job and revenue opportunities Newer business models Emerging markets with limited terrestrial infrastructure Technological advancements	Significant cost toward infrastructure investment Capacity-limited backhaul Geopolitical tensions Spectrum congestion Costs for user equipment Competition from terrestrial networks Cybersecurity risks

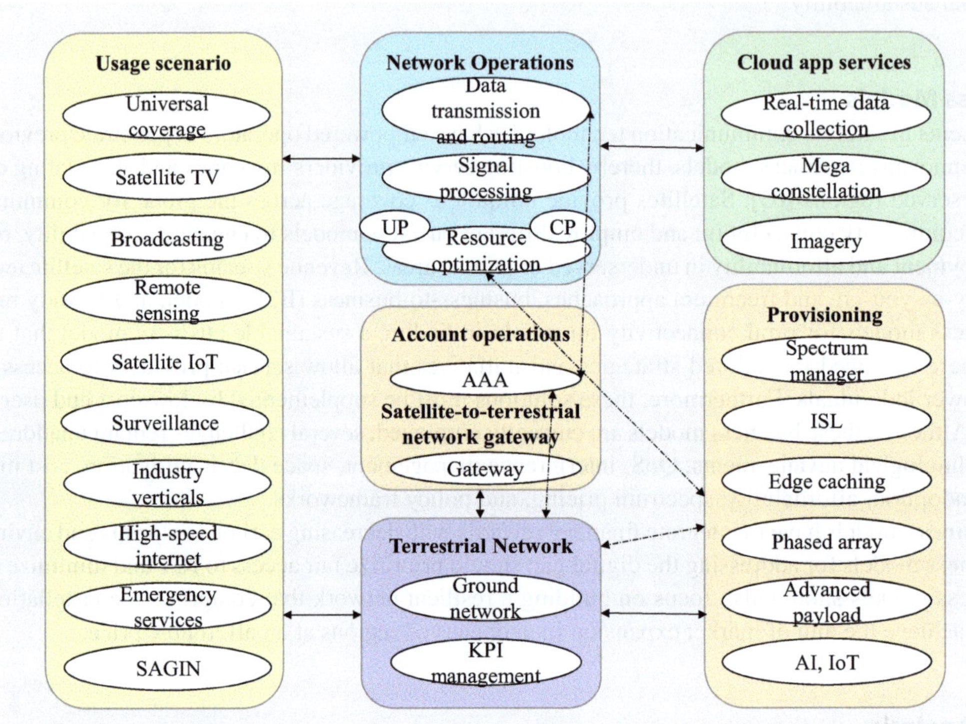

Fig. 4.8 Satellite for digital divide–VNC

The VNC helps in understanding the technical, economic, and strategic requirements to ensure successful deployment. The potential drivers for a satellite in bridging the digital gap are:

- **Satellite Operators**: Companies can play a key role in driving satellite deployments by providing necessary satellite infrastructure and services.
- **Ground Station Providers**: These organizations maintain the link between the satellite and the end users, either directly or through terrestrial networks.
- **Telecommunication Companies**: Telecommunication operators can leverage satellite backhaul to extend coverage in underserved regions. Additional use cases include traffic offloading, broadcast services, and the deployment of resilient disaster recovery networks.

- **Technology Providers**: These stakeholders help in enhancing the technology to provide the hardware and software necessary for efficient and reliable satellite operations.
- **Equipment Manufacturers**: They play a critical role in designing and producing reliable, cost-effective components for satellite communication and end user devices. Their innovations support signal integrity, device compatibility, and power efficiency; ultimately enhancing market accessibility and enabling widespread adoption of satellite communication services in remote and underserved regions.

Collaboration, strategic partnerships, and alignment of interests among the stakeholders are essential to drive cohesive action, ensure desired outcomes, and maximize the impact of the initiative. This helps to drive innovation and knowledge in underserved regions, and the end users could benefit from information, education, healthcare, and economic opportunities facilitated by satellite Internet.

4.3.1.4 Channel Modeling Using Stochastic Geometry

The link budget for satellite signal propagation is better understood using stochastic modeling, which helps in analyzing the characteristics of transmitter and receiver stations, factors affecting the propagation, and impact of noise and interference on the QoS [66, 67]. The noise power at very high frequencies consists of only quantum noise and is modeled as:

$$\begin{aligned} N &= N_o B \\ &= kTB \end{aligned} \tag{4.1}$$

where N is the noise power, k is the Boltzmann constant (1.38×10^{-23} J/K), T is the absolute temperature in K, and B is the noise bandwidth in Hz. The simple link budget to help design a proper link is shown below:

$$\text{Received Power (dBm)} = \text{transmitted Power (dBm)} + \text{gains (dB)} - \text{losses (dB)} \tag{4.2}$$

1. **Uplink: Earth to Space**: The power, P_R, required for transmission is estimated using:

$$P_{R,uplink} = \tfrac{P_T}{4\pi R^2} G_T A_{eff} \tag{4.3}$$

   where P_T is the transmit power in dB, A_{eff} is the effective antenna aperture, R is the distance between the earth's ground terminal and satellite of interest, and G_T is the transmit antenna gain in dB.
2. **Transmission Losses**: Three main factors account for transmission losses: FSPL, atmospheric losses, and pointing losses [68]. The FSPL, L_{FS} in dB, is given by:

$$\begin{aligned} L_{\text{FS}} &= 10\log\left(\frac{4\pi f d}{c}\right)^2 \\ &= 32.4 + 20\log f + 20\log d \end{aligned} \tag{4.4}$$

   where d is the distance between the transmitter and receiver in km, and f is the frequency of operation in MHz. Additionally, there are some other losses determine the satellite link budget, such as L_{AML} is antenna misalignment loss in dB, L_{AA} is the atmospheric absorption loss in dB, and L_{PL} is the polarization mismatch loss and other cabling as well as connector's loss in the transmitters and receivers, which is calculated in dB.
3. **Downlink Calculation**: The received power level at the destination determines the quality of information and is governed by the level of signal power received [68]. The received power for the downlink, $P_{R,\text{downlink}}$ is calculated as:

$$\begin{aligned} P_{R,\text{downlink}} &= P_{\text{sat}} + G_{\text{sat}} - L_{\text{FS}} - L_{\text{AML}} - L_{PL} - L_{AA} + G_R \\ P_{R,\text{downlink}} &= EIRP + G_R - L \end{aligned} \tag{4.5}$$

   where, P_{sat} is the satellite transmitter power in dBW, G_{sat} is the satellite antenna gain in dBi, G_R is the receiver antenna gain in dBi, and L is the total transmission losses.

4. **Total Link Budget**: The total link budget for a communication link is:

$$P_{R,\text{total}} = P_{R,\text{uplink}} + P_{R,\text{downlink}} \tag{4.6}$$

This considers all losses and gains across both uplink and downlink paths.

5. **Link Performance Parameters**: Several parameters need to be monitored to estimate the satellite links and their performance. This is dependent on several factors and the configurations of the radio components. If the received signal power is equal to or less than the total noise levels N, then the signal goes undetected [68]. The performance of the satellite links is directly dependent on the signal to noise ratio (SNR) values and is expressed in dB:

$$SINR = P_{R,total} - N \tag{4.7}$$

4.3.1.5 Techno-Economic Assessment

The TCO for the satellite varies as per the purpose it is designed for. However, the widely used cost models are discussed below [2, 69].

- **Generic Cost Models**: For a satellite deployment, CAPEX, C_c includes all upfront costs related to satellite deployment, such as $C_{\text{satellite}}$ includes satellite manufacturing (such as bus cost (bus performance), payload cost (application of interest)), testing cost, and launch costs include purchasing launch stations; and C_{ground} includes developmental cost toward ground stations, gateways, and user terminals. Let $C_{\text{operating}}$ be the operating expenditure per year. There are most costs involved in $C_{\text{satellite}}$,

$$C_C = C_{\text{satellite}} + C_{\text{ground}} + C_{launch} \tag{4.8}$$

  Similarly, the OPEX, C_o includes costs toward annual operations, such as satellite operations and maintenance, ground station operations, and insurance and regulatory compliance. Furthermore, it also includes the cost of personnel, operations of sensors with transmission capabilities, and data analysis tools.

$$C_O = C_{\text{operating}} + C_{maintenance} + C_{insurance} + C_{regulatory} + C_{dataAnalysis} + C_{sensors} \tag{4.9}$$

  The TCO, C_T, of the satellite over its lifetime is estimated as shown below:

$$C_T = C_c + C_o \times T_{\text{lifetime}} \tag{4.10}$$

- **Cost toward Integrating Satellite in 4G and 5G Networks**: The cost model changes when we introduce enhanced technologies, involving MEC and edge, to the site. These are incorporated in the RAN, satellite terminal, and all edge components. The quantity of equipment required hinges on the site's dimensioning process, such as the bit rate to be delivered, and the number of eNodeB and gNodeB is determined by the bitrates, B_{r_s}, and coverage area, A, to be serviced by the site. These are mathematically modeled as shown below [2]:

$$\begin{aligned} B_{r_s} &= N_u B_{r_u} \\ N_{eNb} &= max(A/Cov_{enb}, B_{r_s}/BW_{enb})) \\ N_{gNb} &= max(A/Cov_{gnb}, B_{r_s}/BW_{gnb}) \\ C_{c,RAN} &= N_{enb}C_{enb} + N_{gnb}C_{gnb} + C_T + C_{inst} \\ C_{o,RAN} &= Site_{rental} + (P_{enbs} + P_{gnbs} + P_{ST}) \times C_{watt} + M \end{aligned} \tag{4.11}$$

  where B_{r_u} is the bitrate per user, N_u is the number of users, Cov_{enb} and Cov_{gnb} are the coverage area offered by eNodeB and gNodeB respectively, BW_{enb} and BW_{gnb} are the eNodeB and gNodeB bandwidth respectively, N_{enb} and N_{gnb} are the number of eNodeB and gNodeB required to provide the minimum B_{r_s} Mbps, C_T is the cost of the tower, C_{inst} is the cost of hardware installations, $Site_{rental}$ is the cost of renting the site for deploying eNodeBs, gNodeBs, and satellite terminals, P_{enbs}, P_{gnbs}, and P_{ST} power consumption of eNodeB, gNodeB, and satellite terminals per year, C_{watt} is the cost of the energy, and M is the ongoing maintenance costs.

The OPEX of the satellite networks (cost of satellite capacity, maintenance, and power consumption) and is represented as:

$$\begin{aligned} C_{osat,S} &= 12(B_{r_s} + B_{r_T}) \times C_{sat,Mbps} \\ C_{o,sat} &= C_{sat,S} + P_{sat,Gat} \times C_{watt} + M \end{aligned} \tag{4.12}$$

where B_{r_T} is the bitrate per site for the traffic control and the overhead, $C_{sat,Mbps}$ is the cost of 1 Mbps per month via satellite link, $C_{sat,S}$ is the cost of satellite capacity for a site S with caching data on the edge, and $P_{sat,Gat}$ is the power consumption of satellite gateway per year.

The cost of a 4G or 5G core is calculated by adding the costs of physical infrastructure (servers, storage, switches, and interfaces), network virtualization, and orchestration software, proprietary software licensing, network maintenance, the energy required to run the core, and upgrades to be supported. Core costs typically range from 8 to 12 % of the total CAPEX. Additional costs involved in the integration of satellite networks with terrestrial networks are:

$$\begin{aligned} C_{inst} &= 15\% \times C_{HW} \\ M &= 10\% \times C_c \\ C_{overhead} &= 22\% \times (C_c + C_o) \end{aligned} \tag{4.13}$$

where C_c and C_o is the sum of all the costs toward CAPEX and OPEX for the proposed solution.

When edge caching is employed to minimize the latency, the cost for the entire MEC deployment in this model is estimated based on the data to be cached at the edge, and the number of servers, as well as the storage equipment required. The additional cost toward MEC deployment is modeled as:

$$\begin{aligned} C_{MEC} &= C_{c_{MEC}} + C_{o_{MEC}} \\ C_{c_{MEC}} &= N_s \times C_s + N_{stg} \times C_{stg} + \tfrac{C_{MEC,software}}{N} \end{aligned} \tag{4.14}$$

where $C_{c_{MEC}}$ is the CAPEX toward MEC which is a sum of product of N_s number of servers and its associated cost, C_s, and N_s number of storage devices and its associated cost, C_s, $C_{MEC,software}$ cost of MEC software management, and N is the number of sites to be served by the GW.

Therefore, the modified TCO for the lifetime of the project is given as:

$$C_T = C_C + \textstyle\sum_{i=1}^{T_{lifetime}} C_o(i) + C_{overhead} + C_{MEC} \tag{4.15}$$

The annual revenue, R, for satellites with Z as the ARPU for a subscription-based service, is computed as shown below:

$$R = N_{\text{users}} \times Z \tag{4.16}$$

The payback period or the break-even time, T_{payback}, is the time required to recover the initial investment, and is calculated as shown below:

$$T_{\text{payback}} = \tfrac{C_c}{R - C_o} \tag{4.17}$$

The NPV evaluates the project's profitability over its lifetime at the discount rate (as a fraction), r, and the year index t, as shown below:

$$\text{NPV} = \textstyle\sum_{t=1}^{T_{\text{lifetime}}} \tfrac{R - C_o}{(1+r)^t} - C_c \tag{4.18}$$

while the ROI, μ measures the financial return relative to the initial investment for the period of study.

$$\mu = \tfrac{R - C_T}{C_T} \times 100\% \tag{4.19}$$

These cost models would help us determine the techno-economic viability of satellite networks for bridging the digital divide in rural communities.

4.3.2 Environmental Frameworks

The advancement in the space layer is driving interest in exploring space for different applications. This would lead to a significant increase in the frequency of space launches, deployment of satellites, and end-of-life (EOL) of satellites, in the coming years. There is a need to determine the environmental impact of satellites from the building to launch, operations to EOL. There are a vast number of parameters that determine the environmental sustainability of satellite systems. The above section offered a mathematical representation of the financial impact of energy consumption on overall costs. However, the environmental effects of satellites and associated technologies, including energy usage, carbon emissions, and space debris, need to be analyzed.

4.3.2.1 Energy Consumption

1. **Subsystem Power Requirements**: The Satellite consists of various subsystems that help in satellite operations, and includes battery, solar panels, thermal control system, payload system, and communication subsystem. The total power consumption of a satellite, P_{total} is the sum of its subsystems [70, 71]:

$$P_{\text{total}} = P_{\text{payload}} + P_{\text{TT\&C}} + P_{\text{thermal}} + P_{\text{ADCS}} + P_{\text{propulsion}} + P_{\text{other}} \tag{4.20}$$

   where P_{payload} is the power needed for mission equipment, $P_{\text{TT\&C}}$ is the power required for TT&C, P_{thermal} is the power for thermal control subsystem, P_{ADCS} is the power required for altitude determination and control system, $P_{\text{propulsion}}$ is the power required for propulsion systems to maintain stability and adjust orbits, and P_{other} for other miscellaneous factors.
2. **Energy Consumption over Time**: To estimate the energy usage by the satellites over a time period from t_1 to t_2, say orbital period or a mission time, is given as:

$$E_{\text{total}} = \int_{t_1}^{t_2} P_{\text{total}}(t)\, dt \tag{4.21}$$

   For steady-state operations:

$$E_{\text{total}} = P_{\text{total}} \times \Delta t \tag{4.22}$$

3. **Battery Sizing**: The required battery capacity, C_{battery} with an efficiency η_{battery}, to support satellite operations during eclipse periods [72], is given as:

$$C_{\text{battery}} = \frac{E_{\text{eclipse}}}{\eta_{\text{battery}}} \tag{4.23}$$

   where the energy needed during the eclipse E_{eclipse}, is estimated as

$$E_{\text{eclipse}} = P_{\text{total}} \times t_{\text{eclipse}} \tag{4.24}$$

   where t_{eclipse} is the duration of eclipse
4. **Solar Array Sizing**: The satellite is powered by solar power [73, 74], and the solar array power, P_{solar}, required to meet the power demands expected during the lifetime of the satellite is modeled as:

$$P_{\text{solar}} = \frac{P_{\text{total}}}{\eta_{\text{solar}} \times \eta_{\text{conversion}}} \tag{4.25}$$

   where
5. **Thermal Control Power**: Satellite operations generate significant heat during the lifetime, which needs to be dissipated to control the temperatures [75]. The thermal control power equation helps calculate the energy needed to manage excess heat in spacecraft systems, especially the energy absorbed from the sun, which needs to be radiated back into space. The power required to control the thermal control subsystem is modeled as:

$$P_{\text{thermal}} = \epsilon \times \sigma \times A \times T^4 \tag{4.26}$$

   where, ϵ is the emissivity of the surface, $\sigma = 5.67 \times 10^{-8}$ is the Stefan-Boltzmann constant, A is radiating surface area, and T is temperature in Kelvin.

6. **Communication Subsystem Power**: This estimates the power required for the efficient data transmission and reception between the satellites and GCS or among the satellites [76], and is calculated as:

$$P_{\text{comm}} = \frac{EIRP \times \lambda^2}{4\pi R^2 \times G_{\text{receiver}}} \tag{4.27}$$

4.3.2.2 Carbon Emissions

The total lifecycle carbon emissions for a satellite are expressed as [77]:

$$CE_{\text{total}} = CE_{\text{launch}} + CE_{\text{manufacturing}} + CE_{\text{operations}} \tag{4.28}$$

where, CE_{total} represents the total emissions (in kg CO_2), CE_{launch} is the emissions from the satellite launch phase (in kg CO_2), $CE_{\text{manufacturing}}$ is the emissions during manufacturing phase (in kg CO_2), and $CE_{\text{operations}}$ is the emissions from satellite operations phase (in kg CO_2).

1. **Carbon emissions from launch**: The carbon emissions generated during the satellite launch are modeled as [78]:

$$CE_{\text{launch}} = F_{\text{fuel}} \times c_{\text{fuel}} \tag{4.29}$$

where, F_{fuel} is mass of the fuel burned (in kg), and c_{fuel} is the carbon emission factor of the fuel (in (kg CO_2/kg fuel).

2. **Carbon emissions from manufacturing**: The emissions from satellite manufacturing are expressed as [79]:

$$CE_{\text{manufacturing}} = \textstyle\sum_{i=1}^{n} M_i \times c_i \tag{4.30}$$

where, M_i is the mass of material i used in the satellite (in kg), and c_i is the carbon emission factor for material i (in kg CO_2/kg material).

3. **Operational carbon emissions**: The emissions during satellite operations are modeled as [80]:

$$CE_{\text{operations}} = P_{\text{ground}} \times t_{\text{lifespan}} \times c_{\text{energy}} \tag{4.31}$$

where, P_{ground} is the power consumption of ground stations (in kW), t_{lifespan} is the satellite operational lifespan (in hours), and c_{energy} is the carbon intensity of electricity (in kg CO_2/kWh).

4.4 Summary

This chapter focused on comprehending satellites as a technology enabler to bridge the digital divide. These discussions address the ongoing advancements in the satellite industry–encompassing enabling technologies and architectures, and explain why satellites are gaining traction as a means to connect the unconnected. Satellite's role in tackling this global challenge depends on how effectively the strategies related to accessibility, availability, and affordability interplay to lower TCO, while leveraging economic and environmental models to build a sustainable and scalable satellite network that is community-focused and inclusive for all. This network would also integrate with the terrestrial network to offer ubiquitous coverage and reliable access. However, without deliberate efforts, it could become a technology that would widen the disparity further.

Exercises

4.1 A satellite operates on a frequency band with a bandwidth of 500 MHz. The spectral efficiency of its modulation scheme is 3.5 bps/Hz. Calculate:

(a) The maximum achievable data rate of the satellite link.
(b) The percentage reduction in throughput if the spectral efficiency drops to 2.8 bps/Hz due to interference.

4.2 A LEO satellite constellation consists of 120 satellites. Each satellite costs $5 million to build, and launching one satellite costs $10 million. Additionally, the operational cost for the constellation is $50 million annually. Calculate:

(a) The total deployment cost for the constellation.
(b) The total cost over a 5-year period, including operational expenses.

4.3 A satellite transmits with an EIRP of 50 dBW. The ground receiver has an antenna gain of 30 dB, and the free-space path loss is 200 dB. Calculate:

(a) The received signal power at the ground station.
(b) Discuss how the result would change if the path loss increases by 10 dB.

4.4 A satellite system operates on both Ku-band (12 GHz) and Ka-band (30 GHz). Evaluate:

(a) The advantages and disadvantages of using Ku-band versus Ka-band for rural connectivity.
(b) Real-world examples of satellite systems utilizing these spectrum bands.

4.5 A satellite's power consumption during transmission is 5 kW, and it achieves a data rate of 10 Gbps. Calculate:

(a) The energy efficiency of the satellite communication in bits per joule.
(b) The impact on efficiency if the power consumption increases to 6 kW while maintaining the same data rate.

4.6 Design a business model canvas for deploying satellite internet services in rural areas. Include:

(a) Key revenue streams and cost components.
(b) Potential partnerships with terrestrial network providers.

4.7 The carbon emissions during the launch of a single satellite are 200,000 kg CO_2. Calculate:

(a) The total emissions for launching a constellation of 100 satellites.
(b) The reduction in emissions if the emissions per launch are reduced by 15%.

4.8 Perform a SWOT analysis for integrating satellite networks with terrestrial networks for rural connectivity.

4.9 In a satellite network, the distribution of ground stations follows a Poisson point process with an average density of 5 stations per 100 km^2.

$P(N(A) = k) = \frac{(\lambda|A|)^k e^{-\lambda|A|}}{k!}$ where $N(A)$ is the number of points in area A, λ is the density of point, and k number of events. Then, Calculate:

(a) The probability of finding at least 3 ground stations in a 50 km^2 area.
(b) The expected number of ground stations in a 200 km^2 area.

4.10 Evaluate the techno-economic feasibility of using satellites for rural connectivity by analyzing:

(a) Deployment cost per Mbps for satellite networks versus terrestrial networks.
(b) Long-term sustainability in terms of operational costs and environmental impact.

4.11 A satellite system operates at a frequency of 12 GHz, and the distance between the satellite and the ground station is 40,000 km. Calculate:

(a) FSPL
(b) Discuss how the path loss would change if the operating frequency increases to 20 GHz.

4.12 A satellite transmits with a EIRP of 50 dBW. The FSPL is 200 dB, and the receiver antenna gain is 30 dB. Calculate:

(a) The received signal power at the ground station.
(b) Comment on the impact of increasing the receiver antenna gain by 5 dB on the received signal power.

4.13 A satellite link has the following parameters:

- Bandwidth $(B) = 20$ MHz,
- Noise temperature $(T) = 290$ K,
- Boltzmann constant $(k) = 1.38 \times 10^{-23}$ J/K,
- Received signal power (P_{received}) = -100 dBW.

Calculate:

(a) The noise power (P_{noise}) using: $P_{\text{noise}} = k \times T \times B$, and convert the result to dBW.
(b) The Signal-to-Noise Ratio (SNR) in dB using: $SNR = P_{\text{received}} - P_{\text{noise}}$.

4.14 A Ku-band satellite link operates at 14 GHz with a slant path length of 10 km through rain. The specific attenuation due to rain is 0.25 dB/km. Calculate:

(a) The total rain attenuation along the slant path.
(b) How rain attenuation would affect the link budget if it increases by 20%.

4.15 A satellite communication link has the following parameters:

- Received power $(P_{\text{received}}) = -90$ dBW,
- Noise power $(P_{\text{noise}}) = -100$ dBW,
- Required SNR = 10 dB.

Calculate:

(a) The actual SNR of the link.
(b) The link margin (difference between actual and required SNR).
(c) Discuss the implications of having a negative link margin.

References

1. L. Toka, M. Konrad, A. Pekar, G. Biczók, Integrating the skies for 6G: techno-economic considerations of lEO, HAPS, and UAV technologies. IEEE Commun. Mag. **62**(11), 44–51 (2024)
2. A. Chiha, M. Van der Wee, D. Colle, S. Verbrugge, Techno-economic viability of integrating satellite communication in 4G networks to bridge the broadband digital divide. Telecommun. Policy **44**(3), 101874 (2020)
3. Y. Shaengchart, N. Bhumpenpein, Bridging the digital divide in rural Thailand: understanding potential factors influencing starlink's satellite internet adoption. Soc. Sci. Humanit. Open **11**, 101355 (2025)
4. B.G. Evans, *Satellite Communication Systems*, vol. 38 (IET, 1999)
5. O. Kodheli, E. Lagunas, N. Maturo, S.K. Sharma, B. Shankar, J.F.M. Montoya, J.C.M. Duncan, D. Spano, S. Chatzinotas, S. Kisseleff et al., Satellite communications in the new space era: a survey and future challenges. IEEE Commun. Surv. Tutor. **23**(1), 70–109 (2020)
6. G. Varrall, *5G and Satellite Spectrum, Standards, and Scale* (Artech House, 2018)
7. A. Lacono, Space explained: what is the satellite spectrum used for? Accessed 29 Oct 2024
8. P. Harati, E. Rosello, I. Dan, E.R. Bammidi, J. Eisenbeis, A. Tessmann, D. Schwantuschke, R. Henneberger, I. Kallfass, E-band downlink wireless data transmission for future satellite communication, in *Topical Workshop on Internet of Space (TWIOS)* (IEEE, 2017), pp. 1–4
9. U.J. Lewark, J. Antes, J. Walheim, J. Timmermann, T. Zwick, I. Kallfass, Link budget analysis for future E-band gigabit satellite communication links (71–76 and 81–84 GHz). CEAS Space J. **4**, 41–46 (2013)
10. P. Ferrazzoli, L. Guerriero, J.P. Wigneron, Simulating L-band emission of forests in view of future satellite applications. IEEE Trans. Geosci. Remote Sens. **40**(12), 2700–2708 (2002)

11. Y. Kawakami, T. Sato, Y. Otsu, S. Hama, Study on S-band mobile communications systems using a geostationary satellite, in *17th AIAA International Communications Satellite Systems Conference and Exhibit* (1998), p. 1212
12. A. Pastukh, V. Tikhvinskiy, S. Dymkova, O. Varlamov, Challenges of using the L-band and S-band for direct-to-cellular satellite 5G–6G NTN systems. Technologies **11**(4), 110 (2023)
13. M. Irsigler, G.W. Hein, A. Schmitz-Peiffer, Use of C-band frequencies for satellite navigation: benefits and drawbacks. GPS Solutions **8**, 119–139 (2004)
14. J. Albuquerque, Satellite operators challenge mobiles' use of C-band. Accessed 29 Oct 2024
15. E.L. Cid, M.G. Sanchez, A.V. Alejos, Wideband analysis of the satellite communication channel at Ku-and X-bands. IEEE Trans. Veh. Technol. **65**(4), 2787–2790 (2015)
16. D. Venugopal, C. Muthugadahalli, K.S. Mohanavelu, K. Narayanan, Ka band satellite communication systems—applications and configurations, in *Proceedings of the International Astronautical Congress (IAC)* (2015), pp. 12–16
17. T. Rossi, M. De Sanctis, M. Ruggieri, C. Riva, L. Luini, G. Codispoti, E. Russo, G. Parca, Satellite communication and propagation experiments through the alphasat Q/V band Aldo Paraboni technology demonstration payload. IEEE Aerosp. Electron. Syst. Mag. **31**(3), 18–27 (2016)
18. Viasat, Satellite communications: a brief history from sputnik to ViaSat-3. Accessed 29 Oct 2024
19. IMSO, Background IMSO. Accessed 29 Oct 2024
20. M. Centenaro, C.E. Costa, F. Granelli, C. Sacchi, L. Vangelista, A survey on technologies, standards and open challenges in satellite IoT. IEEE Commun. Surv. Tutor. **23**(3), 1693–1720 (2021)
21. L. Bartamian, D. Martin, P. Anderson, *Communication Satellites* (American Institute of Aeronautics and Astronautics Inc., 2007)
22. F. Rinaldi, H.-L. Maattanen, J. Torsner, S. Pizzi, S. Andreev, A. Iera, Y. Koucheryavy, G. Araniti, Non-terrestrial networks in 5G & beyond: a survey. IEEE Access **8**, 165178–165200 (2020)
23. M. Hosseinian, J.P. Choi, S.H. Chang, J. Lee, Review of 5G NTN standards development and technical challenges for satellite integration with the 5G network. IEEE Aerosp. Electron. Syst. Mag. **36**(8), 22–31 (2021)
24. M.M. Azari, S. Solanki, S. Chatzinotas, O. Kodheli, H. Sallouha, A. Colpaert, J.F.M. Montoya, S. Pollin, A. Haqiqatnejad, A. Mostaani et al., Evolution of non-terrestrial networks from 5G to 6G: a survey. IEEE Commun. Surv. Tutor. **24**(4), 2633–2672 (2022)
25. J. Liu, Y. Shi, Z.M. Fadlullah, N. Kato, Space-air-ground integrated network: a survey. IEEE Commun. Surv. Tutor. **20**(4), 2714–2741 (2018)
26. Spacewatch Africa, Cost of satellites. Accessed 29 Oct 2024
27. M. Richharia, L.D. Westbrook, *Satellite Systems for Personal Applications: Concepts and Technology* (John Wiley & Sons, 2011)
28. T. Ahmmed, A. Alidadi, Z. Zhang, A.U. Chaudhry, H. Yanikomeroglu, The digital divide in Canada and the role of LEO satellites in bridging the gap. IEEE Commun. Mag. **60**(6), 24–30 (2022)
29. F. Fourati, M.-S. Alouini, Artificial intelligence for satellite communication: a review. Intell. Converg. Netw. **2**(3), 213–243 (2021)
30. I. Siddique, Digital satellite technology: innovations and applications in the modern era. Eur. J. Adv. Eng. Technol. **11**(6), 39–46 (2024)
31. M. Goswami, H.M. Kwon, K. Pham, J. Lyke, Satellite MIMO digital beam-forming under nonlinear high-power amplifier and keyhole, in *2019 International Conference on Computing, Networking and Communications (ICNC)* (IEEE, 2019), pp. 313–317
32. W. Jiang, Software defined satellite networks: a survey. Digital Commun. Netw. **9**(6), 1243–1264 (2023)
33. N. Friedrich, Welcome to the era of software-defined satellites. Accessed 29 Oct 2024
34. S. Kota, G. Giambene, Satellite 5G: IoT use case for rural areas applications, in *Proceedings of the Eleventh International Conference on Advances in Satellite and Space Communications-SPACOMM* (2019), pp. 24–28
35. S. Alfattani, W. Jaafar, Y. Hmamouche, H. Yanikomeroglu, A. Yongaçoglu, Link budget analysis for reconfigurable smart surfaces in aerial platforms. IEEE Open J. Commun. Soc. **2**, 1980–1995 (2021)
36. M.S.B. Syed, H.M. Attaullah, S. Ali, M.I. Aslam, Wireless communications beyond antennas: the role of reconfigurable intelligent surfaces. Eng. Proc. **32**(1), 10 (2023)
37. J.S. Sidhu, S.K. Joshi, M. Gündoğan, T. Brougham, D. Lowndes, L. Mazzarella, M. Krutzik, S. Mohapatra, D. Dequal, G. Vallone et al., Advances in space quantum communications. IET Quantum Commun. **2**(4), 182–217 (2021)
38. C. Wang, Y. Wang, P. Lian, S. Xue, X. Qian, Yu. Yu Shi, B. Du Jia, J. Liu, B. Tang, Space phased array antenna developments: a perspective on structural design. IEEE Aerosp. Electron. Syst. Mag. **35**(7), 44–63 (2020)
39. T. Chaloun, L. Boccia, E. Arnieri, M. Fischer, V. Valenta, N.J.G. Fonseca, C. Waldschmidt, Electronically steerable antennas for future heterogeneous communication networks: review and perspectives. IEEE J. Microwaves **2**(4), 545–581 (2022)
40. J.S. Bardin, Satellite cyber attack search and destroy, in *Computer and Information Security Handbook* (Elsevier, 2025), pp. 1561–1580
41. M.N Sweeting, Modern small satellites-changing the economics of space. Proc. IEEE **106**(3), 343–361 (2018)
42. R. Karim, P. Malcovati, On-chip-antennas: next milestone in the big world of small satellites–a survey of potentials, challenges, and future directions. IEEE Aerosp. Electron. Syst. Mag. **36**(1), 46–60 (2021)
43. NASA, What are smallsats and cubesats? Accessed 29 Oct 2024
44. V. Velivela, Small satellite constellations: the promise of internet for all. ORF Issue Brief **107**, 1–12 (2015)
45. M.D.S. Akhtar, G. Mathur, O. Kravchenko, M. Rakhra, Connecting the unconnected: bridging the digital divide with affordable satellite through enabled smartphones, in *E3S Web of Conferences*, vol. 453 (EDP Sciences, 2023), p. 01054
46. R. Bassoli, F. Granelli, C. Sacchi, S. Bonafini, F.H.P. Fitzek, Cubesat-based 5G cloud radio access networks: a novel paradigm for on-demand anytime/anywhere connectivity. IEEE Veh. Technol. Mag. **15**(2), 39–47 (2020)
47. N. Saeed, A. Elzanaty, H. Almorad, H. Dahrouj, T.Y. Al-Naffouri, M.S. Alouini, Cubesat communications: recent advances and future challenges. IEEE Commun. Surv. Tutor. **22**(3), 1839–1862 (2020)
48. X. Luo, H.H. Chen, Q. Guo, LEO/VLEO satellite communications in 6G and beyond networks–technologies, applications and challenges. IEEE Netw. (2024)
49. G. Giambene, S. Kota, P. Pillai, Satellite-5G integration: a network perspective. IEEE Netw. **32**(5), 25–31 (2018)
50. S. Chen, S. Sun, S. Kang, System integration of terrestrial mobile communication and satellite communication–the trends, challenges and key technologies in B5G and 6G. China Commun. **17**(12), 156–171 (2020)
51. S. Chen, Y.-C. Liang, S. Sun, S. Kang, W. Cheng, M. Peng, Vision, requirements, and technology trend of 6G: How to tackle the challenges of system coverage, capacity, user data-rate and movement speed. IEEE Wirel. Commun. **27**(2), 218–228 (2020)

52. S.B.R. Tirmizi, Y. Chen, S. Lakshminarayana, W. Feng, A.A. Khuwaja, Hybrid satellite–terrestrial networks toward 6G: key technologies and open issues. Sensors **22**(21), 8544 (2022)
53. N2YO, Objects currently in orbit. Accessed 29 Oct 2024
54. N.A. Testoedov, S.N. Karutin, Space geodesy, communications, and navigation: history of the development, state, and prospects. Her. Russ. Acad. Sci. **91**(6), 647–655 (2021)
55. S. Janson, 25 years of small satellites, in *Small Satellite Conference* (2011)
56. Z. Kassas, M. Neinavaie, J. Khalife, N. Khairallah, S. Kozhaya, J. Haidar-Ahmad, Z. Shadram, *Navigation with Starlink Satellites* (UC Irvine, Enter LEO on the GNSS stage, 2021)
57. IARU, Satellites. Accessed 29 Oct 2024
58. A. Wulff, Amateur radio satellites, in *Beginning Radio Communications: Radio Projects and Theory* (2019), pp. 173–194
59. OneWeb, OneWeb. Accessed 29 Oct 2024
60. Y. Henri, The OneWeb satellite system, in *Handbook of Small Satellites: Technology, Design, Manufacture, Applications, Economics and Regulation* (Springer, 2020), pp. 1091–1100
61. A.K. Maini, *Military Satellites* (Wiley Telecom, 2018)
62. B. Jaksic, D. Miljkovic, V. Maksimovic, M. Petrovic, B. Gvozdic, Satellite television transmission in the world-broadcasting systems and standards. Acta Sci. Technol., 42 (2020)
63. Global Data, What's the difference between KPI and KVI ? Accessed 29 Oct 2024
64. Y. Wang, X. Ding, G. Zhang, A novel dynamic spectrum-sharing method for GEO and LEO satellite networks. IEEE Access **8**, 147895–147906 (2020)
65. N.A. Ochuba, D.O. Olutimehin, O.G. Odunaiya, O.T. Soyombo, Sustainable business models in satellite telecommunications. Eng. Sci. Technol. J. **5**(3), 1047–1059 (2024)
66. Intelsat, Satellite link budget. Accessed 29 Oct 2024
67. J.N. Pelton, S. Madry, S. Camacho-Lara, *Handbook of Satellite Applications* (Springer Publishing Company, Incorporated, 2017)
68. S.G. Marquina, Study and simulation of communication links in a leo satellite constellation based on link budget calculations. B.S. thesis, Universitat Politècnica de Catalunya, 2022
69. S. Nakasuka, Micro-satellite project mission and cost model—guidance book. Accessed 29 Oct 2024
70. NASA, Spacecraft electrical power systems (2018)
71. A.O.O. Esho, T.D. Iluyomade, T.M. Olatunde, O.P. Igbinenikaro, A comprehensive review of energy-efficient design in satellite communication systems. Int. J. Eng. Res. Updates **6**(2), 13–25 (2024)
72. Satsearch, Satellite batteries for cubesats and nanosats (2021)
73. Sparkwing, Satellite solar panels (2022)
74. Osborne Engineering, Spacecraft solar array design (2021)
75. NASA, Thermal control systems for satellites (2021)
76. Tutorials Point, Satellite communication power and antenna subsystems (2022)
77. NASA, Carbon emissions tracking for satellites (2023)
78. Cosmos Magazine, Greenhouse gas emissions of spacex rockets (2022)
79. Climatiq, Materials and manufacturing emission factors (2023)
80. MIT, Greenhouse gas emissions impacting satellite operations (2023)

5 Air Networks: UAVs, NTFPs, and HAPS

What are air networks? Air networks are aerial communication networks that operate within the Earth's atmosphere [1]. The main objectives of air networks, when compared to space networks, are to assist military operations and the disaster recovery phase, along with providing regional coverage for a specific use case, location, or application, and offering a temporary communication setup in remote and disaster-affected areas.

An **airborne communication system** comprises **airborne platforms**, which include various systems, vehicles, or structures equipped with communication technologies. It includes technologies such as HAPS (e.g., HIBSs), UAVs, NTFPs, balloons, and aircraft communication systems [1]. With recent advancements in R&D and military applications of airborne communication, their usage has been growing globally. Air networks are emerging as a promising solution for connecting the underserved communities and expanding the connectivity to offer higher network performance, driven by a growing market demand. Nevertheless, the technologies are still quite in an early developmental stage, and further development is required before the technology can globally bridge the digital divide to support wide-scale commercialization as well as large-scale spectrum allocation, monitoring, and optimization.

5.1 State-of-the-Art for Air Layer

Air networks offer wider coverage compared to terrestrial networks, and higher throughput and lower latency compared to space networks. Before going deeper into the topic in this chapter, let's broadly define what the three key technologies of air networks are.

- **UAV and drones**: It is a remotely controlled or fully autonomous aircraft, which was initially built for military purposes, and is now being widely used for applications, such as communication, research, photography, surveillance, and secure communication. It has been studied since World War I.
- **NTFP**: It is an unmanned flying platform that is in quasi-stationary positions, which can move horizontally and vertically to offer cellular communication and has a tether connected to the ground station that provides it with continuous power and data. It overcomes the power and data constraints that exist in UAVs and HAPS. It has been explored since the twentieth century.
- **HAPS**: A long-endurance aerial system that flies in the stratosphere layer of the Earth's atmosphere, often solar powered, and is widely used for communication, surveillance, sensing, and monitoring services.

Air networks may play a basic intermediary role between space and terrestrial networks to facilitate smooth communication. As relay stations, HAPS, and UAVs obtain signals from satellites and transfer them to ground stations, and vice versa through the dynamical reconfiguring of the communication links to suit the needs and environmental factors around them; therefore, it is resilient, flexible, indispensable, and seamless. Airborne relay stations reduce the latency by shortening the communication path and extending coverage to distant or underserved areas. With ongoing advancements in wireless communications, airborne systems will become essential for connectivity across all levels.

S. Koratagere Anantha Kumar, *Wireless Broadband Access to Bridge the Digital Divide*,
Textbooks in Telecommunication Engineering, https://doi.org/10.1007/978-3-032-09261-8_5

5.1.1 UAVs and Drones

Unmanned systems are autonomous platforms that can be easily programmed to perform missions with and without the intervention of a pilot; they can be UAVs, unmanned ground vehicles (UGVs), unmanned underwater vehicles, unmanned surface vehicles, or unmanned spacecraft [2, 3]. In this section, we are going to explore civil applications for UAVs with a special focus on bridging the digital divide. UAVs have attracted growing interest because of their ease of deployment, low cost, high mobility, and hovering capability, making them ideal for real-time applications.

An emerging use case for drones will be UAV-assisted wireless communication with a flying BSs to offer coverage and connectivity over the wireless network. UAVs might act as a low-cost temporary aerial wireless BS in dealing with outages in communication infrastructure during emergencies, or equipment malfunctions in conjunction with catering to high user demand. Some of the primary reasons for the popularity of UAV-assisted wireless communication include increasing endurance, excellent maneuverability, and hovering of these drones for a range of applications; additional factors include falling costs, size, and weight [4].

The fundamental networking architecture of UAV-aided communication networks is depicted in Fig. 5.1 [5]. The UAVs need to support a combination of several systems that enable flight, navigation, communication, and data collection. There are two connection links in the architecture: the control and non-payload communications (CNPC) link and the data link [5].

- **CNPC**: All UAVs that provide high reliability, low latency, minimal data requirements, and secure two-way communications can be operated safely owing to the CNPC link. UAVs and satellites use CNPC links to transmit safety-critical information to GCS. There are three different kinds of information exchanged using the primary CNPC connection, including:

  - common and control from GCS to UAVs,
  - aircraft status report from UAVs to GCS, and
  - sense-and-avoid information among the UAVs.

  The primary CNPC linkages are also essential when human involvement is required. The UAVs are also contacted when they are near or within controlled airspace, for example, the airport, using air traffic controller (ATC) links. The links for CNPC must be operated on protected spectrum L-band (960–977 MHz) and C-band (5030–5091 MHz).

  The secondary CNPC link is a supplementary communication link between the UAV and the satellite to ensure the safe and efficient operation of UAVs. It is always in standby mode and acts as a backup communication link when the primary CNPC link fails or becomes unreliable. CNPC links should employ superior high-security controls to avoid *ghost control*

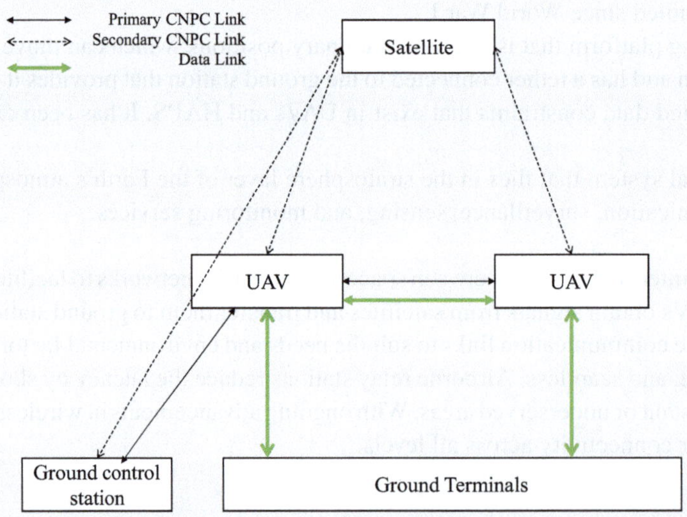

Fig. 5.1 Basic networking architecture of UAVs-aided communication networks

scenario to avoid unauthorized agents controlling the UAVs. The secondary CNPC link is critical in applications such as beyond visual line of sight (BVLOS) operations (in scenarios such as long-distance, autonomous, and direct LOS is not possible), urban operations, disaster management, and military applications. Integrating a secondary CNPC link in UAVs presents many other challenges, such as cost, complexity, spectrum allocation, and power consumption.

- **Data link**: The data link in UAVs is a critical communication bridge between the UAV and its GCS. It ensures the seamless exchange of information required for navigation, control, payload operation, and data transmission. A reliable data link is fundamental for UAVs, whether operating in visual line-of-sight (VLOS) or BVLOS. The key components of a UAV data link include

  - GCS that contains transceivers to communicate with UAVs and interfaces with the operators via software to control and monitor the UAV.
  - UAV onboard communication system for sending and receiving data.
  - communication medium that depends on range and application and includes technologies such as RF systems, satellite communication, 4G, and 5G networks.
  - protocols and standards to ensure interoperability and security for UAVs communication. The most commonly used standards in UAVs are IEEE 802.11 for Wi-Fi or specialized military protocols.

5.1.1.1 Spectrum Bands

Each country has authorized a unique set of frequency bands for UAVs operations depending upon the regional spectrum availability. Currently, there isn't a single standard universal spectrum band for UAVs, however, the most commonly adopted spectrum bands for this technology are 900 MHz, 2.4 GHz, and 5.8 GHz, as they offer a balance between range, data transfer rates, and interference. UAVs operates on various frequency bands serving different purposes in each band [6], such as

- **900 MHz**: This band is often used for long-range communication, integration with terrestrial networks, and telemetry. It provides a good range and penetration through obstacles [7]. This band supports mesh networks for long range (LoRa) and Meshtastic.
- **2.4 GHz**: This is the Industrial, Scientific, and Medical (ISM) band, which is the universal unlicensed spectrum, commonly used for video and control signal transmission for small UAVs [8]. This band is widely used for consumer drones, offering a balance between range and data rates.
- **5.8 GHz**: This frequency range supports high-bandwidth applications and is part of the ISM spectrum range for small UAVs [8]. Hence, the operators in this band need to ensure compliance with ISM band regulations and operating standards [6].
- **UHF**: This band is used for long-range communications and control signal transmissions for commercial and military use cases [9]. It provides a reliable and secure communication link. This band is used for UAV-to-UAV communication and UAVs-to-ground communication [10].
- **L** and **S bands**: These frequency bands are used for UAVs while they are being integrated with the satellite networks and BVLOS operations [11]. L-band operates at 1–2 GHz and is used for radio, telecommunication, and GPS as well. UAVs would help in establishing PTP, PTMP, tactical communication, mesh networks, and support data links [12]. UAVs operating on S-band to transmit video data from UAVs to GCS, either through PTP or PTMP communication links [13, 14].

The regulatory body in each country would ensure that UAVs are operated safely and without causing any interference to other communication technologies in all layers (space, air, and ground) while aligning with international standards to facilitate the scalability of the technology.

5.1.1.2 Applications

Some of the notable applications of UAVs that help in bridging the digital divide are summarized below

- UAVs for ubiquitous and seamless coverage by acting as an aerial BSs. This would be the case in scenarios when there is a natural disaster, an emergency, remote areas, or large events. By hovering at optimal altitude, UAV can ensure reliable and secure communication to end users by offering the range and data rates for the intended use case. The common types of UAVs suitable for these scenarios are fixed wing, Balloon, VTOL, Viking aircraft, Tethered Helikite balloon, and the FALCOM Quadrotor.

- In areas where deploying a terrestrial network is a challenge or impossible, UAVs could potentially serve as a network gateway since it can support the operational frequency bands of end user devices. UAVs can be rapidly deployed to establish emergency communication networks. In remote or disaster-stricken areas, UAVs would act as a temporary backbone network to provide connectivity to the communication network as well as the Internet. UAVs can even provide vital connectivity for rescue teams and affected populations when ground-based infrastructure has been compromised. This ensures that effective coordination can be achieved by emergency services and people can communicate with their loved ones.
- UAVs can provide wireless coverage in rural and remote areas where building a traditional telecommunication network is challenging and not affordable for the end users. In this scenario UAVs could act as relay nodes to provide wireless broadband connectivity between two or more places without having a direct LOS communication link. By using UAVs, operators can increase their range and offer connectivity to underserved communities. With the help of advanced relay techniques, the throughput offered by this system can be maximized to ensure that end users are receiving higher data rates.
- UAVs helps in gathering delay-tolerant data from a large number of distributed wireless devices such as IoT sensors, for applications related to farming, energy management, soil assessment, and other activities.
- The other use of UAVs is for testing and optimization of wireless networks, especially for gathering data on solutions for bridging the digital divide. They are flown over the different areas, measuring signal strength to find dead zones and blind spots. This will, in turn, help network operators improve coverage and performance.
- Cellular-connected UAVs are making low-latency, ultra-reliable communications with ultra-high throughput a reality, guaranteeing coverage for all users.

5.1.1.3 Timeline

Since the early twentieth century until the early 2000s, the UAVs were developed for mostly military applications and surveillance, such as deployment in hostile territory to reduce pilot losses [5]. By the 2010s, drones were accessible to the general public and numerous civilian and commercial use cases were being explored, especially research for UAV-to-ground and UAV-to-UAV communications [15]. The advent of 4G LTE by the mid-2010s provided a basis for UAV extended LTE support for aerial vehicles. By the end of the 2010s, UAVs began offering support for several use cases, including weather monitoring, logistics, delivery, communication, traffic control, emergency search, rescue, and relaying. These drones were designed to be smaller than HAPS, faster to deploy, lightweight ($\leq$25 kg), and miniature in size. The key focus area for designing UAVs was to improve connectivity, manage interference, and ensure secure communication.

In the early 2020s, the research direction steered toward improvements in UAVs communication with the advent of 5G, especially focusing mainly on latency, data rates, and better reliability [15]. The demand for 5G and UAV integration steadily increased. To accomplish the capabilities offered by the SAGIN architecture, the current and upcoming research projects concentrate on creating 3D networks that combine UAV-UAV, UAV-ground, and UAV-satellite communication.

5.1.2 NTFPs

With the research advancing to beyond 5G (B5G) and 6G, there is a need to explore how to connect the unconnected. Some of the upcoming technologies to bridge these gaps are LEO satellites, UAVs, and HAPS. However, each of these technologies has some shortcomings, such as backhaul, high cost, delay, and slow deployment, that need to be overcome before they can be widely available [16, 17]. To overcome these shortcomings, the stakeholders are exploring a newer technology called NTFPs, which can integrate with terrestrial networks and NTN in future 3GPP releases; which are expected to provide ubiquitous coverage and have a ground tether unit that continuously supplies the platform with data, backhaul capacity, endurance, persistence, security, and power. NTFPs are relatively cost-efficient to operate compared to other air, space, and ground layer technologies [18]. Some of the key use cases where NTFPs are explored include communication, energy harvesting, entertainment, science, research, public safety, disaster management, government, and defense [17].

Airborne platforms are vehicles that can fly in the air by opposing gravity with either a static lift or a dynamic lift. There are three key categories of NTFPs:

- **Lighter-than-air (LTA)**: Balloons and blimps use static lift, also called aerostatic lifts, use low-density LTA gas such as helium to rise in the air. The difference in the density of the air inside and the air enveloping the outside of the aerostat produces buoyancy according to Archimedes' principle. Spherical **Balloons** are the most often used NTFPs because they are easy to manufacture, deploy and deploy, and available at a lower cost compared to other NTFP. They can reach a maximum altitude of 600–700 m, carry a maximum payload of around 50 kg, and sustain wind speeds of around 20–40 kmph [17].

  Blimps, also known as a streamlined aerostat, are high-performance LTA platforms that can carry a heavy payload, sustain higher speeds, and stay aloft at higher altitudes. Depending on the size, altitude, and maximum payload, blimps are classified as [19]:

  - *Tactical Class*: These blimps are compact and can be deployed rapidly. Their envelope size ranges from 12–17 m, carries a payload of up to 27 kg, and reaches an altitude of 300 m. They can stay aloft for 7 days, withstand winds of up to 100 kmph, and are mostly employed for border surveillance.
  - *Operations Class*: These blimps have a medium-sized envelope of 22–28 m, carry a payload of up to 200 kg, and reach an altitude of up to 1 km. They can stay aloft for 14 days, withstand winds of up to 100 kmph, and are mostly employed for maritime surveillance and border surveillance.
  - *Strategic Class*: These blimps have a large-sized envelope of 71–74 m, carry a payload of up to 2300 kg, and reach an altitude of up to 4.6 km. They can stay aloft for 30 days, withstand winds of up to 166 kmph, and are mostly employed for long surveillance and monitoring missions to cover a large area.

  Buoyant airborne turbines (BATs) are wind turbines that can reach an altitude of 600 m, where the wind speeds are twice that at the ground level. These LTA platforms are harvesting energy from the wind and transmitting it to the ground stations using the tether that holds them in place. BATs have helium gas in their envelopes.
- **Heavier-than-air (HTA)**: Airplanes and UAVs use dynamic lift, and are called aerodynes. Their upward lift is caused by the relative motion between the HTA platforms and the air using Bernoulli's principle. The most popular HTA NTFPs are [20]:

  - *tUAVs*: These are UAVs with a physical tether to supply them power and data from the ground [21]. They can reach an altitude of 200 m, carry a payload of up to 15 kg, and stay aloft for 2–4 days due to motor overheating challenges [17]. These platforms are also equipped with battery packs as a backup solution when the tether is damaged or there is a power cut.
  - *Airborne Turbine Kite*: These are wind turbines that can harvest energy from the air at altitudes up to 4600 m [22]; since wind speed is higher at higher altitudes. The electric energy generated could be transmitted to the ground for harvesting the energy.

- **Hybrid Category**: This uses both static lift and dynamic lift, and commonly used categories are

  - *Helikites*: This design is a combination of portmanteau helium and kites and is composed of an oblate-spheroid balloon filled with helium gas to provide static lift, and a kite structure to provide dynamic lift [23]. It combines the benefits of both lift types, offering a smaller size and the ability to reach higher altitudes. Helikites can stay in the air for 14 days at an altitude of 1.5 km, and carry a payload of 23 kg.
  - *Hybrid Airships*: These are airships that uses 40% dynamic lift, and 60% static lift [24]. These hybrid airships can land and take off anywhere from a large open space or a flat field. These airships are primarily built for passenger travel, heavy cargo delivery, and military logistics, and are being explored for photography, advertisement, and aerial broadcasting. These can reach up to 6 km in altitude and carry up to 60,000 kg.

Some of the key components of NTFPs are shown in Fig. 5.2. **Envelope** of NTFPs contains gas that allows the platform to soar and stay aloft and could be designed either as spherical, fish-shaped, streamline (blimps) form, and often designed using synthetic material, and additionally use laminates to prevent degradation and abrasion due to ultraviolet exposure [17]. **Lifting gas**, either helium or hydrogen, is filled within the envelope to offer it static lift and has a lower density than air [25]. The **payload** is the weight that NTFPs carries to serve the use case of interest, supporting system, as well as operational payload.

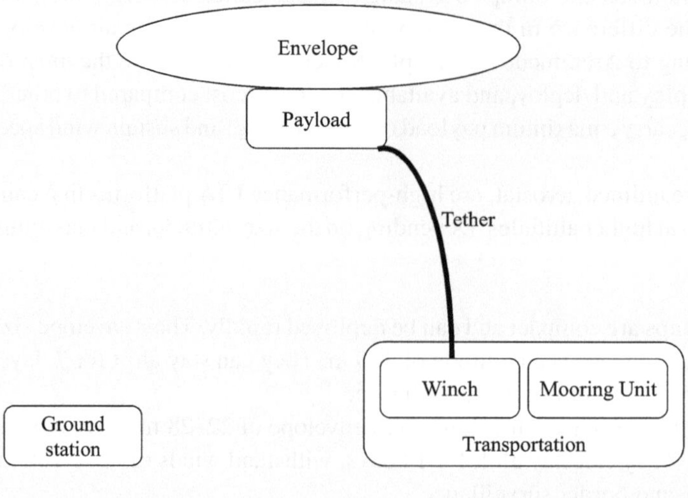

Fig. 5.2 The key components of NTFPs

The **tether** is rolled around a winch at one extremity while the other is connected to the envelope or the shell to provide a power supply to the platform through a power line, and data through the optical fiber, withstanding the harsh operating conditions. The **Mooring stations** system inflates the aerostat envelope before launch and deflates it during maintenance and after flight. **Anchor point or unit** refers to the unit into which the platform is anchored for painting while aloft. **GCS** serves as an operations control station and performs functions such as altitude control of the platforms, control and monitor platforms and equipment, store and process the data (videos and images) related to the application of interest. The components of the NTFPs need to be **transported** to the location of interest via different modes of transportation (truck or ship).

5.1.2.1 Use Cases and Applications

NTFPs is planned to cater to a wide range of applications and use cases related to government, defense, entertainment, science, research, environment sensing, and energy harvesting. NTFPs uses that would help in bridging the digital divide includes [17]:

- **Public safety and disaster relief**: NTFP also offer disaster recovery communication services to help with decision-making by providing communication links between the manager and the crew, among the first responders during the disaster to assess and communicate the situation, aiding in crowd management and riot control, and alerting vehicles to events in their area [26–28].
- **Communication**: NTFPs would reap the benefits of NTN technologies and other air layer technologies to provide over a large area and high throughput to the end users. This would help in bridging the digital divide across the globe at an affordable price [29, 30]. NTFPs could also act as transmitters or relays depending upon the requirements of the use case, and provide coverage along the coastline for sailors, fishermen, and personnel on floating structures [31].
- **Science, research, and environment**: NTFPs are equipped with sensors to collect data for remote sensing use cases to detect natural disasters and changes in the terrain conditions [32, 33]. They can also be used for research and teaching students about the changes that can only be visible from an aerial view. NTFPs are also used by meteorological data recording and aerial survey to gather data such as atmospheric conditions, temperature, wind speed, air pressure, humidity, visibility, pressure formation, seismic activities, and ocean currents [34, 35]. NTFPs helps agriculture and farming by monitoring crop health and combating farming fraud to decrease misleading tactics [36, 37].
- **Energy harvester**: As NTFPs are aloft at high altitudes, they are able to harvest wind and solar energy and transmit it back to the ground to generate power from renewable sources [38, 39].

5.1.2.2 Timeline

The development of tethered flying platforms has seen significant advancement over the recent years. These technologies were initially built for military purposes and then extended for commercial use cases. The tethered flying platforms for communication have undergone a truly revolutionary path over the years, from simple concepts to sophisticated systems integrated with the modern communication network. Even in the early twentieth century, the first concepts of using tethered balloons and kites for basic signal transmission had evolved [40]. The initial versions of this technology offered a rudimentary type of communication, which was primarily built for military communication and observational usage. During World War II, tethered balloons were used widely for military surveillance and communication as they provided a stable platform for transmitting signals and tracking the enemy movements [41, 42].

The late twentieth century was marked by important technical advancements in the creation of materials that allowed the development of advanced tethered platforms [43]. Tethered helicopters and drones appeared in the 1970s and 1980s, which offered more stable communication links, and they were reliable [42, 44].

The commercial applications for tethered platforms began in the early twenty-first century via tethered drones for instant communication during large-scale events or disaster relief [45, 46]. As tethered flying platforms gained popularity, integration with existing terrestrial and NTN systems emerged as a cost-effective and flexible solution [17]. NTFPs provides reliable power and data supply, making it a promising candidate for integration with 6G systems [47, 48]. This integration is intended to bridge the digital gap at an affordable cost while also providing seamless network access.

5.1.3 HAPS and HIBS

A HAPS is a high-altitude pseudo satellite, also known as an atmospheric satellite, offering long endurance at high altitudes to offer observation and communication services. This technology has been regarded as a cost-efficient solution for providing network access to rural or remote areas. Both industry and academia are paying attention to HAPS because of its potential to offer affordable Internet services to a wide geographic area. For instance, one HAPS in the stratosphere at a 20 km altitude can provide connectivity services to a tens-of-kilometers-radius area [49]. Since these satellites are much lower than the space satellites (LEO, MEO, and GEO), they are capable of providing direct connectivity to smartphones [50]. HAPS are expected to provide coverage, capacity, and high QoS both in rural and urban areas.

HAPS can stay at a quasi-stationary position, which would help it provide ubiquitous connectivity [51]. HAPS communication technology uses UAVs, such as balloons, blimps, airships, and other aircraft. Since HAPS has a smaller footprint compared to LEO satellites, it can support higher throughput, low latency (0.13 to 0.33 ms), and would return to earth once it reaches EOL [51]. They are capable of providing Internet backhaul and cellular communication services (4G, 5G, and 6G). Additionally, HAPS are ideal solutions for disaster recovery, as they offer terrestrial network capabilities to users during a disaster or emergency scenario without the need for special equipment.

The HAPS network architecture is shown in Fig. 5.3 [50]. The stratospheric layer's HAPS is situated above the altitude of commercial aviation traffic and is almost free from weather disturbances, including lightning and thunderstorms. The lack of clouds permits higher energy absorption from solar energy [51]. HAPS comprises two types of platforms–aerostatic and aerodynamic, as discussed in Sect. 5.1.2.

HAPS supports two-way communication, which functions as a network node in the stratosphere. It enables bidirectional communication between the users and the wider network infrastructure. ITU radio regulations (RR) defines HAPS as radio stations located on an object at an altitude of 20–50 km and located at a specific, relative point relative to the Earth [52]. A HAPS system comprises of two primary components [51]:

- *Non-terrestrial part* that constitutes all airborne or space-based components necessary for HAPS operations, such as flight control subsystem, energy management subsystem, and communications payload subsystem. Additionally, the non-terrestrial part of HAPS would also consist of aerospace-based communication nodes to interact with space and air layer technologies to form a part of a mega-constellation.
- *Terrestrial part* that includes the ground-based equipment essential for HAPS operations, such as ground station, communication gateway, and terrestrial networks, say 5G and 6G [53].

HAPS technology that operates in IMT spectrum bands and provides mobile broadband access to standard smartphone services is known as HIBS. While HIBS can help close the digital divide, it needs to be vigilant not to interfere with existing

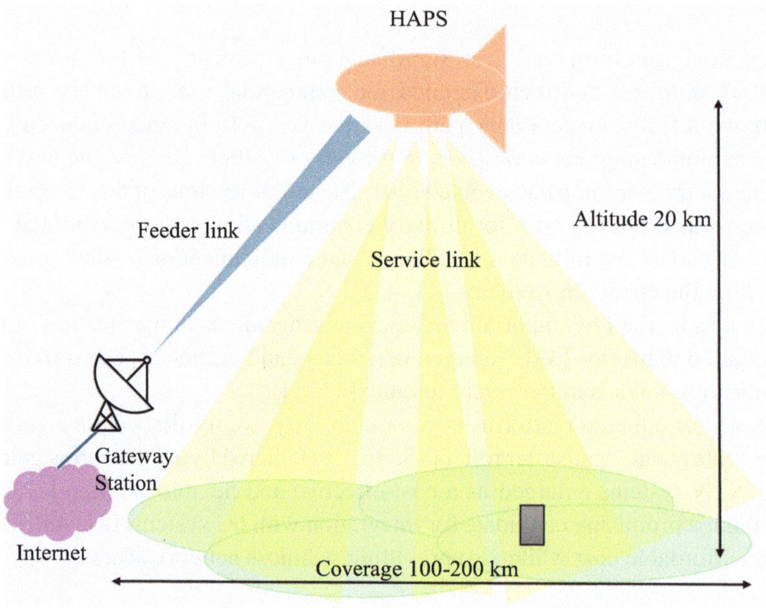

Fig. 5.3 HAPS system architecture

services and safeguard incumbent users (satellites, terrestrial networks, aviation, and weather services) from interference since these technologies also operate in the IMT spectrum bands.

5.1.3.1 Spectrum Bands

Since 1997, ITU has been making efforts to integrate and support HAPS nodes in the communication network [51]. HAPS operates in different frequency bands to ensure the flexibility, efficiency, and ability to address myriad use cases that it needs to serve. The HAPS frequency bands are regulated by the ITU and national regulators at the country level. Based on the technical investigation, the bandwidth size of 396–2969 MHz and 324–1505 MHz is required for ground-to-HAPS links and HAPS-to-ground links, respectively. The primary frequency bands used for HAPS communication are

- **L-band**: This ranges between 1 and 2 GHz, which is shared with the terrestrial IMT systems and is suitable for applications that require low data rate, but long-range communication.
- **S-band**: This ranges between 2 and 4 GHz, which is shared with the terrestrial IMT systems and is suitable for applications that require moderate data rates and stable coverage.
- **C-band**: This ranges from 4 to 8 GHz and is used for long-distance communication with moderate data rates.
- **K-band**: This includes portions of 21.4–22 GHz and 24.25–27.5 GHz bands, which have been allocated for fixed services in Region 2, including the Americas, Greenland, and some Pacific Islands.
- **Ka-band**: This band includes 31–31.3 GHz and 38–39.5 GHZ for worldwide usage to support high data rates and greater throughput. These bands were allocated for HAPS usage during the World Radio Conference (WRC)-19 when they revised the regulatory framework for HAPS and non-geostationary satellite systems.
- **V-band**: This includes confirmed existing HAPS bands, such as 47.2–47.5 GHz and 47.9–48.2 GHz for worldwide usage and is suitable for high-capacity backhaul and broadband services.
- **mmWaves**: This supports frequency bands over 30 GHz for high-throughput communication using smaller antennas, compact array sizes, and reduced interference. However, the challenges in terms of higher path loss need to be addressed.

The 37–43.5 GHz and 66–71 GHz bands were also determined to be the worldwide bands for HAPS [51, 54]. Since these bands also include incumbent users that are supporting critical infrastructure, the ITU regulations specify operating limits. Key parameters include maximum transmit EIRP, antenna radiation pattern, power flux density (PFD) limits, the distance between radio astronomy stations, and the nadir of a HAPS platform.

5.1.3.2 Use Cases and Applications

The most widely explored application for HAPS is its role in communication connectivity. HAPS can provide ultra-high speed last-mile connectivity in regions where the traditional network is not feasible and is underserved. Since the system is compatible with existing end user devices of the terrestrial networks, it can be deployed in disaster-hit regions to restore telecommunication services quickly. Furthermore, it can also act as a relay network and serve as a backhaul for next-generation communication systems. HAPS also facilitate support for IoT using different standards such as oneM2M, Sensus, and LoRaWAN, to enable seamless data transmission across vast areas supporting IoT applications for industry verticals.

With high definition (HD) cameras and sensors on board, it can also conduct timely disaster assessments to ensure proper resource allocation and prioritize rescue operations. HAPS integrated with an early warning system can monitor environmental changes in anticipation of natural disasters and offer warnings in advance to lower their impact. HAPS will also be supporting commercial broadband services to address user needs for connectivity, which are growing by the day. They are meant to enhance connectivity in an urban or suburban environment and offload traffic during peak periods while complementing conventional cellular networks. HAPS will provide seamless connectivity in areas where broadband has typically not been reliable, such as in remote mining sites or oil rigs, ensuring their operations do not come to a standstill.

HAPS could potentially become a techno-economically feasible technology in underserved regions, when it acts as an aerial BS to extend mobile and broadband services to underserved regions. HAPS would enable communities to access education, healthcare, governance, and employment opportunities, thereby giving rise to economic development and a digitally inclusive society.

5.1.3.3 Timeline

The idea of using high-altitude balloons or platforms for telecommunication was introduced in the 1970s to provide large area coverage [55]; however, the progress was hindered by technological limitations [56, 57]. NASA began exploring the concepts of using high-altitude platforms for communication via tethered balloons and solar-powered platforms to get connectivity in remote areas [58–60]. The studies related to the HAPS communication system began in the 1990s with the significant improvement in the UAV technology [61].

Emerging prototypes such as Airbus solar-powered UAV called the Zephyr demonstrated the feasibility of long-endurance platforms [62–64]. Additionally, ITU included HAPS in its recommendations for frequency allocations and its role in the telecommunication sector [65–67]. Several organizations formed the HAPS alliance and invested in HAPS systems to deliver Internet connectivity [68, 69]. This encouraged ITU and 3GPP to focus their efforts toward integrating HAPS in the global standards for 4G and 5G [70–72]. With the advancements of the solar array, the HAPS were capable of staying aloft for weeks. HAPS alliance launched collaborative initiatives to develop scalable solutions. Additionally, HAPS became part of B5G and 6G communication systems for extending rural and remote area connectivity to bridge the digital divide [73, 74]. Furthermore, several countries started spectrum allocation regulations for HAPS, which enabled large-scale deployment. Some of the key trials are Airbus's Zephyr staying aloft for 64 days, showcasing the long-endurance capabilities of the platform [75], and HAPSMobile Inc. (a joint venture between Softbank and AeroVironment) trialed HAPS-based 5G connectivity through a solar-powered UAV named Sunglider [50, 76]. Though Google's Loon has showcased a potential for completing a 100-day flight with its balloon-type UAV [77], it was shut down in 2021 due to a lack of a clear business case and techno-economic feasibility, along with limited adoption [78].

HAPS are expected to play a key role in 6G and SAGIN architecture to provide ubiquitous coverage, especially in remote areas [79]. Government and private entities aim to deploy HAPS to bridge the digital access gap. Advancements in energy systems, materials, and manufacturing costs are expected to lower the overall cost, which would improve their techno-economic feasibility and make it a commercially viable solution [1, 53].

5.2 Worldwide Air Layer Deployments

5.2.1 UAVs

5.2.1.1 Deployment Across the Globe

The global deployment of UAVs has been expanding rapidly, which is being driven by technological advancements and increased demand for UAVs in sectors such as agriculture, construction, infrastructure inspection, search and rescue operations, delivery services, and wireless broadband. The UAVs can play versatile roles and is an invaluable tool for businesses aiming to optimize operations and reduce costs. The UAVs market is contributing $30.2 billion in 2024 and is anticipated to reach $48.5 billion by 2029, at a compound annual growth rate (CAGR) of 9.9% between 2024 and 2029 [80].

Geographically, North America is the largest market for commercial UAVs and accounts for about 40% of the global market. This region offers a boost to the UAVs market by formulating a strong regulatory framework and technological infrastructure to encourage widespread adoption of this technology. Several countries, such as the USA, China, India, the UK, France, Germany, Israel, the UAE, Russia, Ukraine, and Turkey, are investing significantly in these technologies to advance their military combat capabilities, as well as focusing on innovating civilian use cases of these technologies. UAV is a global phenomenon, and countries are investing heavily in this technology.

As the technology is still in an early stage for civilian use cases, there is a need to understand the test trials and projects funded in these areas. UAVs are gaining traction worldwide with various projects and test trials showcasing their potential to bridge the digital divide. One notable initiative is the *Marktschwalbe Project* in Germany, which highlights the benefit of being connected to the Internet in rural areas [81]. The drones here are used for delivering goods to rural areas by connecting them to the nearby urban center, and the drones have a carrying capacity of 3 kg with a radius of 12 km.

The SkyLiTE Project in Finland aims to provide LTE connectivity using low-altitude UAV networks [82]. In this project, UAVs functions as an aerial BS to expand cellular network coverage to regions with minimal or no infrastructure, offering digital connection. This model's major use cases are remote healthcare and education, intending to provide dependable and cost-effective connection solutions to neglected regions.

Several other projects are being explored to study the feasibility of using LTE networks for UAV connectivity in rural areas, such as BVLOS drone connectivity in Finland for logistics to effectively support low-capacity command and controls for UAVs and use parallel connections to ensure seamless coverage [83, 84]. Similarly, the project by Northern Arizona University in the USA is experimenting with the use of extended cellular connectivity for data transfer in fields with high QoS and reliable connectivity. Similar initiatives are launched in India as well, such as *Digital Sky Platform* to provide a single-window service for drone operations and regulations across the country [85]. This would support different use cases and shorten the time taken to test and deploy UAVs Internet connectivity to rural and remote areas.

5.2.1.2 Challenges and Potential Solutions

There are several problems while deploying UAV-based wireless communication systems, especially when it is integrated with space layer technologies and 3D placement of UAVs [3]. When using UAVs to provide wireless broadband for rural deployments and other regions, the following challenges need to be addressed:

- UAVs have a limited flight time and payload capabilities mainly because of battery power requirements, which restrict their operation time. There is a need to charge or swap the batteries, which leads to a disruption in the service provided. There are several ongoing research projects ranging from minimizing the total power required of UAVs while supporting the user demands to creating improved battery capabilities such as wireless charging and large capacity batteries. Another key problem with increasing the payload to integrate UAVs with future networks is the increase in energy consumption, which could potentially be addressed using solar-powered UAVs, fuel cells, and energy-efficient communication protocols.
- Several studies have underlined the problem of UAV placement in terms of the number of consumers to be served and the revenue gained from delivering the service. Furthermore, UAVs flying at high altitude faces increased path loss, restricting the coverage area with high QoS. Furthermore, employing several UAVs to achieve a wide range may result in additional delay, leading to increased cost and poor QoS for end users. Potential solutions for these scenarios include smart beamforming to offer a user-centric network, adaptive frequency selection by monitoring the propagation channel to serve the applications of interest and minimize path loss, and optimizing the aerial relay path to ensure that the signal is transmitted in the shortest possible route to minimize the introduction of additional delay in the signal propagation.
- There is a substantial cost required for building, launching, and operating a UAV at a certain altitude to serve a specific region of interest. There is a need to optimize the number of UAVs required to provide continuous coverage, especially in remote and rural areas. Studies have shown that it is a nondeterministic polynomial time (NP)-complete problem in determining the minimum number of UAVs required in providing the continuous coverage, considering the need for battery requirements (either replace or recharge). The results obtained should be analyzed from a cost point of view to study the techno-economic feasibility of the proposed number of UAVs to serve the region of interest.
- As the number of UAVs in the air layer increases, there is a need to manage and optimize the position, frequency of operation, interference, co-channel interference, and other resources. As spectrum is a scarce resource, UAVs compete with space and ground layer using the concept of spectrum reuse. However, without efficient spectrum planning and management,

this could lead to co-channel interference, leading to poor QoS or loss of data packets. Additionally, high-speed UAVs would introduce Doppler frequency shifts affecting the signal quality. This could potentially be mitigated using AI-driven spectrum-based allocation, spectrum sensing, directional antennas, and cognitive radio technologies.
- UAVs are vulnerable to jamming and spoofing attacks that can disrupt communication. There is a need to prevent unauthorized access to the data being transmitted and also prevent the data from being hacked. The network security should also focus on the cybersecurity for the GCS as well. Secure communication, strong encryption protocols, quantum security, anti-jamming protocols, and user cybersecurity awareness play a crucial role in the scalability of UAVs in bridging the digital divide.
- Scalability and coordination of UAV networks pose additional challenges for the operator, for example, a wide-scale network with thousands of UAVs would require proper management of networks in order not to face congestion or communication disruption. Additionally, routing and handover between the UAVs for seamless connectivity should also be without disruption. There is a need to optimize the orbital path of UAVs to avoid a collision that would result in space debris or communication disruption. There are several ongoing research projects to overcome challenges in terms of AI-driven network control, SDN for UAVs, intelligent swarm management, obstacle detection and avoidance, collision avoidance algorithms, drone swarm, and path optimization.

5.2.2 NTFPs

5.2.2.1 Deployment Across the Globe

NTFPs has become popular globally, with several countries and projects exploring the potential and understanding the business model for using it. The global tethered flying platform market including NTFPs is expected to grow from $ 237.08 million in 2023 to $ 349.48 million by 2030, at a CAGR of 5.7% which is fueled by the growth in demand for aerial communication, surveillance, and monitoring use cases [86]. This technology supports versatile use cases and is faster to deploy.

Several academia and industry organizations have been exploring tethered flying platforms and drones for use cases and applications discussed in Sect. 5.1.2.1. For example, Hooverfly technologies in the USA has developed advanced tethered drones for communication, surveillance, and emergency response [87], while King Abdullah University of Science and Technology has been exploring the integration of NTFPs into the 6G ecosystem for enhancing global connectivity and seamless communication [88]. Meanwhile, Institut National Polytechnique de Toulouse has assessed the economic feasibility and performance of NTFPs in B5G and 6G wireless systems [89].

MNOs worldwide are testing NTFPs to determine its influence on their operations and the telecommunications sector in general. NTT Docomo in Japan has been experimenting with using tethered drones to improve network coverage during huge events and catastrophes, ensuring a stable and secure network in situations where standard networks are insufficient [90]. Deutsche Telekom in Germany has been conducting pilot trials using tethered drones to bridge the digital divide by enhancing the coverage in rural and remote areas [91]. Meanwhile, British Telecom Group in the UK has been investigating the simplicity of establishing communication links for emergency response and disaster assistance using NTFPs [92]. Meanwhile, China Mobile has been investing in tethered drones to improve network resilience, enable large-scale gatherings, and emergency response [93]. In India, Reliance Jio is testing tethered drones to deliver seamless coverage and excellent QoS in crowded metropolitan areas and during huge gatherings [94].

5.2.2.2 Challenges and Potential Solutions

There are challenges in terms of mobility, tethered constraints, optimal placement, and intervention operations that need to be resolved before NTFPs becomes widely available. NTFPs has limited mobility compared to the satellites, but higher mobility compared to the terrestrial networks. They cannot move beyond the radius of the tether's length. Furthermore, the tether itself could cause challenges in the operations when damaged. And this could cause significant damage to critical operations. They can optimize their position only at a certain distance that is permissible by the tether length and position of the anchor.

To overcome these challenges, free-flying platforms were designed that would have a wireless backhaul system using mmWave, RF, and FSO communication links. The free-flying platform should have a low speed to maintain LOS; otherwise, if the target receiver is NLOS, then it can optimize its location to get the target receivers within LOS. However, free-flying platform systems have identical issues as satellites, UAVs, and HAPS. They offer high data speeds and mobility but are prone to calibration errors, pointing inaccuracies, positioning challenges, and are also cost-intensive.

5.2.3 HAPS

5.2.3.1 Deployment Across the Globe

The deployment of HAPS has gained significant momentum in recent years. HAPS has been part of initiatives ranging from research studies to pilot trials to large-scale commercial deployments. There is a steady interest from the stakeholders, which is driving the technology further.

Several countries, such as Japan, the USA, Peru, Puerto Rico, India, China, and EU countries are exploring the use of HAPS for extending broadband to rural and remote regions under the digital divide initiatives [95, 96]. These projects involve collaboration between multiple stakeholders, such as governments, research organizations, businesses, companies, and NGOs. HAPS commercial deployments are gaining traction on a global scale. In Africa, HAPS aims to connect 600 million individuals who are living in underserved regions, and this is being led by Airbus, who plan to roll out Zephyr HAPS to bridge the digital divide [97]. Indonesia has an extensive network of 17,000 islands, making it difficult to deploy traditional networks to bridge the digital divide [98]. In this scenario, HAPS are expected to extend the coverage to another 30% of the population residing in remote areas. Similarly, in Brazil and Latin America, HAPS has demonstrated its capabilities by expanding the coverage by 7,500 km^2 with each platform, catering to the underserved population [72, 99].

Globally, the HAPS market is expected to grow to $4 billion by 2030, which would be driven by investments from tech giants and government-backed initiatives. The demand is also fueled by the potential use cases and applications supported by HAPS. For instance, Zephyr HAPS has integrated with IoT devices for environmental monitoring in the UK [100]. For IoT applications, HAPS can support up to 1 million sensors over a vast area. Similarly, South Korea and Japan are integrating HAPS with the 5G network to expand their urban and rural network coverage [101]. On the other hand, the USA has prioritized surveillance and disaster management use cases, with HAPS being deployed to establish emergency communication networks in areas affected by hurricanes, earthquakes, wildfires, and floods [102, 103].

5.2.3.2 Challenges and Potential Solutions

There are several challenges that need to be addressed while employing HAPS, especially when integrating with other technologies. One of the major hurdles is designing the mMIMO and ultra mMIMO architecture for HAPS that is energy-efficient when operating in the mmWave and THz communication bands. Furthermore, there is pilot contamination, inter-beam interference, and inter-user, along with inter-beam correlation due to a large number of beams and the footprint of each beam [73]. Potential solutions for this include applying analog or hybrid beamforming techniques to minimize the number of RF chains, user grouping, and increasing antenna element spacing.

There are several challenges in integrating space, air, and terrestrial networks that need to support dynamic resource allocation in different technologies, dynamically switch frequency bands when changing the technology, and support seamless integration. A potential solution could include developing a rapidly decentralized architecture that can be implemented using GPUs, ML, and AI that would help in optimizing the resource allocation and improve performance efficiency.

For HAPS to provide high throughput to the users, there is a need for a high-capacity backhaul link between the HAPS node and the network core, which is sensitive to pointing errors and changes in weather conditions. The frequency band for operation should be capable of withstanding these challenges, and the ground station transceivers should be robust to minimize these errors. Since HAPS constitute a large number of terrestrial and aerial vehicles, the power consumption is limited by onboard energy. This could be addressed by using renewable energies on board, mid-air refueling, replacing batteries regularly, powering via tether, and developing energy-efficient protocols and technologies.

Another critical challenge that needs to be tackled by HAPS is jittering, which occurs due to physical disturbances in the stratosphere layer. Jittering refers to the inconsistency in the position or timing of the signal or system. This would impact coherence time, beam alignment, user disconnection, variation in the footprints, randomness to GNSS, and poor inter-HAPS communication. This could be addressed by integrating stale channel state information (CSI) techniques to solve channel estimation problems, deploying efficient jitter-ware beam tracking, efficient handover algorithms, predicting HAPS location based on weather conditions, and using ground reference beacons to accurately measure the exact position of the HAPS.

There is a need to improve the security aspects of the HAPS, which is vulnerable to wireless channel attacks. Malicious participation in federated learning (FL) or centralized server risks makes the system vulnerable. Implementing ML-based anomaly detection in networks and ensuring local model parameters remain private through FL techniques can safeguard against these threats.

In conclusion, addressing the challenges with using HAPS would promote wide-scale usage of this technology. The potential solutions would ensure HAPS are reliable, secure, scalable, and serve the future connectivity demands globally.

5.3 Air-Layer: Accessibility, Affordability, and Climate

Next, there is a need to understand the air networks in terms of accessibility, affordability, and climate sustainability.

5.3.1 UAVs

5.3.1.1 Economic Frameworks

The capabilities offered by UAVs play a crucial role in society and are attracting the attention of stakeholders, for building innovative business models and use cases that would boost the UAVs solutions in the market. For UAVs to effectively bridge the digital divide, it must be both cost-effective and affordable for end users. Their potential lies in the ability to seamlessly integrate with terrestrial networks and NTNs, enabling extended connectivity in underserved regions. Some key economic frameworks for UAVs are discussed here.

Business Models: Advancement in the UAVs has revolutionized the use cases supported by the technology, which is driving the technology to solve digital divide challenges. This would help empower the unconnected and boost the economies worldwide. UAVs over a wide range of services, and the common business models include drone-as-a-service, commercial telecommunication network, and B2B services. Furthermore, UAVs is capable of integrating with the satellites to extend the limited coverage of terrestrial networks to provide seamless coverage. Another possibility is to build a mesh network using a network of UAVs to act as a relay network to reach a wider area. The potential revenue streams for communication network business models are subscription-based access, data-as-a-service, and service fees; that is highly dependent on who the consumers are for the services offered by UAVs.

SWOT Analysis: UAVs has the potential to bridge the digital divide by offering cost-effective, scalable, and reliable connectivity solutions. However, the widespread usage of UAVs for providing wireless services would depend on the regulatory restrictions, policies supporting the use cases, energy limitations, market competition from other solutions, and the business model. By leveraging the collaboration with stakeholders and advancement in the integration with terrestrial networks and NTN, UAVs would play a vital role in closing this gap. Table 5.1 shows the SWOT analysis for the UAV in bridging the digital divide globally.

Channel Modeling Using Stochastic Geometry: Path loss models for UAVs vary based on the environment, altitude, and communication link [3]. The path loss models for UAVs are classified into different categories as shown in Fig. 5.4 [3].

There are 4 common communication links in the UAV communication systems that are mainly developed for urban environments and can be extended to rural scenarios with appropriate modifications, such as lower building height, diffraction over terrain, long-distance signal propagation, and increased probability of LOS communication. The mathematical modeling of the different path loss models is given below, from [3, 104]:

Table 5.1 SWOT analysis of UAVs

Strengths	Weaknesses
Cost-effective deployment Rapid deployment and mobility Scalable based on the use case of interest Compatible with advanced telecommunication technologies such as 4G, 5G, and 6G Smaller carbon footprint Higher data rates	Limited coverage compared to satellites Battery constraints reduce their flight time and performance Signal interference with the terrestrial networks and sensitivity to weather conditions High maintenance costs due to regular upgrades Stricter airspace and spectrum regulation as critical infrastructure that operates on these bands Power and backhaul data
Opportunities	**Threats**
Bridging the digital divide in remote and rural areas Capable of providing emergency communication networks during disasters and emergencies Integration with space technologies to offer seamless connectivity Support future networks, AI, and energy storage to improve efficiency and lower cost	Competition from LEO satellites that are capable of offering wider coverage Risk involved in air space debris collision, equipment failure, and harsh operating conditions of UAVs UAV networks are vulnerable to hacking, data breaches, and jamming The initial CAPEX is significant in a scenario where demand is not

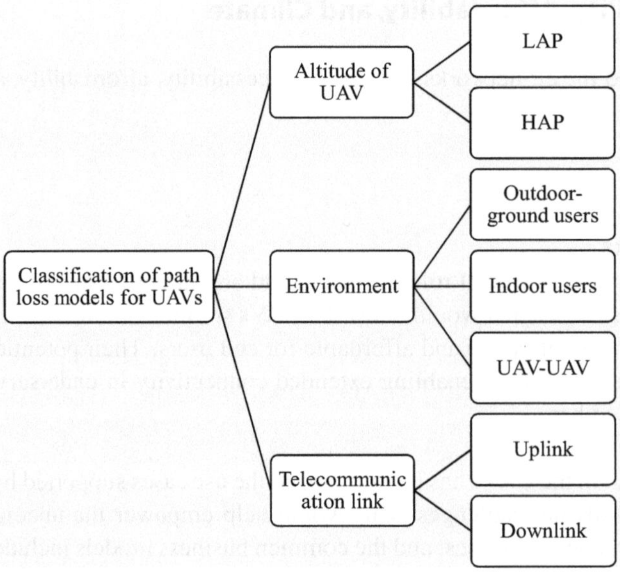

Fig. 5.4 UAV–classification of path loss models

1. ***Air-to-Ground for LAP***: This model assumes that all users are located outdoors and the location of each user can be represented by an outdoor 2D point in an urban environment [105]. This assumption helps in building a simpler model but has limited applicability for indoor user scenarios and focuses more on the LOS and near LOS, as there are only 3% of receivers that have deep fading conditions resulting from consecutive reflection and diffraction. Hence, for the UAV at height h and with a coverage radius R, path loss model, $L(h, R)$, using air-to-ground communication link using LOS and near LOS is given as [3]:
$$\begin{aligned} L(h, R) &= P(LOS)L_{LOS} + P(NLOS)L_{NLOS} \\ s.t., & \\ P(LOS) &= \frac{1}{1+\alpha \exp(-\beta[\frac{180}{\pi}\theta - \alpha])} \\ L_{LOS} &= 20\log(\frac{4\pi f_c d}{c} + \varsigma_{LOS}) \\ L_{NLOS} &= 20\log(\frac{4\pi f_c d}{c} + \varsigma_{NLOS}) \\ P(NLOS) &= 1 - P(LOS) \end{aligned} \tag{5.1}$$
where $P(LOS)$ and $P(NLOS)$ are the probability of having LOS connection link at an elevation angle θ, and the probability of having NLOS connection link, L_{LOS} and L_{NLOS} are the average path loss for LOS and NLOS paths, α and β are constants which depend on the propagation environment, d is the distance between the UAV and the ground user, c is the speed of the light, and ς_{LOS} and ς_{NLOS} are the average additional losses which depends on the environment.
2. ***Outdoor-to-Indoor Path Loss Model for LAP***: To overcome the shortcomings of the above model, the ITU-certified outdoor-to-indoor path loss model, $L_{o,i}$, to provide wireless connectivity for indoor users using UAV [106] in urban areas, and is shown below
$$\begin{aligned} L_{o,i} &= L_F + L_B + L_I \\ &= (w\log_{10} d_{out} + w\log_{10} f + g_1) + (g_2 + g_3(1-\cos\theta_i)^2) + (g_4 d_{in}) \end{aligned} \tag{5.2}$$
where L_f, L_B, and L_I are path loss due to free space path loss, building penetration loss, and indoor loss, d_{out} is the distance between the UAV and the indoor user, θ_i is the incident angle, d_{in} is the distance of the user inside the building to the exterior toward the direction of propagation, and f is the carrier frequency. Furthermore, w, g_1, g_2, g_3, and g_4 are constants dependent on the propagation environment.
3. ***Cellular-to-UAV Path Loss Model for LAP***: When the signal is being transmitted from cellular BS to UAV, the path loss model is dependent on depression angle θ between the UAV and the cellular BS, and the terrestrial coverage beneath the UAV. This model is based on extensive field experiments involving the collection of both terrestrial and aerial coverage signal strengths. The cellular-to-UAV path loss, $L(d, \theta)$, is modeled as shown below

$$\begin{aligned} L(d,\theta) &= L_{ter}(d) + \eta(\theta) + \chi(\theta) \\ &= 10\alpha log(d) + A(\theta - \theta_0)\exp^{(-\frac{\theta-\theta_0}{B})} + \eta_0 + N(0, a\theta + \sigma_o) \end{aligned} \tag{5.3}$$

where $L_{ter}(d)$ is the terrestrial path loss at a given distance d from the cellular BS to the UAV, $\eta(\theta)$ is the excess aerial path loss, $\chi(\theta)$ is the Gaussian random variable with an angle dependent on the standard deviation $\sigma_{UAV}(\theta)$ representing the shadow component, and α is the path loss component. Furthermore, A, B, θ_0, and η_0 are fitting parameters.

4. ***Air-to-Ground for HAP***: For providing mobile communication using HAP would operate between 2 and 6 GHz bands. The base formula of $L(h, r)$ remains the same, but the $P(LOS)$, $P(NLOS)$, L_{LOS}, and L_{NLOS} would vary depending upon the propagation environment [107]. With time, the model has been enhanced to suit the study from the view of parameters of interest, considering the environments [104]. For example, in the case where the focus is on the elevation angle to calculate the path link losses in an air-to-ground communication scenario, then it is modeled as

$$L(h,\theta) = 20\log_{10}(\frac{\Delta h}{\sin\theta}) + 20\log_{10} f - 27.55 \tag{5.4}$$

where $\Delta h = h_{UAV} - h_{GCS}$ is the difference in the height of the UAV and the height of the GCS and is calculated in meters, f is the frequency in MHz. This model could be applied directly to the rural scenarios.

If the log-distance path loss model is used, then the path loss can be calculated as

$$\begin{aligned} L &= L_0 + 10n\log_{10}(\frac{d}{d_0}) + X_\sigma \\ L_0 &= 10\log_{10}(\frac{4\pi d_0}{\lambda}) \end{aligned} \tag{5.5}$$

where d_0 is the reference distance and it is generally set to 1 m, n is the path loss exponent ($n \in (2,6)$, where 2 is for free space and 6 is for ultra-dense areas), λ is the wavelength of frequency, and X_σ is the random variable which presents variations in the free space path loss model that follows a normal distribution with a standard deviation σ.

5. ***UAV-to-UAV Path Loss Model***: Generally, UAV-to-UAV communication link is dominated by LOS, therefore, the FSPL model would be applied in this scenario [3, 104]. This is mathematically represented as

$$L_{FSPL} = 20\log_{10}(\frac{4\pi d}{\lambda}) \tag{5.6}$$

where d is the distance between the UAVs, and λ is wavelength of the carrier frequency.

6. **Weather Conditions**: The signal gets impacted by the weather conditions, such as attenuation due to rain, fog, and signal delay due to atmospheric conditions. The path loss equation is modified as given below

$$\begin{aligned} L_{rain} &= \gamma R^\alpha d \\ L_{fog} &= kd \end{aligned} \tag{5.7}$$

where γ is the rain attenuation coefficient, R is the rain rate (mm/hr), α is the frequency-dependent exponent, k specific attenuation due to fog (dB/km), and d distance between the transmitter and the receiver.

Substituting these parameters in the standard link budget equation, discussed in Sect. 2.5.4.1, helps in modeling the link budget for the UAVs.

Techno-economic Assessment: A TEA model for UAVs deployment analyzes the technical feasibility and economic viability of using UAVs for specific applications, such as bridging the digital divide [1]. The CAPEX, C_c, and OPEX, C_o, for UAVs network is shown below

$$C_c = C_{UAV} + C_{infra} + C_{software} + C_{spectrum} \tag{5.8}$$

where C_{UAV} is the cost toward N UAVs that are required to support the desired use case and accounts for procurement, sensors, and communication equipment, C_{infra} is the cost toward the GCS, such as charging stations, control hubs, and receivers, $C_{software}$ is the cost incurred toward the navigation, fleet management, AI-based decision-making software, and $C_{spectrum}$ is the cost toward spectrum licensing and regulatory fees.

$$\begin{aligned} C_{o,t} &= C_{energy,t} + C_{maintenance,t} + C_{personnel,t} + C_{data,t} + C_{insurnace,t} \\ C_o &= \sum_{t=1}^{T} \frac{C_{o,t}}{(1+r)^t} \end{aligned} \tag{5.9}$$

where $C_{energy,t}$ is the cost toward energy (battery charging, fuel cells) for a duration t years, $C_{maintenance,t}$ is the cost of ongoing maintenance, repair, and replacement, $C_{personnel,t}$ is the salaries of personnel involved in the operations of UAVs, $C_{data,t}$ is the cost toward data handling, edge computing, cloud processing, and communication networks, and $C_{insurnace,t}$ is the cost toward insurance that would cover the risk and liability. Additionally, r is the discount rate, and T is the project duration.

The revenue for the UAVs is dependent on the use case that it would serve. It would have income streams from different sources, such as providing broadband services in rural areas, and disaster recovery zones, or selling remote sensing data that is requested by users for the application of their interest. This could be mathematically modeled as

$$R_t = PD_t \tag{5.10}$$

where P is the price per unit (per user, or per MB of data), and D_t is the demand for the year t, which could be quantized as either the number of users or data volume.

5.3.1.2 Environmental Frameworks

Mathematical modeling for the environmental framework for UAVs focuses on quantifying and mitigating the environmental factors along with improving the performance and efficiency of UAVs.

Energy Consumption: The total energy consumption for the UAV, E_{total} can be modeled as [108]:

$$\begin{aligned} E_{total} &= E_{hover} + E_{travel} + E_{compensate} \\ &= P_{hover}t_{hover} + P_{travel}t_{travel} + P_{compensate}t_{compensate} \end{aligned} \tag{5.11}$$

where E is the energy, P is the power required per hour, and t is the duration of the activity in hour. The key activities that consume energy for UAVs are staying aloft or hovering, traveling to the desired position, and energy required toward counteracting weather conditions, such as wind and turbulence.

Carbon Emissions: The total carbon footprint, CE, for a UAVs network can be expressed as

$$CE = CE_p + CE_t + CE_c + CE_o \tag{5.12}$$

where CE_p is the carbon emissions of the production phase, CE_t is the carbon emissions during the transportation phase, CE_c is the carbon emissions during the construction phase, and CE_o is the carbon emissions during the operations phase. For example, to ensure sustainability and study the environmental impact, it is important to understand the operations of UAV in terms of carbon footprint/ emissions analysis. The total carbon emission of operations, CE_o is modeled as [108]:

$$CE_o = \int_0^T m_{fuel}(t) c_{CO_2} dt \tag{5.13}$$

where $m_{fuel}(t)$ is the fuel consumption rate, and c_{CO_2} is the emission per unit of fuel consumed. For more details calculations, refer [108].

Other environmental aspects that need to be considered for UAVs include noise pollution, flight path optimization, wildlife preservation, recyclable materials, minimizing debris, compliance with national and international environmental policies, clearance for sensitive zones, weather resilience, low carbon emissions using renewable energy sources, lifecycle management, and environmental conservation.

5.3.2 NTFPs

5.3.2.1 Economic Frameworks

Economic Modeling: There are several pricing strategies for NTFPs, which are dependent on the potential end user service plans, which could be a pay-per-use model, subscription model, leasing model, and freemium model. The key services that attract revenue for the NTFPs are **service-as-a-platform (SaaP)**, where NTFPs is expected to offer key services such as logistics, connectivity, remote sensing, mapping, and infrastructure examination. Other potential revenue streams are **data-as-a-service (DaaS)**, which leverages value from selling data collected from NTFPs for applications related to agriculture, urban planning, site inspection, and environmental research [17]. This data provides valuable insights for decision-makers in formulating strategies.

The newer applications for NTFPs are **vertical integration** of industries using a single platform. They can support two or more applications using a single NTFP, for example, act as an aerial BSs to provide last-mile connectivity in remote areas while serving other use cases such as agriculture monitoring applications [17]. The additional revenue stream for NTFPs is **market segmentation**, where they would offer customized solutions for each specific industry, such as military, government, logistics, and airports. There is a need to develop a collaborative business ecosystem with all the stakeholders involved for encouraging a wide-scale adoption of NTFPs. This would promote a shared risk model, ensuring adoption and driving innovation, scalability, and sustainability of the technology [88].

SWOT Analysis: Table 5.2 shows the SWOT analysis of NTFPs in bridging the digital divide.

It shows that the multilayered uses for connectivity, logistics, and surveillance, and their ability to reach remote or underserved areas, are all factors that constitute strengths while providing innovative solutions when traditional infrastructure fails. Some of the key weaknesses of this technology include the high capital cost of establishment, limited flight range, and technical challenges such as tether constraints and payload capacity. The opportunities for NTFPs lie in providing services to agriculture, disaster management, and urban planning industries, along with recent developments in autonomous technology and integration with renewable energy [17, 88]. The key threats to this technology are competition from space layer technologies and other air layer technologies, policy constraints, security issues, and environmental challenges.

Table 5.2 SWOT analysis of NTFPs

Strengths	Weaknesses
Rapid deployment for emergencies Cost-effective than building terrestrial networks in the same scenario Flexible coverage Energy efficient with the usage of renewable energy sources Scalable for large gatherings	Limited by the high-density traffic generated in the coverage area Requires skilled personnel to operate and run the network High power consumption if renewable sources of energy are not available The QoS is dependent on the weather conditions Lack of uniformity in the spectrum licensing norms for this technology across the globe
Opportunities	**Threats**
To bridge the digital divide in rural and remote areas Rapid deployment during emergencies and disaster recovery Integration with the terrestrial network and SAGIN would make it a valuable technology and be easily supported by end user devices Potential for large-scale poisson point process (PPP) Potential solution for enhancing the coverage and QoS	Competitive technologies in the same frequency bands Rival technologies such as LEO satellites, UAVs, HAPS Concerns about the environmental impact of using NTFPs High energy consumption model Significantly high investment would make it not a lucrative option if the ROI is poor

Channel Modeling Using Stochastic Geometry: The ITU proposed a model for estimating NTFP LOS probability in an urban environment between a transmitter h_{Tx} and a receiver, h_{Rx} at an elevation θ, and is modeled as [17, 109, 110]:

$$P(LOS, \theta) = \frac{1}{1+a_1 \exp(-b_1(\theta - a_1))} \tag{5.14}$$

where a_1 and b_1 are functions of the propagation environment variables p_α, p_β, and p_γ. Similarly, for a NLOS propagation environment, the probability of NLOS is modeled as

$$P(NLOS, \theta) = 1 - P(LOS, \theta) \tag{5.15}$$

Typically, there is a smaller path loss exponent α_{PL} in rural environments and a greater path loss exponent α_{PL} in urban environments, and this supports the development of a realistic channel model for signal propagation. The path loss equation is modified as

$$PL(\alpha, \theta, d) = d^{-\alpha_{PL}^{(\theta)}}[P_{loss,LOS}P(LOS, \theta) + P_{loss,NLOS}P(NLOS, \theta)] \tag{5.16}$$

where $P_{loss,LOS}$ and $P_{loss,NLOS}$ is the mean additional losses for LOS and NLOS transmission for a distance d between the transmitter and the receiver.

Techno-economic Assessment: The CAPEX and OPEX for the NTFPs depends on several factors which determine the TEA feasibility.

CAPEX: The initial cost includes costs toward the platform acquisition C_{pa}, tethering equipment C_{te}, communication hardware C_{ch}, BS integrating with the NTFPs equipment $C_{BS,I}$, ground support C_{gs}, tether C_{tether}, anchor C_a, and power supply infrastructure C_{ps}; and is mathematically modeled as

$$C_c = C_{pa} + C_{te} + C_{ch} + C_{BS,I} + C_{gs} + C_{tether} + C_a + C_{ps} \tag{5.17}$$

OPEX: The ongoing cost of operations and maintenance, $C_{o,t}$ for the tth year, includes regular servicing maintenance and replacement costs $C_{m,t}$, energy costs $C_{energy,t}$, spectrum $C_{spectrum,t}$ and personnel costs $C_{personnel,t}$; and is mathematically modeled as

$$\begin{aligned} C_{o,t} &= C_{m,t} + C_{energy,t} + C_{spectrum,t} + C_{personnel,t} \\ C_{energy,t} &= P_{total} H_{op} Z_{energy,buy} \end{aligned} \tag{5.18}$$

where P_{total} is the total power consumption from the grid (watts), H_{op} is the operational hours, and $Z_{energy,buy}$ is the cost of electricity per kWh purchased from the grid.

TCO: To estimate the TCO with a discount rate r over a period of T years, the model is expressed as

$$C = C_c + \sum_{t=1}^{T} \frac{C_{o,t}}{(1+r)^t} \tag{5.19}$$

Revenue: There are different revenue streams for NTFPs as discussed earlier, hence, it is required to understand the potentially all use cases that would generate revenue for the service provided: subscription ($Z_{n,t}$ is the price per user n connectivity per month and N_t is the number of users for the year t, pay-as-you-go ($Z_{data,t}$ is the price per GB of data downloaded and U_t is the data downloaded), data package ($Z_{s,t}$ is the price for emergency events or special events, and E_t is the occurrence of it for the year t), and energy generation ($Z_{energy,sell}$ is the price per unit energy when sold to the grid and $U_{energy,t}$ is the units of power harvested from the system). This is expressed as

$$R_t = 12(\sum_{n=1}^{N} Z_{n,t})N_t + Z_{data,t}U_t + Z_{s,t}E_t + Z_{energy,sell}U_{energy,t} \tag{5.20}$$

5.3.2.2 Environmental Frameworks

An environmental framework for NTFPs involves a comprehensive assessment of the entire lifecycle, starting from manufacturing to EOL. The results from these studies help in understanding the framework and the policy context to promote sustainable technologies that will help to achieve their SDGs for regulators and policymakers.

Energy Consumption: In the TEA modeling for NTFPs, the total energy consumption is represented as E_{total} (in kWh), and is shown below

$$E_{total} = E_{platform} + E_{payload} + E_{tether} \tag{5.21}$$

which is a sum of energy consumption by the platform for flight stabilization and tether management $E_{platform}$, power required for the communication equipment $E_{payload}$, and power losses in the tether cable due to resistances E_{tether} [17].

If the site uses renewable energy sources, then the energy harvested from it, $E_{renewable}$, could be subtracted from the total power consumption, and it could be modified as shown below

$$E_{total,grid} = E_{total} - E_{renewable} \tag{5.22}$$

Carbon Emissions: The total carbon footprint, CE, for a NTFPs network can be expressed as

$$CE = CE_p + CE_t + CE_c + CE_o \tag{5.23}$$

where CE_p is the carbon emissions of the production phase, CE_t is the carbon emissions during the transportation phase, CE_c is the carbon emissions during the construction phase, and CE_o is the carbon emissions during the operations phase.

For NTFPs, the carbon emissions of operations, CE_o, depend on the source of the energy drawn, and are mathematically modeled as

$$CE_o = E_{total,grid} \times c_{intensity} \tag{5.24}$$

where $c_{intensity}$ is the carbon intensity of the electricity from the grid (in kg CO_2/kWh) [17, 111]. If the batteries with $\eta_{efficiency}$ as the battery efficiency are used to support the E_{load} network load supported by the batteries during the peak hours to lower the OPEX, then it is modeled as

$$C_{batteryEmissions} = \frac{E_{load}}{\eta_{efficiency}} c_{intensity} \tag{5.25}$$

5.3.3 HAPS

5.3.3.1 Economic Frameworks

Economic modeling: The pricing models for HAPS mirror those used in UAVs and NTFPs, and the cost sharing frameworks include usage-based pricing, freemium, tiered model, and community collaborating model [112]. The key revenue-generating business models for HAPS are collaborating under poisson point process (PPP) for selling **connectivity-as-a-service (CaaS)** to TSPs. Additionally, HAPSs are capable of supporting **multi-tenants** on the same platforms, which would help in improving the revenues and resource utilization. HAPS ecosystem integrated with terrestrial and space layer communication systems to support end-to-end solutions for different applications. HAPS could also be used for enhancing digital literacy to bridge the digital access gap across the globe.

SWOT analysis: Table 5.3 shows the SWOT analysis for HAPS in bridging the digital divide. Aligning the strengths, addressing the weaknesses, and working toward the technology advancement with the threats in consideration would enable HAPS to unlock new opportunities to bridge the digital divide and open new markets beyond global connectivity.

Channel modeling using Stochastic geometry: To understand the communication channel characteristics for HAPS in rural areas, a stochastic model incorporates probabilistic elements to represent the variation in environmental factors, which can be simulated and analyzed to understand the system performance under realistic conditions. The received signal power, P_r for HAPS is derived from [50, 113, 114], is expressed as

Table 5.3 SWOT analysis of HAPS

SWOT Analysis	
Strengths	Weaknesses
High flexibility, throughput, and adaptability, plus low latency Cost-effective for deploying with renewable energy sources in rural and remote areas Access remote or hazardous locations Offer a wide coverage area (up to 200 km in diameter) HAPS can be launched faster than satellites, or by deploying a terrestrial network	Limited battery life and payload capacity during adverse conditions when renewable energy can't be harvested Susceptibility to adverse weather conditions Significant CAPEX for development, launch, and deployment Shorter lifespan compared to the satellites Technical challenges such as jittering, positioning issues, and atmospheric losses
Opportunities	Threats
Expanding to bridge the digital divide in underserved areas Partnerships with industries and governments would speed up the rollout of HAPS Technological advancements in IoT sensors, 5G, 6G, edge computing, and AI. Direct connectivity with smartphones, which allows for rapid connectivity restoration during disaster recovery Gaining popularity in continents that are facing an acute digital divide challenge.	Competition from alternative technologies, such as LEO satellites, UAVs, and NTFPs. Risk of cybersecurity threats and data interception. Regulatory restrictions in certain regions. Interference to the incumbent users, both satellites and terrestrial networks. Vulnerable to extreme climatic conditions Complex regulation due to airspace permissions and spectrum sharing challenges

$$\begin{aligned}
P_r &= P_r + G_t + G_r - L_{FSPL} - L_{shadow} \\
&\quad - L_{fading} - L_{atmosphere} \\
L_{FSPL} &= 20\log_{10} d + 20\log_{10} f + 20\log_{10} \tfrac{4\pi}{c} \\
L_{shadow} &= X_\sigma \\
X_\sigma &= N(0, \sigma^2) \\
L_{fading,LOS} &= 10\log_{10}(\tfrac{K}{K+1} + \tfrac{1}{K+1}|h|^2) \\
L_{fading,NLOS} &= -10\log_{10}|h|^2 \\
L_{atmosphere} &= \alpha d
\end{aligned} \tag{5.26}$$

where L_{FSPL} is the deterministic FSPL component which is dependent on the distance d, frequency f, and the speed of the light c; L_{shadow} is the random shadowing loss which follows a log-normal distribution with standard deviation as σ (dependent on the environmental conditions), and arises from large-scale environmental obstructions; $L_{fading,LOS}$ is the multipath fading due to the beamforming and signal reflections from multiple sources. In rural and remote areas, the signal would mostly have a LOS with the receiver with Rician K-factor representing the ratio of LOS to scattered signal power, and h is the channel coefficient which is a function of the complex normal distribution function, $h \sim CN(\mu, \sigma^2)$; and $L_{atmosphere}$ is the atmospheric absorption losses that are modeled dependent on the α which is the specific attenuation factor (dB/km), determined by ITU-R models for rain, humidity, and gases.

For the uth UE that is receiving the signal from the cth channel of hth HAPS, the mathematical modeling for its characteristics are described below [50]:

$$\begin{aligned}
R_{u,c,h} &= P_t + G_{u,c,h} - L_{u,h} + G_{UE} \\
L_{u,h} &= 20log_{10}(\tfrac{4\pi r_{u,h}}{\lambda})
\end{aligned} \tag{5.27}$$

where P_t is the BS transmission power, $G_{u,c,h}$ is the gain of the cth cell in h^{th} HAPS, $L_{u,h}$ is the FSPL between the uth UE and h^{th} HAPS, $r_{u,h}$ is the distance between the uth UE and h^{th} HAPS, and λ is wavelength of the signal. For estimating the SNR, $\gamma_{SNR,u}$ of the signal u^{th} UE that is receiving the signal from the cth channel of hth HAPS, it is mathematically modeled as [50]:

$$\gamma_{SNR,u} = 10log_{10}(\frac{10^{R_{u,c,h}/10}}{\sigma^2}) \tag{5.28}$$

where σ^2 is the noise power.

Techno-economic Assessment: There are different factor tat determine the cost model for HAPS, which contribute to its lifecycle. As these are newer technologies, a detailed cost model is not yet available.

Cost: The technical parameters that determine the cost and functionality of HAPS are payload capacity, coverage area, energy efficiency, and flight duration. altitude stability, latency, throughput, reliability, and interoperability. The high-level framework for HAPS cost model is expressed below

$$\begin{aligned} C &= C_c + C_o \\ C_c &= C_{manufacturing} + C_{deployment} + C_{groundSupport} \\ C_{o,t} &= C_{operations} + C_{maintenance} \end{aligned} \tag{5.29}$$

where C is the TCO for a HAPS network, C_c is the CAPEX, and $C_{o,t}$ is the OPEX for the year t.

$C_{manufacturing}$ accounts for costs spent R&D toward HAPS, to design structural design, aerodynamics, energy systems, payload integrations, and communication capabilities. It also includes building and testing the prototypes in a simulated real-world scenario to study their performance. In a later stage, this cost also involves the cost spent toward the manufacturing, such as material, fabrication, and assembly of HAPS.

$C_{deployment}$ is the cost toward the procurement and deployment of launch pads, which are used for launching HAPS into the stratosphere. It also included the costs associated with obtaining permission and licensed access for the operational spectrum, airspace regulations, and compliance with national and international aviation laws. There is also a need for ground infrastructure that would control HAPS, provide power, and act as a gateway for the network when connecting with the terrestrial network. $C_{groundSupport}$ is the cost spent toward setting up GCS for HAPS operations and monitoring.

$C_{operations}$ is the cost for powering HAPS, the cost of operating the communication payload, antenna, and other mission-specific equipment, and integrating with the existing terrestrial and space networks for seamless connectivity. $C_{maintenance}$ accounts for the cost spent toward routine maintenance, flight system checks, energy components, and payload functionality. It also considers the costs toward software upgrades, replacement of parts, and enhancing the capabilities of the HAPS that are aloft.

Revenue: While discussing the economic models and business models for HAPS, we discussed different revenue streams from CaaS R_{CaaS}, multi-tenant infrastructure sharing R_{NHN}, integrations with other technologies $R_{Integration}$, advertisement $R_{advertise}$, industry-specific applications $R_{collaboration}$, and direct subscription from the N_t end users R_{users} paying an average ARPU of Z_t for the year t. Then, the total revenue R is modeled as,

$$\begin{aligned} R_t &= R_{CaaS} + R_{NHN} + R_{Integration} + R_{advertise} + R_{collaboration} + 12Z_tN_t \\ R &= \sum_{t=1}^{T} R_t \end{aligned} \tag{5.30}$$

5.3.3.2 Environmental Frameworks

A thorough evaluation of the complete lifetime, from production to EOL, is part of an environmental framework for HAPS [115].

- **Energy Consumption**: In the TEA modeling for HAPSs, the total energy consumption is represented as E_{total} (in kWh), and is shown below

$$E_{total} = E_{propulsion} + E_{payload} + E_{support} \tag{5.31}$$

which is a sum of energy consumption by the platform for flight stabilization, altitude, and path $E_{propulsion}$, energy required for the communication equipment and data processing $E_{payload}$, and energy for onboard systems $E_{support}$.

If the site uses renewable energy sources, then the energy harvested from it, $E_{renewable}$, could be subtracted from the total power consumption, and it could be modified as shown below

$$E_{total} = E_{total} - E_{renewable} \tag{5.32}$$

- **Carbon Emissions**: The total carbon footprint, CE, for a HAPS network can be expressed as

$$CE = CE_p + CE_t + CE_c + CE_o \tag{5.33}$$

where CE_p is the carbon emissions of the production phase, CE_t is the carbon emissions during the transportation phase, CE_c is the carbon emissions during the construction phase, and CE_o is the carbon emissions during the operations phase. The carbon emissions of the operations stage depend on the source of the energy drawn, and are mathematically modeled as

$$CE_o = \frac{E_{total}}{\eta_{efficiency}} \times c_{intensity} \tag{5.34}$$

where $c_{intensity}$ is the carbon intensity of the electricity from the energy source(in kg CO_2/kWh), the energy source that is used to support the network has an $\eta_{efficiency}$ efficiency.

5.4 Summary

This chapter focused on appreciating the air communication networks-UAVs, NTFPs, and HAPS, as a technology and its role in bridging the digital divide. It offers capabilities which are a mix of benefits offered by space and ground networks, making it lucrative to pursue as an emerging solution in bridging the digital divide. The chapter focuses on technology, architecture, global deployments, enabling technologies, and applications that are driving the market as a solution for connecting the unconnected. Once the technology reaches a mature stage, it would become more affordable for commercial deployments and support wide-scale commercialization across different spectrum bands. However, without focused efforts in developing this technology to serve the underserved market, it is hard to lower the overall TCO, and build a sustainable air network leveraging the economic and environmental modeling. The three main technologies offer different capabilities and are suitable for various deployment scenarios depending on the market needs.

Exercises

5.1 A UAV operates at an altitude of 500 meters and has a field of view angle (θ) of 60 degrees. Calculate:

(a) The radius of the circular coverage area on the ground.
(b) The total area covered by the UAV on the ground.

Use the formula:

$$R = h \cdot \tan\left(\frac{\theta}{2}\right),$$

where R is the radius, h is the altitude, and θ is the field of view angle.

5.2 A rural connectivity project requires deploying a fleet of 10 UAVs. Each UAV costs $20,000 to manufacture and $5,000 annually for maintenance. Calculate:

(a) The initial deployment cost for the UAV fleet.
(b) The total cost of operating the fleet for 5 years, including maintenance expenses.

5.3 An NTFP operates at an altitude of 200 meters and transmits at a power of 20 dBm. The receiver sensitivity is -90 dBm, and the FSPL with d is the communication range, and $f = 2.4\,\text{GHz}$. Calculate the maximum communication range (d) in meters.

5.4 A HAPS platform has a solar array with an efficiency of 22% and an available solar irradiance of $1,000\,\text{W/m}^2$. The platform requires 10 kW of power to operate. Calculate the minimum solar panel area required to sustain the HAPS platform. Use the formula:

$P_{\text{solar}} = \eta_{\text{solar}} \cdot A \cdot I$, where η_{solar} is the solar efficiency, A is the panel area, and I is the solar irradiance.

5.5 A HAPS platform transmits with an EIRP of 60 dBW to a ground station located 30 km away. The receiver antenna gain is 25 dB, and the free-space path loss is given as 145 dB. Calculate the received signal power at the ground station.

5.6 A UAV fleet of 20 drones emits 50 kg of CO_2 per drone annually. Calculate:

(a) The total annual carbon emissions of the fleet.
(b) The total emissions over a 10-year period.

5.7 A UAV communication system operates with a bandwidth of 10 MHz and achieves a data rate of 50 Mbps. Calculate:

(a) The spectral efficiency in bps/Hz.
(b) Discuss the impact of increasing the bandwidth to 20 MHz while maintaining the same data rate.

5.8 An NTFP system is deployed in a flood-prone region for emergency connectivity. Evaluate:

(a) The advantages of using NTFPs over traditional terrestrial networks in disaster scenarios.
(b) The challenges of maintaining NTFP deployments during prolonged disasters.

5.9 A UAV uses 500 W of power and operates for 6 hours per day. Calculate:

(a) The daily energy consumption of the UAV.
(b) The total energy consumption for a fleet of 10 UAVs operating for a month (30 days).

5.10 Analyze the techno-economic feasibility of using HAPS for rural connectivity by discussing:

(a) The cost per Mbps compared to terrestrial networks.
(b) The long-term benefits of HAPS in addressing rural digital divides.

References

1. L. Toka, M. Konrad, A. Pekar, G. Biczók, Integrating the skies for 6G: Techno-economic considerations of LEO, HAPS, and UAV technologies. IEEE Commun. Mag. **62**(11), 44–51 (2024)
2. A. Koubâa, A. Allouch, M. Alajlan, Y. Javed, A. Belghith, M. Khalgui, Micro air vehicle link (mavlink) in a nutshell: A survey. IEEE Access **7**, 87658–87680 (2019)
3. H. Shakhatreh, A.H. Sawalmeh, A. Al-Fuqaha, Z. Dou, E. Almaita, I. Khalil, N.S. Othman, A. Khreishah, M. Guizani, Unmanned aerial vehicles (UAVs): a survey on civil applications and key research challenges. IEEE Access **7**, 48572–48634 (2019)
4. K. Shafafi, E.N. Almeida, A. Coelho, H. Fontes, M. Ricardo, R. Campos, UAV-assisted wireless communications: an experimental analysis of A2G and G2A channels, in *International Conference on Simulation Tools and Techniques* (Springer, 2023), pp. 243–256
5. Y. Zeng, R. Zhang, T.J. Lim, Wireless communications with unmanned aerial vehicles: opportunities and challenges. IEEE Commun. Mag. **54**(5), 36–42 (2016)
6. M.A. Jasim, H. Shakhatreh, N. Siasi, A.H. Sawalmeh, A. Aldalbahi, A. Al-Fuqaha, A survey on spectrum management for unmanned aerial vehicles (UAVs). IEEE Access **10**, 11443–11499 (2021)
7. Aerocorner. What frequency do drones use? Accessed 29 Oct 2024
8. P. Stoica, S. Basak, C. Molder, B. Scheers, Review of counter-UAV solutions based on the detection of remote control communication, in *2020 13th International Conference on Communications (COMM)* (IEEE, 2020), pp. 233–238
9. C. Ge, D. Zhai, Y. Jiang, R. Zhang, X. Yang, B. Li, X. Tang, Pathloss and airframe shadowing loss of air-to-ground UAV channel in the airport area at UHF-and L-Band. IEEE Trans. Veh. Technol. **72**(6), 8094–8098 (2023)
10. Z.H. Firouzeh, M. Moradian, A. Safari-Hajat-Aghaei, H. Mir-Mohammad-Sadeghi, Design and implementation of ground station antennas for UAV data radio link in UHF band, in *2006 2nd International Conference on Information & Communication Technologies*, vol. 2. (IEEE, 2006), pp. 2195–2200
11. B. Shang, X. Li, C. Li, Z. Li, Coverage in cooperative LEO satellite networks. J. Commun. Inf. Netw. **8**(4), 329–340 (2023)
12. Md. Mehedi Farhad, A. Manavi Alam, S. Biswas, M. Abdus Shahid Rafi, A.C. Gurbuz, M. Kurum, Sdr-based dual polarized L-Band microwave radiometer operating from small UAS platforms. IEEE J. Select. Top. Appl. Earth Obs. Remote Sens. (2024)
13. A. Jenerowicz, K. Siok, M. Woroszkiewicz, A. Orych, The fusion of satellite and UAV data: simulation of high spatial resolution band, in *Remote Sensing for Agriculture, Ecosystems, and Hydrology XIX*, vol. 10421. (SPIE, 2017), pp. 364–375
14. Y. Lyu, Y. He, Z. Liang, W. Wang, Yu. Junyi, D. Shi, Measurement-Based Tapped Delay line channel modeling for fixed-wing unmanned aerial vehicle air-to-ground communications at S-Band. Drones **8**(9), 492 (2024)
15. T. Poongodi, A. Saini, G. Dhuriya, V. Gupta, Historical perspectives and introduction to UAV cellular communications, in *Unmanned Aerial Vehicle Cellular Communications* (Springer, 2022), pp. 1–23

16. W. Saad, M. Bennis, M. Chen, A vision of 6G wireless systems: applications, trends, technologies, and open research problems. IEEE Network **34**(3), 134–142 (2019)
17. B.E.Y. Belmekki, M.-S. Alouini, Unleashing the potential of networked tethered flying platforms: prospects, challenges, and applications. IEEE Open J. Veh. Technol. **3**, 278–320 (2022)
18. G. Araniti, A. Iera, S. Pizzi, F. Rinaldi, Toward 6G non-terrestrial networks. IEEE Netw. **36**(1), 113–120 (2021)
19. TCOM. Global leader in persistent surveillance aerostats ISR solutions. Accessed 29 Oct 2024
20. C.L Nickol, M.D Guynn, L.L Kohout, T.A. Ozoroski, High altitude long endurance uav analysis of alternatives and technology requirements development. Technical report, NASA, 2007
21. M.N. Marques, S.A. Magalhães, F.N. Dos Santos, H.S. Mendonça, Tethered unmanned aerial vehicles—a systematic review. Robotics **12**(4), 117 (2023)
22. A. Cherubini, A. Papini, R. Vertechy, M. Fontana, Airborne wind energy systems: a review of the technologies. Renew. Sustain. Energy Rev. **51**, 1461–1476 (2015)
23. S.J. Maeng, O. Ozdemir, İ. Güvenç, M.L. Sichitiu, M. Mushi, R. Dutta, M. Ghosh, SDR-Based 5G NR C-Band I/Q monitoring and surveillance in urban area using a helikite, in *2023 IEEE International Conference on Industrial Technology (ICIT)* (IEEE, 2023), pp. 1–6
24. M. Manikandan, R.S. Pant, Research and advancements in hybrid airships—a review. Prog. Aerosp. Sci. **127**, 100741 (2021)
25. G.A. Khoury, *Airship Technology*, vol. 10. (Cambridge University Press, 2012)
26. S.H. Alsamhi, F.A. Almalki, O. Ma, M.S. Ansari, M.C. Angelides, Performance optimization of tethered balloon technology for public safety and emergency communications. Telecommun. Syst. **75**, 235–244 (2020)
27. S.H. Alsamhi, M.S. Ansari, N.S. Rajput, Disaster coverage predication for the emerging tethered balloon technology: capability for preparedness, detection, mitigation, and response. Disaster Med. Public Health Prep. **12**(2), 222–231 (2018)
28. S.H. Alsamhi, M.S. Ansari, O. Ma, F. Almalki, S.K. Gupta, Tethered balloon technology in design solutions for rescue and relief team emergency communication services. Disaster Med. Public Health Prepar. **13**(2), 203–210 (2019)
29. D. Espinoza, D. Reed, Wireless technologies and policies for connecting rural areas in emerging countries: a case study in rural Peru. Digital Policy, Regul. Gov. **20**(5), 479–511 (2018)
30. P.K. Chopra, R. Manchanda, R. Mehrotra, S. Jain, A new topology for telecom and broad band services in spars, remote and hilly areas. WSEAS Trans. Commun. **10**(9), 273–286 (2011)
31. S. Guan, J. Wang, C. Jiang, R. Duan, Y. Ren, T.Q.S. Quek. MagicNet: the maritime giant cellular network. IEEE Commun. Mag. **59**(3), 117–123 (2021)
32. K.G. Knapp, M.L. Jensen, B.B. Balsley, J.A. Bognar, S.J. Oltmans, T.W. Smith, J.W. Birks, Vertical profiling using a complementary kite and tethered balloon platform at Ferryland Downs, Newfoundland, Canada: observation of a dry, ozone-rich plume in the free troposphere. J. Geophys. Res. Atmos. **103**(D11), 13389–13397 (1998)
33. A. Hill, Y. Rowan, M.M Kersel, Mapping with aerial photographs: recording the past, the present, and the invisible at Marj Rabba, Israel. Near Eastern Archaeol. **77**(3), 182–186 (2014)
34. B. Zhang, L. Tang, M. Roemer, Probabilistic weather forecasting analysis for unmanned aerial vehicle path planning. J. Guid. Control. Dyn. **37**(1), 309–312 (2014)
35. H.-J. Koo, M. Belorid, H.J. Hwang, M.-H. Kim, B.-Y. Kim, J.W. Cha, Y.H. Lee, J. Baek, J.-W. Jung, S.-K. Seo, Development and case study of unmanned aerial vehicles (UAVs) for weather modification experiments. Atmosphere **34**(1), 35–53 (2024)
36. A.K. Singh, S. Gupta, Agriculture and the use of unmanned aerial vehicles (UAVs) current practices and prospects. Adv. Aerial Sens. Imag. 87–108 (2024)
37. S. Despoudi, K. Spanaki, O. Rodriguez-Espindola, E.D. Zamani, S. Despoudi, K. Spanaki, O. Rodriguez-Espindola, E.D. Zamani, Technological applications of agricultural 4.0 supply chains, in *Agricultural Supply Chains and Industry 4.0: Technological Advance for Sustainability* (2021), pp. 29–36
38. Y. Jiang, M. Lv, C. Wang, X. Meng, S. Ouyang, G. Wang, Layout optimization of stratospheric balloon solar array based on energy production. Energy **229**, 120636 (2021)
39. B.-G. Chae, O. Hyun-Ung, Development of power plant trio cansat platform powered by solar, wind and piezo energy harvesting systems. Int. J. Aeronaut. Space Sci. **22**(6), 1464–1480 (2021)
40. E. Young, Tethered balloons-present and future, in *2nd Aerodynamic Deceleration Systems Conference* (1968), p. 941
41. N. Smith, A history of first world war technology in 11 obiects. Eng. Technol. **13**(11/12), 34–37 (2018)
42. R.A. Grossnick, *Kite Balloons to Airships–: The Navy's Lighter-than-air Experience*. Deputy Chief of Naval Operations (Air Warfare, 1987)
43. R. Hallion, *Taking Flight: Inventing the Aerial Age, from Antiquity Through the First World War* (Oxford University Press, 2003)
44. J.F. Patton, Tethered aerostats for RADAR platforms, in *Aerospace Research Central* (1975)
45. L.A. Vierling, M. Fersdahl, X. Chen, Z. Li, P. Zimmerman, The short wave aerostat-mounted imager (SWAMI): a novel platform for acquiring remotely sensed data from a tethered balloon. Remote Sens. Environ. **103**(3), 255–264 (2006)
46. D. Grace, M. Mohorcic, M. Oodo, M.H. Capstick, M. Bobbio Pallavicini, M. Lalovic, CAPANINA—communications from aerial platform networks delivering broadband information for all, in Proceedings of the 14th IST Mobile and Wireless and Communications Summit (2005)
47. M. Mohorcic, D. Grace, G. Kandus, T. Tozer, Broadband communications from aerial platform networks, in *IST Mobile Summit* (2004)
48. A.M. Nazar, M.Y. Selim, A.E. Kamal, Mounting RIS panels on tethered and untethered UAVs: a survey. Arab. J. Sci. Eng. **49**(3), 2857–2885 (2024)
49. Y. Zhang, M.A. Kishk, M.-S. Alouini, HAP-enabled communications in rural areas: when diverse services meet inadequate communication infrastructures. IEEE Open J. Commun. Soc. (2023)
50. Y. Shibata, N. Kanazawa, M. Konishi, K. Hoshino, Y. Ohta, A. Nagate, System design of gigabit HAPS mobile communications. IEEE Access **8**, 157995–158007 (2020)
51. G.K. Kurt, M.G. Khoshkholgh, S. Alfattani, A. Ibrahim, T.S.J. Darwish, Md.S. Alam, H. Yanikomeroglu, A. Yongacoglu, A vision and framework for the high altitude platform station (HAPS) networks of the future. IEEE Commun. Surv. Tutor. **23**(2), 729–779 (2021)
52. International Telecommunication Union (ITU). High-altitude platform systems (haps). Accessed 7 Feb 2025
53. D. Renga, M. Meo, Can high altitude platform stations make 6G sustainable? IEEE Commun. Mag. **60**(9), 75–80 (2022)

54. S. Stellatou, C. Erotokritou, High-altitude platform stations (HAPS); regulatory obstacles blocking their deployment, in *2024 International Conference on Unmanned Aircraft Systems (ICUAS)* (IEEE, 2024), pp. 363–369
55. A.S. Carten, *An Investigation of the Applicability of High Altitude, Lighter-Than-Air (LTA) Vehicles to the Tactical Communications Relay Problem*, vol. 74. Air Force Cambridge Research Laboratories, Air Force Systems Command, United ... (1974)
56. M. Quinlan, Balloons and the awareness of a new age. Eighteenth Century **14**(3), 221 (1973)
57. S. Wishna, C. Hamby, D. Reed, NASA balloon: aircraft ranging, data and voice experiment. Technical report, NASA, 1972
58. J. Willis, Using the STD bus for encoder/decoder systems aboard a high-altitude balloon platform, in *International Telemetering Conference Proceedings* (International Foundation for Telemetering, 1983)
59. W.J. Vogel, E.K. Smith, Theory and measurements of propagation for satellite to land mobile communication at UHF, in *35th IEEE Vehicular Technology Conference*, vol. 35. (IEEE, 1985), pp. 218–226
60. C.T. Force, The NASA tracking and data acquisition networks their history and their future. Acta Astronautica **18**, 231–240 (1988)
61. D. Stuchlik, K. Tillery, The NASA long duration balloon project, in *International Balloon Technology Conference* (1991), p. 3680
62. J. Gavan, S. Tapuchi, D. Grace, Concepts and main applications of high-altitude-platform radio relays. URSI Radio Sci Bull. **2009**(330), 20–31 (2009)
63. A. Rapinett, *Zephyr: A High Altitude Long Endurance Unmanned Air Vehicle* (Department of Physics, University of Surrey, Doctor, 2009)
64. V. Cleave, Solar power: a flight to remember. Nature **451**(7181) (2008)
65. E. Cianca, M. Lucente, T. Rossi, C. Stallo, M. Ruggieri, E. Morelli et al., EHF channel sounding for telecommunications applications via HAPs and balloons. Memorie della Societa Astronomica Italiana **79**, 932 (2008)
66. T.C. Tozer, D. Grace, High-altitude platforms for wireless communications. Electron. Commun. Eng. J. **13**(3), 127–137 (2001)
67. A.K. Widiawan, R. Tafazolli, High altitude platform station (HAPS): a review of new infrastructure development for future wireless communications. Wirel. Pers. Commun. **42**, 387–404 (2007)
68. E. Setiawan, The potential use of high altitude platform station in rural telecommunication infrastructure, in *2018 International Conference on ICT for Rural Development (IC-ICTRuDev)* (IEEE, 2018), pp. 35–37
69. T. Mahapol, Design and optimisation of a power efficient high-altitude platform for equatorial operation, in *2018 IEEE Transportation Electrification Conference and Expo, Asia-Pacific (ITEC Asia-Pacific)* (IEEE, 2018), pp. 1–8
70. A. Anttonen, P. Ruuska, M. Kiviranta, 3GPP nonterrestrial networks: a concise review and look ahead. *VTT Technical Research Centre of Finland* (2019)
71. Z. Yang, *Coexistence, deployment and business models of heterogeneous Wireless Systems incorporating High Altitude Platforms*. Ph.D thesis, Blekinge Institute of Technology, 2013
72. F.A. d'Oliveira, F. Cristovão Lourenço de Melo, T.C. Devezas, High-altitude platforms—present situation and technology trends. J. Aerosp. Technol. Manag. **8**(3), 249–262 (2016)
73. O. Abbasi, A. Yadav, H. Yanikomeroglu, N.-D.D. Jao, G. Senarath, P. Zhu, HAPS for 6G networks: potential use cases, open challenges, and possible solutions. IEEE Wirel. Commun. **31**(3), 324–331 (2024)
74. K.N. Kuteesa, C.U. Akpuokwe, C.A. Udeh, Theoretical perspectives on digital divide and ICT access: comparative study of rural communities in Africa and the United States. Comput. Sci. IT Res. J. **5**(4), 839–849 (2024)
75. R.D. Breeje, A.D. Lange, J. Boterman, M. Visser, L. Meijer, M. Stopel, I. Proimadis, H. de Man, Prototype test results of a high-altitude pseudo-satellite laser communication terminal, in *2023 IEEE International Conference on Space Optical Systems and Applications (ICSOS)* (IEEE, 2023), pp. 36–41
76. V. Hill, J. Mukherjee, D. Lisoski, B. Danowsky, S. Haviland, In-flight stability analysis and envelope clearance of the sunglider Solar HALE UAS, in *AIAA AVIATION 2021 FORUM* (2021), p. 2796
77. M. Trancossi, *High altitude platform system airship for telecommunication and border monitoring design and physical model* (Technical report, SAE Technical Paper, 2020)
78. P. Serrano, M. Gramaglia, F. Mancini, L. Chiaraviglio, G. Bianchi, Balloons in the sky: unveiling the characteristics and trade-offs of the google loon service. IEEE Trans. Mob. Comput. **22**(6), 3165–3178 (2021)
79. C. Wang, M. Pang, T. Wu, F. Gao, L. Zhao, J. Chen, W. Wang, D. Wang, Z. Zhang, P. Zhang, Resilient massive access for SAGIN: a deep reinforcement learning approach. IEEE J. Select. Areas Commun. (2024)
80. MarketsandMarkets. Unmanned aerial vehicles (UAV) market report. Accessed 29 Oct 2024
81. Beyond Vision. The Marktschwalbe project: addressing rural connectivity problems. Accessed 29 Oct 2024
82. K. Sundaresan, E. Chai, A. Chakraborty, S. Rangarajan, SkyLiTE: end-to-end design of low-altitude UAV networks for providing LTE connectivity (2018). arXiv:1802.06042
83. Flymarshall. Unlocking BVLOS Drone Connectivity in Rural Areas: A Comprehensive Study on LTE network viability. Accessed 29 Oct 2024
84. North European Logistics Institute. Unlocking BVLOS Drone Connectivity in Rural Areas: A Comprehensive Study on LTE network viability. Accessed 29 Oct 2024
85. Ministry of Civil Aviation. Civil aviation regulatory landscape of Indian drone ecosystem. Accessed 29 Oct 2024
86. Maximize Market Research. Tethered drone market-industry analysis and forecast 2030. Accessed 04 Feb 2025
87. Hoverfly Technologies. Hoverfly technologies-tethered drones. Accessed 04 Feb 2025
88. King Abdullah University of Science and Technology (KAUST). Unleashing the potential of networked tethered flying platforms. Accessed 04 Feb 2025
89. Institut National Polytechnique de Toulouse. Unleashing the potential of networked tethered flying platforms for B5G/6G. Accessed 04 Feb 2025
90. NTT Docomo. Ntt docomo-tethered drones for network coverage. Accessed 04 Feb 2025
91. Deutsche Telekom. Deutsche telekom-tethered drones for rural coverage. Accessed 04 Feb 2025
92. BT Group. Bt group-tethered drones for emergency response. Accessed 04 Feb 2025
93. China Mobile. China mobile-tethered drones for network resilience. Accessed 04 Feb 2025
94. Reliance Jio. Reliance jio-tethered drones for urban coverage. Accessed 04 Feb 2025

95. H. Omote, S. Kimura, H.-Y. Lin, A. Sato, HAPS propagation loss model for urban and suburban environments, in *2020 International Symposium on Antennas and Propagation (ISAP)* (IEEE, 2021), pp. 681–682
96. GSMA. High altitude platform systems (HAPS) whitepaper 2021. Accessed 7 Feb 2025
97. Airbus. Zephyr-the world's leading solar-electric stratospheric UAS. Accessed 07 Feb 2025
98. Y.M. Putro, R.A. Nugraha, T.R. Nugraha, R. Christiawan, A. Bodhihanna, J.P. Ong, M.R. Testarosa, Legal issues pertaining to high altitude platform station implementation in Indonesia as an Archipelagic state. Udayana J. Law Culture **7**(2), 144–164 (2023)
99. BNamericas. We are really trying to establish a new connectivity industry here. Accessed 7 Feb 2025
100. Sierra Nevada Corporation (SNC). UK stratospheric HAPS project enters next phase. Accessed 7 February 2025
101. SoftBank Corporation. HAPS: High altitude platform station-non-terrestrial network solution. Accessed 7 Feb 2025
102. Chanakya Forum. High altitude platform systems: Imperative players in surveillance, communications, and kill chains. Accessed 7 Feb 2025
103. J.D. Deaton, High altitude platforms for disaster recovery: capabilities, strategies, and techniques for emergency telecommunications. EURASIP J. Wirel. Commun. Netw. **2008**, 1–8 (2008)
104. D. Aláez, M. Celaya-Echarri, L. Azpilicueta, J. Villadangos, UAVradio: radio link path loss estimation for UAVs. SoftwareX **25**, 101628 (2024)
105. A. Al-Hourani, S. Kandeepan, A. Jamalipour, Modeling air-to-ground path loss for low altitude platforms in urban environments, in *Global Communications Conference* (IEEE, 2014), pp. 2898–2904
106. H. Shakhatreh, A. Khreishah, B. Ji, Providing wireless coverage to high-rise buildings using UAVs, in *International Conference on Communications (ICC)* (IEEE, 2017), pp. 1–6
107. J. Holis, P. Pechac, Elevation dependent shadowing model for mobile communications via high altitude platforms in built-up areas. IEEE Trans. Antennas Propag. **56**(4), 1078–1084 (2008)
108. M.A. Figliozzi, Lifecycle modeling and assessment of unmanned aerial vehicles (Drones) CO2e emissions. Transp. Res. Part D Transp. Environ. **57**, 251–261 (2017)
109. P. Series, Propagation data and prediction methods required for the design of terrestrial broadband radio access systems operating in a frequency range from 3 to 60 GHz, in *Recommendation ITU-R* (2013), pp. 1410–1415
110. A. Al-Hourani, S. Kandeepan, S. Lardner, Optimal LAP altitude for maximum coverage. IEEE Wirel. Commun. Lett. **3**(6), 569–572 (2014)
111. D.S. Vohra, Enhancing the total airborne duration for a quadcopter/hexacopter using tethered power cable/mechanism, in *2024 15th International Conference on Computing Communication and Networking Technologies (ICCCNT)* (2024), pp. 1–4
112. M.Z. Farooqi, J. Laimio, J. Manner, 5G in the air: energy efficient future 5G high altitude platform stations, in *2025 IEEE Conference on Technologies for Sustainability (SusTech)* (IEEE, 2025), pp. 1–8
113. S. Alfattani, W. Jaafar, Y. Hmamouche, H. Yanikomeroglu, A. Yongaçoglu, Link budget analysis for reconfigurable smart surfaces in aerial platforms. IEEE Open J. Commun. Soc. **2**, 1980–1995 (2021)
114. X. Liu, H. Liu, C. Liu, Y. Luo, Time-varying communication channel high altitude platform station link budget and channel modeling. Information **9**(9), 210 (2018)
115. M.A. Solazzo, L. Travascio, A. Vozella, Proposed life cycle assessment approach to estimate the environmental impact due to high altitude platform systems. J. Phys. Conf. Ser. **2716** (IOP Publishing, 2024), pp. 012038

Seamless Connectivity: Wi-Fi

6

The IEEE 802.11 standard for wireless local area networks (WLAN), commercially known as Wi-Fi, has been driving digital connectivity across the globe [1]. Pilot experiments and network deployments have shown that Wi-Fi is a viable way to bridge the digital divide, that is capable of providing free, high-speed Internet in public spaces and educational hubs, enabling Internet connectivity for millions [2]. It is popular due to its affordability, accessibility, and scalability in underserved areas; additionally, it is cost-efficient, offers low latency, ultra-high throughput, and can be deployed quickly. These features make it ideal for deployment as a public Wi-Fi, community networks, and rural network initiatives focusing on connecting the unconnected as well as bridging the usage gap to build an inclusive digital community. Therefore, there is a need to explore the role of Wi-Fi in bridging the digital divide globally, while leveraging economic and environmental modeling to build a sustainable solution.

6.1 State-of-the-Art for Wi-Fi Technology

Wi-Fi is a wireless technology that supports LAN devices and Internet access. Wi-Fi has a shorter range, and the signal QoS is affected by penetration losses, which helps in minimizing the interference in ultra-dense deployments. The two main reasons why Wi-Fi is popular across the globe are that it operates in a license-free spectrum band and is cost-effective [3]. Figure 6.1 shows that the Wi-Fi architecture consists of several components working together to provide seamless and reliable wireless connectivity for end user devices [4]. The key components for a Wi-Fi network include

- **Access Point (AP)**: This is a networking device that allows other Wi-Fi devices to connect to a wired or wireless network [4]. There are two main Wi-Fi operation modes: *infrastructure mode* and *ad hoc mode*. The *infrastructure mode* is commonly used in homes, offices, and public Wi-Fi networks where the devices connect to the network via an AP, as shown in Fig. 6.1. On the contrary, *ad hoc mode* is primarily used for peer-to-peer networking and temporary communication, where devices communicate with each other directly without using an AP. The AP connects to the wired LAN typically via Ethernet cables, and the AP provides wireless connectivity to the end user devices via wireless LAN technology using the Wi-Fi standards [5].
- **Router**: This is a networking device that routes the traffic between the end user devices to support navigation through the Internet [4]. A router connects two or more data lines with different IP addresses. When the data packet is being transmitted, it reads the packet header to determine the destination network address. It uses a routing table and routing policy to transmit the data packet from one network to another until it reaches the destination IP address.
- **Devices**: These are the end user devices that support Wi-Fi technology and connect through the AP and include smartphones, laptops, TV, IoT sensors, earphones, headphones, consoles, tablets, wearable devices, and smart automation. These are the devices that a user interacts with to use the Wi-Fi network.
- **Service set identifier (SSID)**: This is the name of the Wi-Fi network to which the device connects. A wireless router or AP broadcasts SSIDs to nearby devices within its range, allowing them to distinguish it from other networks. To enhance the security of the Wi-Fi network, devices must authenticate using a password before gaining access.

S. Koratagere Anantha Kumar, *Wireless Broadband Access to Bridge the Digital Divide*,
Textbooks in Telecommunication Engineering, https://doi.org/10.1007/978-3-032-09261-8_6

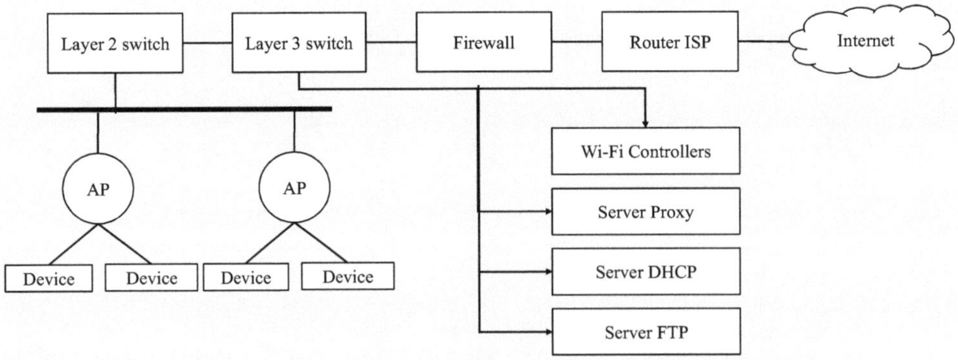

Fig. 6.1 Wi-Fi architecture

- **Firewall**: A firewall filters traffic by port, host, and network segment, which allows it to block incoming query requests at the server level to keep the network safe [6]. A firewall acts as a barrier to block unauthorized access and only allows trusted data to pass through the network. It allows traffic to flow from the end user device to the external network and vice versa [4].
- **Switches**: A switch is a multiport bridge with a buffer that can boost the signal strength, efficiency, and performance. This is a data link layer device that performs an error check before data forwarding. Switch only forwards data that is free from error and sends back packets with errors [7]. There are several types of switches, including Layer 2 and Layer 3 switches. The Layer 2 switches operate in the data link layer of the open system interconnection (OSI) model and are responsible for data forwarding among the devices connected to the same network. It connected the devices by an Ethernet port [4]. Whereas, Layer 3 switches operate in the network layer of the OSI model and route data between different networks, such as large and complex networks. Layer 3 switches perform similar functionalities as routers but are also connected to a firewall, servers, switches, and controllers.
- **Wireless Router**: This is different compared to the router that was discussed earlier. A wireless router, or Wi-Fi router, is a device that includes all the components discussed above, such as a router, an AP, a firewall, switches, and controllers [8]. A Wi-Fi router typically consists of one or more wireless interface controllers, a Gigabit Ethernet port integrated on a SoC, MIMO along with beamforming capabilities, antennas, and a router.
- **Server**: A **server proxy** allows a device to connect to the Internet through an intermediary server. When the device sends a request to the Internet via a server proxy, the proxy would encrypt the packet such that the Internet would understand it as the request from the proxy rather than the device [9].

Service providers and network administrators leverage the advanced capabilities of Wi-Fi to manage complex use cases and support high-density deployments efficiently. Wi-Fi offers higher data and more reliable indoor connectivity at a lower cost [5]. Indeed, Wi-Fi is a popular choice among end users seeking ultra-high-speed data and for service providers to offload their IP data traffic.

6.1.1 Spectrum Bands

Wi-Fi technology operates mainly on three spectrum bands: 2.4, 5, and 6 GHz [10]. Additionally, some other spectrum bands are currently popular for very niche applications using Wi-Fi, such as the 900 MHz and 60 GHz. These are described below and summarized in Table 6.1:

- **2.4 GHz band**: The first Wi-Fi was introduced in this spectrum band, which offers lower speed but better range due to the lower wavelength [11]. This spectrum band is a ISM band; therefore, it is shared with other technologies; hence, this band is more prone to congestion and interference from other devices [12].

Table 6.1 Wi-Fi spectrum bands and their characteristics

Band	Frequency range	Max speed	Range	Interference	Wi-Fi versions
2.4 GHz	2.400–2.483 GHz	1 Gbps	Long	High	Wi-Fi 1–6
5 GHz	5.150–5.850 GHz	9.6 Gbps	Medium	Low	Wi-Fi 3–7
6 GHz	5.925–7.125 GHz	46 Gbps	Short	None (New Band)	Wi-Fi 6E, Wi-Fi 7
900 MHz	902–928 MHz	100 Mbps	Very Long	Medium	Wi-Fi HaLow
60 GHz	57–71 GHz	7 to 40 Gbps	Very Short	No Interference	WiGig, Wi-Fi 7 Future

- **5 GHz band**: The 5 GHz spectrum band was introduced with IEEE 802.11a to offer higher data rates and lower the interference compared to the 2.4 GHz band [13]. The 5 GHz band offers short wavelength, reduced range, weaker penetration through walls, higher data rates, higher bandwidth, and increased channel availability. Unlike the 2.4 GHz band, the 5 GHz band is less crowded, mainly because there are fewer non-Wi-Fi devices. In some countries, this band is opportunistically shared between the Wi-Fi devices and the radar systems [14]. In this scenario, Wi-Fi device is subjected to dynamic frequency selection (DFS) regulations to avoid interfering with the radar systems.
- **6 GHz band**: This is a newer spectrum band added in the last few years to assist with the Wi-Fi capabilities. The 6 GHz band offers significantly higher throughput than 2.4 and 5 GHz combined, making it an ideal solution to complement the advanced terrestrial technologies [10]. This band is exclusively used by Wi-Fi networks in some regions, and other regions are considering the spectrum allocation in this band (either the entire band for unlicensed use, the entire band for licensed use, or a hybrid model that would support both unlicensed and licensed use of the spectrum).
- **900 MHz band**: The 900 MHz spectrum is part of the ISM band, and Wi-Fi in this band is popular for low-power, long-range communication technologies, such as smart meters, industrial IoT, smart agriculture, and remote applications [15]. The 900 MHz band offers long-range, better penetration capabilities through walls and other obstacles on the path of propagation; however, it offers lower data rates.

  In 2017, IEEE 802.11ah was published, called Wi-Fi HaLow, which uses 900 MHz license-exempt bands to provide the extended-range Wi-Fi networks [15]. This technology is designed for battery-powered sensors and supports applications and protocols that require low data rates, higher data rates, a wider coverage range, and low power consumption [16]. The maximum data rates of HaLow are 347 Mbps, and it focuses on reliability and efficiency rather than throughput rates. It can provide a range of up to 1 km and support up to 1000 devices via one access point.
- **60 GHz band**: Another band that has become popular for Wi-Fi networks is the 60 GHz spectrum band [17, 18]. This is known as the mmWave Wi-Fi band. It supports extremely high throughput (multi-gigabit speeds) and a shorter wavelength, resulting in a shorter range. This band includes the IEEE 802.11ad and IEEE 802.11ay standards [19]. The propagation path for the signal in this band is easily blocked by walls and objects; however, it can propagate via reflections from walls, ceilings, floors, and objects.

  Wi-Fi on this 60 GHz band was standardized as WiGig, first introduced with 802.11ad in 2012 ($\leq$7 Gbps), and a WiGig version 2 was released with 802.11ay in 2021 ($\leq$40 Gbps) [20]. It operates in the 57–71 GHz spectrum band, providing low interference and high bandwidth [21]. Unlike the traditional Wi-Fi, modern Wi-Fi standards use beamforming to focus the signal toward intended devices, minimizing power consumption, and improving throughput [22].

6.1.2 Timeline

The evolution of Wi-Fi standards is driven by the advancements of IEEE 802.11 standards, which define the protocols and performance of wireless networking to improve network efficiency, speed, latency, security, and coverage [23]. Over the years, the technology has upgraded in different angles, such as the evolution of physical (PHY) and MAC for wireless communication to support novel transmission technologies supported by Wi-Fi [5]. The brief timeline for the Wi-Fi technology is shown in Table 6.2 [5], and is discussed below

- **Pre-1997**: Around the year 1980, the research on using infrared (IR) rays to design WLAN began in IBM's Rueschlikon Laboratory, Zurich, Switzerland [5, 24]. At that time, wired LANs was popular in large manufacturers and offices, which

Table 6.2 Comparison of Wi-Fi generations (802.11 standards)

Wi-Fi Gen	IEEE Std	Year	Max speed	Bands	Ch. width	Modulation	Key features
Wi-Fi 1	802.11	1997	2 Mbps	2.4 GHz	22 MHz	FHSS/DSSS	First standard
Wi-Fi 2	802.11b	1999	11 Mbps	2.4 GHz	22 MHz	DSSS	Improved range
Wi-Fi 3	802.11a/g	1999/2003	54 Mbps	5 GHz/2.4 GHz	20 MHz	OFDM	Less interference (a)
Wi-Fi 4	802.11n	2009	600 Mbps	2.4/5 GHz	40 MHz	64-QAM	MIMO support
Wi-Fi 5	802.11ac	2014	3.5 Gbps	5 GHz	160 MHz	256-QAM	MU-MIMO, Beamforming
Wi-Fi 6	802.11ax	2019	9.6 Gbps	2.4/5 GHz	160 MHz	1024-QAM	OFDMA, BSS Coloring
Wi-Fi 6E	802.11ax (Ext)	2021	9.6 Gbps	2.4/5/6 GHz	160 MHz	1024-QAM	Adds 6 GHz band
Wi-Fi 7	802.11be	2024	46 Gbps	2.4/5/6 GHz	320 MHz	4096-QAM	Multi-Link Operation

was not feasible when it required dense deployments. This fueled the need for wireless connectivity, along with the need for aesthetically pleasing solutions.
In 1991, NCR Corporation and AT&T developed the precursor to Wi-Fi, called WaveLAN, using DSSS, to provide wireless cash registry systems that could achieve 2 Mbps data rates [5]. Other groups that achieved similar results were Proxim, which resorted to FHSS, while Photonics developed IR solutions for WLAN.

- **1997**: In this year, IEEE introduced the 802.11 standards, which were built on the architecture and learnings of all the R&D and pre-1997 commercial deployments of Wi-Fi. This standard is the first wireless networking standard that offers a speed of 2 Mbps [25]. Either DSSS or FHSS modulation technique could be used by the technology to encode the data. As a result, there was a surge in the development of wireless networking equipment and solutions, and devices with Wi-Fi capabilities were made possible, enabling users to access the Internet without a wired connection.
- **1999**: IEEE 802.11a and 802.11b were developed simultaneously [26–28]. The Wi-Fi alliance was formed to promote the usage of Wi-Fi and ensure interoperability.
The first widely adopted standard was 802.11b, which would operate in the 2.4 GHz band and support up to 11 Mbps data rates with a fallback to the legacy 802.11 standard; it played a crucial role in creating the demand for wireless networking [29]. The IEEE 802.11b used a complex M-Ary orthogonal coding known as complementary code keying (CCK), to achieve higher data rates. This technology uses DSSS modulation to improve signal robustness and reduce interference with other users on the same spectrum band. Depending on the environmental conditions, 802.11b offers an indoor range of up to 30 m and an outdoor range of 90 m [30]. The standardization encouraged the development of devices at a lower cost, making them affordable for consumers and businesses. The devices and users in this band faced challenges such as spectrum congestion, interference from other devices, and security challenges as they used the wired equivalent privacy (WEP) protocol [31].
The IEEE 802.11a would operate in the 5 GHz spectrum band and support speeds up to 54 Mbps. It was primarily adopted in businesses and enterprise environments that required higher performance and reliability. However, the adoption rate was lower due to the higher cost and limited device compatibility compared to IEEE 802.11b, along with a shorter coverage area.
- **2003**: The extension of the 802.11b standard is the IEEE 802.11g standard, which was ratified in 2003 [32]. IEEE 802.11g operates on the same 2.4 GHz spectrum band while offering improved data rates of up to 54 Mbps in the same band. This technology is backward compatible with 802.11b to ensure broader adaptability and a smooth transition [33]. 802.11g standard uses OFDM schemes and supports CCK; it was widely adopted in both consumer and business markets and integrated into various devices. 802.11g supported faster Internet browsing, faster file transfer, and streaming media [34]. However, the key challenges involved managing the limited spectrum resource in the 2.4 GHz band from becoming congested and reducing the interference from other devices.
In 2004, the Wi-Fi alliance was launched Wi-Fi protected access (WPA) to improve the security of Wi-Fi networks. By 2005, the number of public Wi-Fi hotspots started increasing significantly, with some even offering paid access to customers. The local government started contemplating the idea of using municipal Wi-Fi networks to provide city-wide Internet access

to residents and visitors. In 2007, Apple introduced its first iPhone, which integrated Wi-Fi capabilities to highlight the importance of Wi-Fi for mobile access for consumers. [35]. Apple also introduced AirPort Extreme, which supports IEEE 802.11n [36].

- **2009**: The previous advancements paved the path for the standardization of IEEE 802.11n, which offers a significant performance improvement compared to the older generations [37]. 802.11n can support up to 600 Mbps and offer MIMO capabilities for spatial multiplexing to increase the throughput. This standard is also capable of performing spectrum aggregation in regions where the channel is available using channel bonding techniques. The advancements in the MAC protocol improved efficiency by reducing the overhead and increasing the throughput. 802.11n became widely popular in consumer devices [38]. This technology is also reliable, which makes it suitable for bandwidth-intensive applications [39]. 802.11n is also capable of offering and altering the performance based on the channel conditions and environmental factors. This technology is also backward compatible with previous Wi-Fi generations.
- **2013**: Building on 802.11n technology, the subsequent generation of Wi-Fi is 802.11ac, delivers data rates up to 3.5 Gbps with Wave 1 devices (channel width of 80 MHz) and 6.93 Gbps with Wave 2 devices (channel width of 160 MHz), via 256-QAM advanced modulation techniques that encoded more data per signal to offer this increased throughput [40]. Furthermore, 802.11ac is integrating with advanced technological features such as multi user MIMO (MU-MIMO) beamforming. It is backward compatible with previous generations. The improved performance and higher data rates drove the demand for this technology. 802.11ac was widely integrated into Wi-Fi enabled devices and networking equipment for high bandwidth applications. With time, more devices started operating on the 5 GHz band, and they can now experience interference from other devices operating in the same frequency range [41]. The performance of the network is dependent on the operating conditions, MU-MIMO, and the number of antennas.
- **2019**: In 2019, 802.11ax, also known as Wi-Fi 6, was standardized, which is the next-generation wireless standard designed to improve speed, efficiency, and QoS in highly dense environments [42, 43]. It operates in both 2.4 GHz and 5 GHz spectrum bands and supports up to 9.6 Gbps using 1024 QAM modulation to offer improved data rates. Wi-Fi 6 uses OFDM to allow multiple devices to share the channel simultaneously [44, 45]. Wi-Fi 6 became popular in places where there is a need for dense deployments. Some of the other key features of Wi-Fi 6 include target wake time (TWT) in crowded environments, improving the energy efficiency for IoT and other connected devices by scheduling transmissions between the access point and the devices [46]. TWT also helps in extending the battery life of connected devices. Wi-Fi 6 also uses basic service set coloring schemes and beamforming to minimize interference from the nearby devices and enhance the signal strength, range, and reliability [47].
  In 2021, Wi-Fi 6 was extended further to Wi-Fi 6E, which supports the 6 GHz spectrum band, additional bandwidth, and less interference [48]. This spectrum band can offer seven additional 160 MHz channels, which significantly boosts the performance of the Wi-Fi 6E and reduces congestion compared to the bands at which Wi-Fi 6 operates. Wi-Fi 6E supports high-bandwidth applications.
- **2024**: With the advancement of telecommunication technologies, the video traffic generated has gone up exponentially, and the demands surpass the capabilities offered by Wi-Fi 6E [49]. Therefore, 802.11be, also known as Wi-Fi 7, was released, which is capable of offering extremely high throughput (EHT), deterministic low latency, enhanced network capacity that can support a minimum throughput of 30 Gbps, and operating at 2.4 GHz, 5 GHz, and 6 GHz [50, 51]. Some of the unique features of Wi-Fi 7 include multilink operations (MLO) capabilities that are a combined framework to manage spectrum aggregations across different bands, manage multiple links, and optimize resource allocation across the different spectrum bands [52]. The feature of Wi-Fi 7 aggregates up to 320 MHz of spectrum and offers a maximum data rate of 46 Gbps using 4096-QAM OFDM access technology [53]. This technology also supports multi-RUs and uses the WPA3 security protocol. Wi-Fi 7 supports up to 16 spatial streams, preamble puncturing to minimize the interference, and multi-AP coordination to improve the overall network performance [23].
  The primary use case of Wi-Fi 7 includes time-sensitive applications. By 2025 to 2026, the adoption of Wi-Fi 7 is expected to mature [54]. However, several countries are yet to standardize the spectrum allocation of 6 GHz spectrum bands, and this factor would contribute significantly to the widespread adoption of Wi-Fi 7 [55].
- **2028**: The discussions for Wi-Fi 8, possibly named 802.11bn, have begun [56, 57]. Some of the features that Wi-Fi 8 is expected to focus include [58]:

  - AI-optimized networking for bandwidth allocation, congestion control, and energy efficiency,
  - expansion to frequency band above 7 GHz, integration with early 6G development, merging into cellular networks, and existing Wi-Fi generations,
  - speeds in the range of 100 Gbps per user with higher-order modulation (8192-QAM) or beyond,

- ultra-low latency for metaverse, quantum networking, holographic communication, and AR/VR applications, and
- seamless integration with terrestrial networks, especially 6 GHz cellular networks, to offer seamless indoor-outdoor coverage.

The official standard and commercial adoption of Wi-Fi 8 are expected to be done in the year 2030 [59].

6.1.3 Use Cases

As of 2025, there are around 19.5 billion Wi-Fi enabled devices connected to the Internet. This shows us that Wi-Fi has become an integral part of our daily lives and has become an integral part of a range of applications across various sectors. Some of the popular use cases of Wi-Fi enable seamless Internet access for different applications for end user devices ranging from home networking to public hotspots and smart homes to industrial IoT. Public hotspots are convenient for travelers, commuters, and anyone else who needs Internet access.

Access to Wi-Fi in educational institutions is playing a crucial role in empowering communities, particularly in remote and rural areas. It enables students and teachers to collaborate and learn digitally. Wi-Fi is preferred by several rural connectivity solutions across the globe, mainly because of its cost-effectiveness, which reduces the overall network deployment cost. Deploying technologies that require extensive groundwork and installation drives the network cost higher. Wi-Fi Solutions are generally faster to deploy since they operate on the unlicensed spectrum band and provide seamless connectivity with any device. Most of the end user devices are manufactured with integrated Wi-Fi capabilities, making them a versatile solution for bridging the digital divide at an affordable cost.

In rural and remote areas, community networks can be established using Wi-Fi technology to give residents in the region shared Internet access. This would improve people's digital literacy and facilitate communication, information access, and recreational purposes. These programs would create an inclusive society by encouraging cooperation, resource sharing, and affordable connectivity.

6.1.4 Technologies Powering Wi-Fi

A combination of technologies empowers and drives the performance of Wi-Fi networks and will also play a key role in the development of Wi-Fi 8. Some of these technologies powering Wi-Fi include

- **Beamforming**: It is a technology that enhances the performance of the transmitted signal by directing the beams toward the user or a specific device rather than omnidirectional beam broadcasting. By focusing the signal directly on the intended receiver device, the Wi-Fi can minimize the error, transmit higher data rates, and improve network security. This technology further powers Wi-Fi to manage and reduce the interference with other devices operating in the same band. Another important use case of beamforming in Wi-Fi is the ability to co-exist with other technologies, especially 5G NR in the upper 6 GHz band [60].
- **Spectrum Sharing**: Sometimes the Wi-Fi channels are very crowded where multiple devices are competing for the same resources [61]. In such scenarios, spectrum sharing and DSA would play a crucial role in improving the network performance by dynamically changing the operating spectrum band to meet the demand and minimize the interference [62]. The newer generations, such as Wi-Fi 6 and above, offer the capabilities of performing spectrum sharing protocols and algorithms to avoid interfering with the incumbents on the same band [63, 64]. Stakeholders are investigating ways for the different technologies to coexist in the highly anticipated 6 GHz band [65].
- **MEC**: This is a technology that brings the computing resources to the edge of the network and enhances the performance of the Wi-Fi networks while lowering the latency [66]. Wi-Fi connects the IoT devices to the network, while the MEC complements this deployment by processing the data at the edge and sending less data to the central network. This model improves the latency, QoS, energy efficiency, and maximizes the resource utilization [67].
- **MLO**: The main characteristic of Wi-Fi 7 and its upgraded version that aids in transmitting the extreme throughput is MLO [68]. This technology supports multi-link devices simultaneously operating across different spectrum bands, such as 2.4,

Fig. 6.2 Multi-link operations in Wi-Fi 7 and beyond

5, and 6 GHz, as shown in Fig. 6.2 [69, 70]. This feature is quite useful in a scenario where the network is facing heavy congestion and the devices can transmit and receive data over different bands at any point in time.

- **Modulation Techniques**: Some of the key modulation techniques employed by Wi-Fi generations are DSSS, OFDM, FHSS, and adaptive modulation. Based on the availability of the channel bandwidth and propagation environment, Wi-Fi can use different modulation techniques, such as BPSK, QPSK, and QAM, to offer higher speeds. For example, Wi-Fi 7 uses 4096-QAM to deliver 46 Gbps data rates to the users. The latest Wi-Fi standards use adaptive modulation techniques to dynamically change the modulation techniques based on the wireless channel conditions. The primary benefits of this technique allow Wi-Fi to offer seamless connectivity, even in poor channel conditions.
- **AI and ML for QoS Optimization**: Another driving factor, which is empowering Wi-Fi is the applications of AI and ML in this domain. It helps in improving the network's efficiency in terms of power, energy, channel selection, resource allocation, adaptability, reliability, seamless connectivity, managing network traffic, and congestion performance. Furthermore, AI/ML would help in the prediction of network faults and maintenance of the network to prevent service disruption. Furthermore, AI/ML would help in integrating Wi-Fi networks with other terrestrial, air, and space layer technologies. AI/ML has aided further in monitoring KPI, documentation, data cleaning, user behavior analysis, and addressing connectivity issues [71].
- **Wider Bandwidth**: The availability of wider bandwidth plays a crucial role for Wi-Fi technology to be capable of delivering numerous benefits, such as very high QoS, improved network efficiency, and better network congestion management [72]. The wider bandwidth also supports higher data rates and reduces interference between the Wi-Fi sub-channels and nearby devices operating in the same band. Wider bandwidth would also ensure that advanced Wi-Fi generations would offer a reliable, secure, resilient, and low-latency network. Since the indoor devices generate over 80% of traffic, wider bandwidth plays a crucial role in handling the traffic generated.
- **Enhanced Security**: Wireless network security plays a critical role in ensuring that users and their data are safe from cyber threats, unauthorized access, and data breaches. Enhanced network security would involve designing, implementing, and ensuring the security of Wi-Fi networks to preserve the confidentiality, integrity, and authenticity of Wi-Fi networks and their resources. Wi-Fi networks use 4 main types of network security protocols, which include WEP, WPA, WPA2, and WPA3.
  Wi-Fi technology needs to ensure that it is a safe digital environment for IoT ecosystems, online financial transactions, work from home, telehealth, and other applications that rely on it for primary access to the Internet. Furthermore, using Wi-Fi CERTIFIED Passpoint on guest or public networks includes WPA3-enterprise security, which seamlessly authenticates user devices on the network using the credentials provided by the service provider. This is a popular technique that is used in the community hub networks in rural areas to ensure users have access to a safe Internet.

6.1.5 Rural Wi-Fi Architectures

Why would Wi-Fi be discussed so extensively in this chapter? This is mainly because Wi-Fi is one of the key contender technologies to bridge the digital divide. Different deployment strategies are utilized Wi-Fi in providing coverage in rural and remote areas at an affordable price and are shown in Fig. 6.3.

6.1.5.1 Wi-Fi Mesh Networks

A mesh network uses multiple APs nodes that communicate with each other via wireless links. The node that communicates with the Internet is called a gateway node, while the outer node is called a leaf node, and the rest of the nodes between the gateway and a leaf are called repeaters, as shown in Fig. 6.4 [73]. The key features of a mesh network are practically unlimited network expansion, simple networking, and easy maintenance and administration. The backhaul links provide long-distance

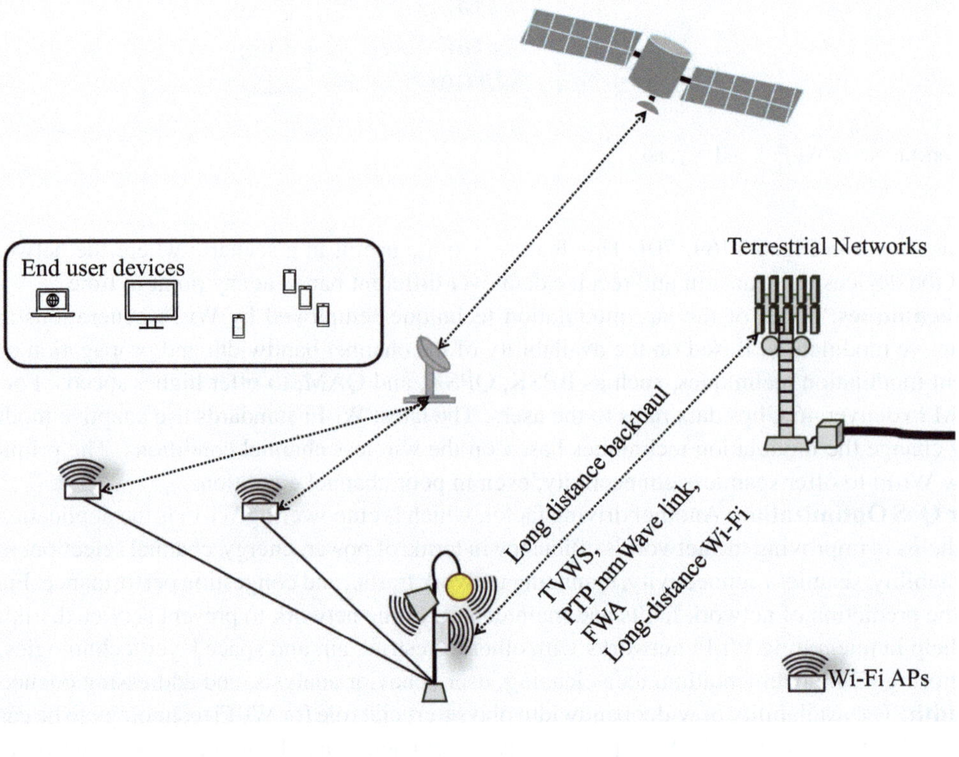

Fig. 6.3 Architecture for different Wi-Fi deployment strategies in rural areas

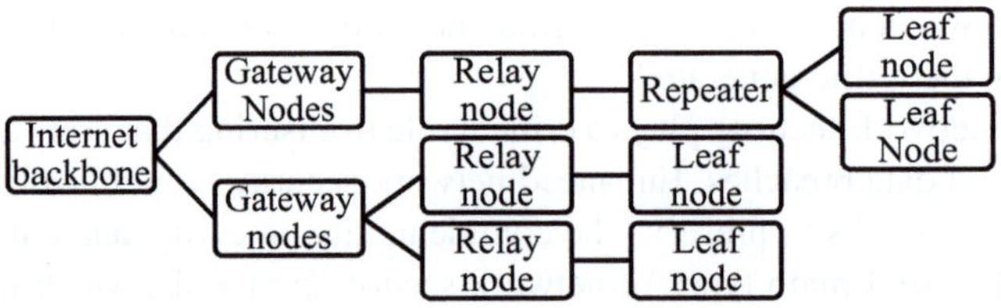

Fig. 6.4 Wi-Fi mesh networks

connectivity between the mesh nodes and the gateway, especially using high-gain directional antennas. For rural areas, solar panels with battery backups are a potential solution to mitigate the inconsistency in power supply in those regions [74]. The key challenges with mesh networks are bandwidth management, power reliability, and extreme weather conditions.

6.1.5.2 TVWS and Wi-Fi

There are several rural pilot trials and test beds that have been exploring the use of TVWS and Wi-Fi together to bridge the digital divide [75]. Since the TVWS signal operates in the UHF bands, it offers a longer range and superior penetration capabilities over Wi-Fi link. These properties of TVWS make it an ideal candidate for acting as a backhaul link for the local Wi-Fi mesh deployed in the community of interest [76, 77]. Some of the key benefits offered by TVWS include cost-effective coverage and backhaul solutions. Though TVWS offers excellent propagation characteristics, it is not compatible with a wide range of devices, as in the case of Wi-Fi. The integration of TVWS and Wi-Fi helps to bridge the benefits offered by both at a lower cost. However, the key challenge with this model is the need for the spectrum licensing of TVWS radios at a recurring duration and the availability of the spectrum. This method has been explored under pilot studies in Africa, Asia, and Europe.

6.1.5.3 Wi-Fi Backhaul with Satellite or Fiber

Another promising approach to bridge the digital divide is the integration of Wi-Fi networks as a last-mile technology and using satellite technology as backhaul [78]. This deployment framework is popular in underserved regions, where it is hard to deploy terrestrial networks. In this deployment strategy, the satellite would transmit the data to the ground station or the Wi-Fi access point directly. Once the gateway node is connected to the Internet via satellite network, the network would be further connected to a local mesh network to provide wide area coverage in remote and rural areas [79]. The proposed integration model combines satellite long-range coverage with Wi-Fis high-throughput performance to deliver affordable, multi-user connectivity [80].

As technology evolves, it holds the greater potential to offer terrestrial network capabilities in rural areas at a lower cost. For a wide-scale adoption of this technology, certain challenges need to be addressed, such as the latency of LEO, MEO, and GEO satellites, link susceptibility to the climatic conditions, cost of provisioning a satellite backhaul capacity, and policy and regulations in supporting these solutions in rural markets [81].

6.1.5.4 Public Wi-Fi

Public Wi-Fi has revolutionized the digital transformation for several industries across the globe. The end user devices would connect to the public Wi-Fi network for different applications. Figure 6.5 shows the high-level architecture model for public Wi-Fi [82, 83]. This model is popular in numerous locations across the globe, and the business model adopted around it varies depending on the deployment location.

6.1.5.5 Community Networks

The community Wi-Fi networks are grassroots initiatives that are designed, built, and managed by the community. This type of deployment is popular in remote and rural areas. The local community is trained by NGOs and organizations to run their networks. This deployment leverages mesh networking to extend gateway connectivity to houses, schools, clinics, and libraries via an open-access model that is cost-effective, spurs local employment, and boosts digital literacy and self-reliance. Governments across the globe are providing subsidies for the development of community networks at an affordable cost.

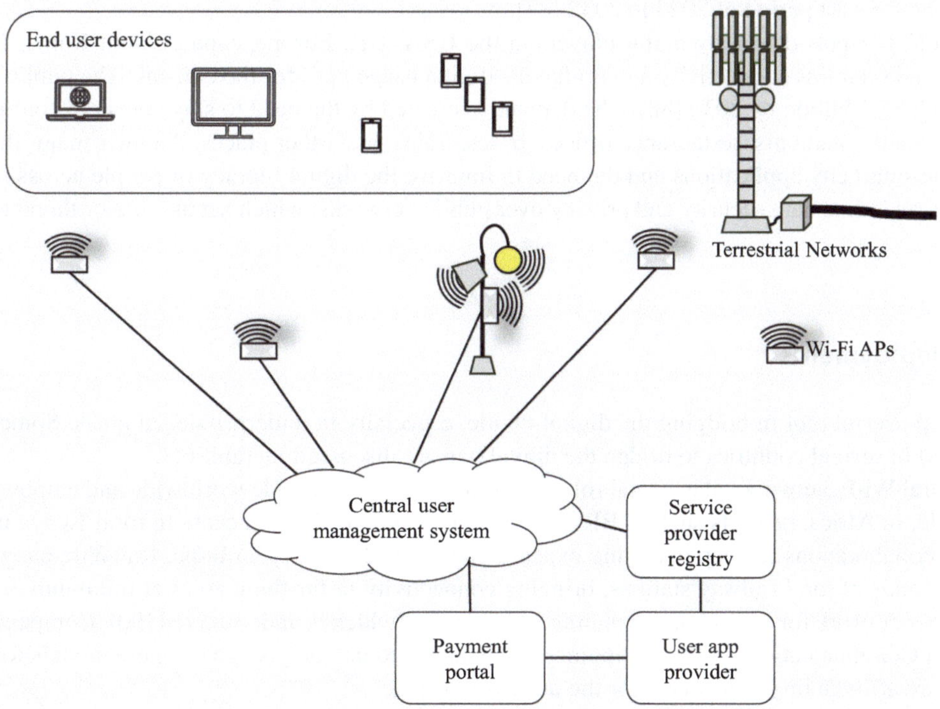

Fig. 6.5 Public Wi-Fi architecture

However, certain key challenges need to be addressed for scaling the network, for example, the cost of connecting to PoP, funding sources, technical expertise, regulatory hurdles for a commercial large-scale network, and QoS.

6.2 Worldwide Wi-Fi Deployments

Wi-Fi plays a key role in the digital transformation that supports businesses of every shape and size. There are around 19.5 billion Wi-Fi enabled devices, which are connected to over a billion Wi-Fi networks across the globe. Wi-Fi deployments are present from homes and on light poles in public areas.

6.2.1 Wi-Fi Deployment Across the Globe

Wi-Fi 6 deployments have been growing rapidly across the globe, with a projected 4.1 billion devices to be shipped in 2024 [84]. The Wi-Fi 6 market share has been growing at a CAGR of 36.8% from 2023 to 2024, bringing it to $23.56 billion. The demand for being connected and the applications supported by the Internet are driving the demand for higher data throughput and the growing bandwidth-intensive applications. By 2028, the market share for Wi-Fi 6 is expected to reach around $76.32 billion at a CAGR of 34.2% [85].

The Wi-Fi alliance has predicted that there would be around 2.1 billion Wi-Fi 7 enabled devices for bandwidth-intensive applications worldwide by 2028. Wi-Fi 7 is expected to be part of 6G ubiquitous coverage to enable better and reliable indoor connectivity in both urban and rural areas [86]. A survey was conducted to assess the deployment strategies of global industry executives, who stated that they are planning on deploying OpenRoaming for seamless integration between Wi-Fi and cellular networks (LTE/5G/6G), enhancing Wi-Fi security, offering seamless roaming on the Wi-Fi networks, and facilitating access across diverse technology networks [86].

With the growing complexity of Wi-Fi deployment, **Wi-Fi-as-a-service** is a newer deployment business model that is becoming popular. In this model, a firm offers a subscription service that integrates software, infrastructure, and hardware and manages the operations into one package that would allow faster rollout of complex Wi-Fi networks for the customized use cases and applications. The global market for this deployment strategy is expected to reach $4.4 billion by 2032, at a CAGR of 14.6% during the forecast period of 2023 to 2032 [87].

The public Wi-Fi hotspots driven by major players in the USA, UK, Europe, Japan, South Korea, India, and Africa to provide inclusive and seamless connectivity and bridge the digital usage gaps for the citizens. The public Wi-Fi market share is expected to reach $5.1 billion by 2030 [88]. The demand is created by the need to stay connected online in places such as airports, shopping malls, stadiums, restaurants, offices, buses, trains, and other places. Another major driving factor for the public Wi-Fi is the smart city applications and the need to improve the digital literacy of people across the globe. However, there are concerns regarding data security and privacy over public networks, which act as a major threat to the growth of this strategy.

6.2.2 Rural Deployments

Wi-Fi has been a powerful tool in bridging the digital divide, especially in underprivileged areas. Some key initiatives and programs launched in several countries to bridge the digital gap are discussed in Table 6.3.

Commercial rural Wi-Fi networks play a vital role in bridging the digital divide worldwide and empowering rural communities. For example, in Africa, projects such as BRCK and Poa, affordable Wi-Fi access in rural Kenya using solar-powered routers and local collaborations to keep operating expenses low [89]. Similarly, in India, RailWire has implemented public Wi-Fi across thousands of rural railway stations, bringing connectivity to far-flung areas at minimum expense. The railway has a dedicated fiber network for control and signaling for the trains, which is underutilized [90]. Companies, such as AirJaldi in India and Smart Communications in the Philippines, have deployed pay-as-you-go business models for Wi-Fi solutions in rural areas to ensure affordability at low cost for the underprivileged.

The architecture shown in Fig. 6.5 can be further extended to support rural private networks and SME Wi-Fi ecosystem. Even developed countries such as Australia offer an Outback Wi-Fi initiative for their citizens via long-range wireless systems

Table 6.3 Global public Wi-Fi deployments initiatives–digital divide and digital inclusion

Country	Initiative details
The United States	The "Internet for All" initiative provide funding to expand broadband and public Wi-Fi access, including in remote and rural areas
India	The "PM-WANI" scheme focuses on connecting the millions in the remote and rural areas using low-cost public Wi-Fi hotspots. They use the marketplace business model where there is a platform that connects Wi-Fi service providers with the end users who would require these services by connecting to the Internet. This is similar to the business model employed by Airbnb and Uber
South Korea	As a developed nation, provides public Wi-Fi hotspots in the entire country to ensure there is inclusivity
Kenya	Initiatives like BRCK's SupaBRCK aim to create a solar-powered hotspot that is self-reliant and generates clean water, portable Wi-Fi hotspots, and other necessary amenities that are sustainable in remote and rural areas
Rwanda	To enhance the digital literacy in the country, it has deployed free Wi-Fi in public areas and public transport under the "Smart Kigali" scheme

using Wi-Fi technology to enhance the digital connectivity for indigenous people and offer inclusive digital transformation [91]. In the UK, TVWS integration with Wi-Fi was explored in connecting the remote rural areas and the Island of Scotland [92].

6.2.3 Challenges in Rural Wi-Fi Deployment

Studies have highlighted that deploying Wi-Fi networks in rural areas are essential in successfully bridging the digital divide. To truly empower the underserved communities, the technology must address certain challenges that are unique to Wi-Fi, and include [93]:

- **Poor performance** is a common issue with Wi-Fi mainly because it operates on a licensed exempt ISM band, which is impacted by interference, congestion, and hardware limitations. The performance would be poor when the indoor Wi-Fi technologies need to support thousands of households in rural areas, especially if the network planning is not optimized. There is also a need to enhance the Wi-Fi capabilities to support long-range PTP and PTMP links when used as a backhaul technology in rural areas.
- **Poor upfront Wi-Fi planning** results in a network whose performance would not reach the expected levels. There is a need to select an optimal location to deploy gateways, repeaters, and leaf nodes and mount them in outdoor sites. This would impact the use cases to be served, costs, and interoperability challenges, especially where mobile networks are offloading to Wi-Fi networks or when a particular community is not able to provide connectivity to their critical infrastructure and educational hubs. Efficient network planning for Wi-Fi networks would also lead to improved QoS and lower interference.
- **Wi-Fi outages** in rural areas are a significant challenge, resulting from challenges in infrastructure limitations, power supply inconsistencies, environmental factors, maintenance issues, outdated equipment, hardware failure, theft, and limited backhaul options. In rural areas, it would be hard to send a technician whenever the network has an outage. The large-scale Wi-Fi networks require ways to be remotely rectified during breakdown to keep the operational costs low and minimize the downtime. Another possibility would be to train the local community in handling such scenarios at an additional cost, requiring efforts to train and upskill the community.
- **Wi-Fi management** is critical for a large-scale Wi-Fi network in rural areas. It would require personnel who are capable of managing the network, allocating resources, optimizing the network performance, ensuring efficient bandwidth utilization, and ensuring the network is secure and stable. There is a lack of on-site IT support, which could be compensated for by deploying cloud resources for network management to enable remote monitoring and troubleshooting. Therefore, there is a need to train the local community to run the networks and perform checks to avoid Wi-Fi outages.
- The initial deployment **cost** could be exorbitant for a wide area network (WAN) when factoring the cost of channel management, interference mitigation, cost of backhaul connectivity, and initial CAPEX. Wi-Fi equipment costs the same for urban and rural scenarios; hence, in a low subscriber density scenario, to improve the techno-economic feasibility of the deployment, there is a need for higher ARPU.
- **Technological Challenges**: Deployments must use outdoor-grade cabling and robust backup power systems to endure harsh environmental conditions and power inconsistencies.

6.3 Wi-Fi: Accessibility, Affordability, and Climate

There is a need to develop innovative business models in these underserved regions to build a sustainable business model due to low population density, network adoption, affordability constraints, and high infrastructure costs.

6.3.1 Economic Frameworks

6.3.1.1 Economic Modeling

Economic modeling for Wi-Fi technology in rural areas would enable a solid understanding of the technical requirements, deployment challenges, and adoption challenges to devise a potential strategy to overcome the problem and help empower the community with digital connectivity. The business models and revenue streams for Wi-Fi for rural deployments could use a freemium model, community-based model, Wi-Fi-as-a-service, pay-as-you-go model, government-led model, and advertisement model. Different pricing models are employed depending on the deployment location and revenue streams.

It is important to select an appropriate and suitable business model that reflects the demand and affordability of the underserved community. When the stakeholders align their interests with the socioeconomic conditions of the community, rural connectivity would thrive in such a scenario. Wi-Fi deployment offers opportunities and challenges that must be well considered. Table 6.4 shows the SWOT analysis for the deployment of Wi-Fi in rural communities.

6.3.1.2 Channel Modeling Using Stochastic Geometry

To understand the Wi-Fi link budget and stochastic modeling associated with it, there is a need to develop a coverage model for the desired area of interest. For a Wi-Fi AP, the path loss in dB, $PL(d)_{Wi-Fi}$, is modeled according to the heuristic model [94]:

$$PL(d)_{Wi-Fi} = PL(d_0) + 10nlog_{10}\frac{d}{d_0} + \Sigma_i L_{Wi} + \Sigma_j L_{Bj} \tag{6.1}$$

where $PL(d_0)$ is the path loss at a reference distance d_0 m in dB according to the distance loss model, d m is the distance between the AP and the receiver, n is the path loss component, $\Sigma_i L_{Wi}$ is the wall loss component along the path, and $\Sigma_j L_{Bj}$ is the interaction loss accounting for the propagation direction losses [94]. Simplifying the equation using the Friis transmission model, the link budget is written as

$$PL(d)_{Wi-Fi} = -27.55 + 20log_{10}f + 20log_{10}d \tag{6.2}$$

Table 6.4 SWOT analysis of Wi-Fi in bridging the digital divide

Strengths	Weaknesses
Cost-effective and rapid deployment, higher data throughout, and support mobility Scalable based on the use case of interest Compatible with advanced telecommunication technologies such as 4G, 5G, 6G, satellites, UAVs, and FWA Smaller carbon footprint Economic growth and social inclusion Subsidized by the government	Limited coverage, interference, and range Lower operating power constraints Sensitivity to weather conditions High setup and maintenance costs due to upgrades Stricter regulations as critical for infrastructure operating on 6 GHz bands Vulnerable to cybersecurity threats, especially in public Wi-Fi deployments
Opportunities	**Threats**
Bridging the digital divide in remote and rural areas Empowering the rural communities and providing inclusivity at an affordable price Integration with other technologies to offer seamless connectivity Collaboration with stakeholders to drive a sustainable model Emerging Wi-Fi generations that are power efficient and powered by renewable energy sources	Limitations in terms of cost and technical capabilities of backhaul Constraints in the operating conditions, especially in higher bands Limited purchasing power in remote and rural areas Competition from other technologies that would offer services in the same price range Lack of sustainable business models in underprivileged regions Lack of appropriate digital skills, and challenges of being able to afford the devices that utilize Wi-Fi

After estimating the link budget for the desired location of interest, next, estimate the Wi-Fi AP's range in km, D_{Wi-Fi}, using the following equation [95]:

$$D_{Wi-Fi} = 10^{(PL(d)^{max}_{Wi-Fi} - 32.44 - 20log_{10} f)/20} \tag{6.3}$$

Since Wi-Fi operates on ISM bands, it is often difficult to model the interference from other devices. Hence, most models use field test values for the assessment [95–97].

6.3.1.3 Techno-Economic Assessment

The TEA for rural broadband deployment of Wi-Fi and any other technology would depend on several factors [98]. The key factors that determine the cost of network deployment are coverage distance, coverage area, and data rate per area, which in turn depend on the independent variables such as antenna beamwidth, Wi-Fi generation, antenna gain, power, bandwidth, frequency, maximum data rate, and price of the equipment. Some studies have shown that the cost of the network and APs is highly dependent on these parameters [98], for example,

$$\begin{aligned} Price = &= 168.34\beta^{-0.362} \\ &= 7.6034P - 122.82 \\ &= 4.083A + 58.051 \end{aligned} \tag{6.4}$$

where β is the antenna beamwidth in °, P is the transmit power requirements in dBm, and A is the intended coverage area on km^2.

A major component of the network costs when deploying comes from the backhaul connectivity. The cost is dependent on the backhaul technology and network architecture. In rural communities, there is also a need to consider alternative power sources, as there would be power inconsistency using renewable energy sources and batteries. Once the technological modeling helps in determining parameters such as the number of APs required for the deployment, energy sources, and backhaul options, then comes into the picture the cost modeling.

- **CAPEX**: The initial CAPEX, C_c for the deployment would include

$$C_c = N_{AP}C_{AP} + C_{power} + C_{backhaul} + C_{install} + C_{spectrum} \tag{6.5}$$

  where N_{AP} is the number of Wi-Fi AP priced at C_{AP}, C_{power} is the cost toward the installation of power sources (renewable and non-renewable), $C_{install}$ is the cost for installation of the network that would account for installation, labor, configuration, wiring, and miscellaneous, and $C_{spectrum}$ is the cost toward the spectrum paid by the operators to the government when they provided advanced coverage capabilities by automating Wi-Fi channel management, which otherwise is license exempt.
- **OPEX**: Similarly, OPEX toward Wi-Fi deployment for year t, $C_{o,t}$, would include costs toward total energy consumption $E_{energy,t}$ in kWh priced at $P_{unit,t}$ per unit, $C_{bandwidth,t}$ is the cost toward the data traffic handled by the network per year, $C_{maintenance,t}$ is the annual network maintenance cost that includes repairs, outages, and breakdown, and $C_{secure,t}$ is the cost toward the additional cybersecurity and access control. This is mathematically modeled as

$$C_{o,t} = E_{energy,t}P_{unit,t} + C_{bandwidth,t} + C_{maintenance,t} + C_{secure,t} \tag{6.6}$$

- **TCO**: To estimate the TCO with a discount rate r over a period of T years, the model is expressed as

$$C = C_c + \sum_{t=1}^{T} \frac{C_{o,t}}{(1+r)^t} \tag{6.7}$$

- **Revenue**: As Wi-Fi is a mature technology, there are several revenue streams that generate income for the service provider, as discussed in Sect. 6.3.1.1. This is mathematically modeled as shown below

$$R_{Wi-Fi,t} = R_{sub,t} + R_{Pay-per-use,t} + R_{ads,t} + R_{partmership,t} \tag{6.8}$$

where $R_{Wi-Fi,t}$ is the revenue for the service provider from all the different streams for the year t that they provision, $R_{sub,t}$ is the revenue from the monthly subscription, $R_{Pay-per-use,t}$ is the revenue from the pay-as-you-go model. The revenue from the freemium can be either the subscription model or pay-as-you-go, or both. $R_{ads,t}$ is the revenue generated through advertisements, and $R_{partnership,t}$ is the revenue from industry verticals, SMEs, and government bodies. Each of these revenues is modeled as

Subscription Revenue: This would depend on number of subscriber $N_{sub,t}$, and the average subscription fee per month Z_t, and is given as

$$R_{sub,t} = 12 N_{sub,t} Z_t \tag{6.9}$$

Pay-as-you-go Revenue: This is estimated based on the number of pay-as-you-go customers $N_{Pay-per-use,t}$, average data consumption $D_{avg,t}$, and cost per GB $Z_{data,t}$, and is modeled as

$$R_{Pay-per-use,t} = N_{Pay-per-use,t} D_{avg,t} Z_{data,t} \tag{6.10}$$

Advertisements: The revenue from advertisements is calculated as the product of the number of ad impressions $V_{imp,t}$ and the cost per 1,000 impressions $C_{imp,t}$, and is shown as

$$R_{ads,t} = V_{imp,t} C_{imp,t} \tag{6.11}$$

Partnerships: The revenue from collaborating with partners and firms is modeled based on the number of partners $N_{partner,t}$ and the average partnership contract value $Z_{partner,t}$, and is modeled as

$$R_{partnership,t} = N_{partner,t} Z_{partner,t} \tag{6.12}$$

NPV: The NPV for this technology is modeled as

$$NPV = \frac{(R_{Wi-Fi} - Co,t)}{C_c}.100\% \tag{6.13}$$

6.3.2 Environmental Frameworks

As Wi-Fi is a popular solution in rural areas, there is a need to understand the energy demands required for supporting the network operations and explore ways to minimize them. Wi-Fi requires regular hardware upgrades that generate several discarded hardware items, which require a proper recycling process and e-waste management. This would also help in lowering the carbon footprint of Wi-Fi [99].

6.3.2.1 Energy Consumption

Wi-Fi is widely deployed across the globe, for example, an enterprise-grade Wi-Fi APs would consume around 5 to 15 W per AP, whereas a low-grade IoT APs would consume around 1 to 3 W per AP. A mesh network and repeater Wi-Fi APs would consume more power than leaf nodes; meanwhile, a traditional AP with wired backhaul is more power efficient. The energy consumption of Wi-Fi AP would also increase further in hard operating conditions and hot weather.

Data usage of end users and the number of simultaneous active users significantly drive the power consumption of Wi-Fi APs. For example, the power consumption of WNIs, P_{WNI}, which is present on end user devices, is modeled as [100]:

$$P_{WNI} = P_{baseline} + (48 - 0.768 \times R_{channel}) r_{data} \tag{6.14}$$

where $P_{baseline}$ is the baseline power corresponding to the power state of WNI, $R_{channel}$ is the uplink channel rate, and r_{data} is the packet rate, which are collected from the phone during the runtime.

The key to reducing the power consumption would be to upgrade the equipment to energy-efficient Wi-Fi generations. But generally, this method is cost-intensive; therefore, alternatively, the solution of exploring renewable energy sources is beneficial. There are several different ways to monitor the energy consumption of the Wi-Fi networks. The empirical model to understand the total power consumption of the Wi-Fi APs, repeaters, and connected devices is mathematically modeled as shown below [100]:

$$\begin{aligned} E_{Wi-Fi} &= E_{idle} + E_{active} \\ &= P_{idle}T_{idle} + P_{active}T_{active} \\ E_{active} &= P_{AP,active}T_{AP}N_{AP} + P_{devices,active}T_{devices}N_{devices} \\ E_{idle} &= P_{AP,idle}(24 - T_{AP})N_{AP} + P_{devices,idle}(24 - T_{devices})N_{devices} \end{aligned} \tag{6.15}$$

where P_{idle} and P_{active} are the power ratings of the device during the idle and active phases of traffic usage, and T_{idle} in hours and T_{active} in hours account for the time the device spends being idle and active, respectively. Additionally, there is energy consumed during the data transfer based on the traffic, E_{data}, which is mathematically modeled as [100]:

$$E_{data} = D_{total} \times EC_{Wi-Fi} \tag{6.16}$$

where D_{total} is the total data transferred (GB), and EC_{Wi-Fi} is the energy consumption per GB of data. Therefore, the total energy consumption of a Wi-Fi network is

$$E_{total} = E_{Wi-Fi} + E_{data} \tag{6.17}$$

If the site uses renewable energy sources, then the energy harvested from it, $E_{renewable}$, could be subtracted from the total power consumption, and it could be modified as shown below

$$E_{total,grid} = E_{total} - E_{renewable} \tag{6.18}$$

6.3.2.2 Carbon Emissions

The total carbon footprint, CE, for a Wi-Fi network can be expressed as

$$CE = CE_p + CE_t + CE_c + CE_o \tag{6.19}$$

where CE_p is the carbon emissions of the production phase, CE_t is the carbon emissions during the transportation phase, CE_c is the carbon emissions during the construction phase, and CE_o is the carbon emissions during the operations phase.

The carbon emissions for Wi-Fi the network CE_o (kg CO_2), is dependent on the energy consumption of Wi-Fi the network and its source and are modeled as [101]:

$$CE_o = E_{total,grid} \times c_{intensity} \tag{6.20}$$

where $c_{intensity}$ is the carbon emission factor in (kg CO_2/kWh). However, this model is enhanced when we model the recycling of Wi-Fi hardware and equipment, and is modeled as follows:

$$\begin{aligned} CE_{Wi-Fi,recycle} &= CE_{Wi-Fi} - CE_{recycle} \\ CE_{recycle} &= \Sigma_{i=1}^{N} M_i \times c_i \end{aligned} \tag{6.21}$$

where $CE_{recycle}$ is the emission savings from material recovery and reduced pollution, which is a product of M_i the mass of the material i recovered in kg, c_i is the emission reduction factor of material i (kg CO_2 saved per kg recycled), and N is the number of elements or recyclable materials recovered.

6.4 Summary

This chapter highlights the usage of Wi-Fi as a technology in bridging the digital divide. It is a popular solution worldwide since it offers high throughput and low latency along with being accessible, affordable, and available; however, the challenge lies in constant hardware upgrades with the newer generation of Wi-Fi driving the cost of deployment. This chapter focuses on global deployments, advanced architecture, key enabling technologies, and specific applications that are driving the connectivity to build a globally digitally inclusive society. The economic and environmental frameworks examined in this chapter help in understanding the impact Wi-Fi would have on the community of interest and hold the potential to address the digital divide. Globally, there is an ongoing focus on spectrum harmonization for different Wi-Fi generations to ensure scalability at an affordable price. The initiatives using Wi-Fi in connecting the unconnected have to ensure efforts are in the direction of building a digitally inclusive society for all.

Exercises

6.1 A Wi-Fi mesh network consists of 5 nodes, each with a coverage radius of 50 m. Assuming circular coverage and no overlap between adjacent nodes, calculate:

(a) The total coverage area of the mesh network.
(b) The percentage increase in total coverage area if an additional 2 nodes are added to the network.

6.2 A rural public Wi-Fi project involves deploying 10 access points, each costing $2,000. Annual operational expenses per access point are $500. Calculate:

(a) The initial deployment cost of the project.
(b) The total cost of deploying and maintaining the network for 5 years.

6.3 A Wi-Fi network operates on a 20 MHz channel with a data rate of 200 Mbps. Calculate:

(a) The spectral efficiency of the network in bps/Hz.
(b) The additional bandwidth required to achieve a data rate of 400 Mbps, assuming the same spectral efficiency.

6.4 A satellite backhaul link provides a data rate of 1 Gbps for a rural Wi-Fi network. Evaluate:

(a) The maximum number of simultaneous users the network can support, assuming each user requires 10 Mbps.
(b) The reduction in the number of users if the effective backhaul data rate reduces to 800 Mbps due to interference.

6.5 A Wi-Fi access point consumes 15 W of power and operates 24 h a day. Calculate:

(a) The daily energy consumption of the access point in kWh.
(b) The total energy consumption for a network of 20 access points over a year.

6.6 The power grid in a rural area has a carbon intensity of 0.5 kg CO_2 per kWh. For a network of 15 access points, each consuming 15 W of power, calculate:

(a) The total annual carbon emissions from the Wi-Fi network.
(b) The reduction in emissions if energy-efficient access points consuming 10 W are deployed instead.

6.7 A Wi-Fi network serves users distributed according to a Poisson Point Process with an average density of 5 users per 100 m^2. Calculate:

(a) The probability of having exactly 3 users in a 50 m^2 area.
(b) The expected number of users in a 200 m^2 area.

6.8 Evaluate the feasibility of deploying TVWS technology to backhaul Wi-Fi networks in a rural area. Discuss:

(a) The advantages of using TVWS for backhauling Wi-Fi in underserved regions.
(b) The technical challenges and regulatory considerations for such a deployment.

6.9 A community Wi-Fi network requires an initial investment of \$10,000 for hardware and \$2,000 annually for maintenance. The network serves 500 users who each pay \$2/month for access. Calculate:

(a) The payback period for recovering the initial investment.
(b) The total revenue generated over a 5-year period.

6.10 Analyze the challenges of deploying public Wi-Fi in rural areas. Discuss:

(a) Technical constraints, including coverage and backhaul availability.
(b) Economic and social barriers to adoption and scalability.

References

1. U. Varshney, The status and future of 802.11-based WLANs. Computer **36**(6), 102–105 (2003)
2. P. Ashok, G. Hallur, Wireless broadband unleashed: charting regulatory pathways for broadband development, in *2024 2nd World Conference on Communication & Computing (WCONF)* (IEEE, 2024), pp. 1–6
3. B. Raman, K. Chebrolu, Experiences in using WiFi for rural internet in India. IEEE Commun. Mag. **45**(1), 104–110 (2007)
4. F.B. Garrido, W. Torres Cerón, F.A. Simanca H, P.E. Carreño H, M.A. Moncada, Design of a model for the implementation of an access control to a Wi-Fi network, in *International Conference on Advanced Engineering Theory and Applications* (Springer, 2019), pp. 185–192
5. K. Pahlavan, P. Krishnamurthy, Evolution and impact of Wi-Fi technology and applications: a historical perspective. Int. J. Wireless Inf. Netw. **28**, 3–19 (2021)
6. The Windows Club. How to set up a router firewall. Accessed on 27 Feb 2025
7. GeeksforGeeks. Network devices: Hub, repeater, bridge, switch, router, gateways. Accessed on 27 Feb 2025
8. Computer Tech Info. What is a router? definition, functions, uses, types, examples. Accessed on 27 Feb 2025
9. N. Xhemajli, Z. Tafa, Mobile proxy in public WiFi networks: a tool against MITM attacks, in *2024 13th Mediterranean Conference on Embedded Computing (MECO)* (IEEE, 2024), pp. 1–5
10. G. Naik, J.-M. Park, J. Ashdown, W. Lehr, Next generation wi-Fi and 5G NR-U in the 6 GHz bands: opportunities and challenges. IEEE Access **8**, 153027–153056 (2020)
11. M. Zhang, R.S. Wolff, Using Wi-Fi for cost-effective broadband wireless access in rural and remote areas, in *2004 IEEE Wireless Communications and Networking Conference (IEEE Cat. No. 04TH8733)*, vol. 3 (IEEE, 2004), pp. 1347–1352
12. I. Dolińska, M. Jakubowski, A. Masiukiewicz, Interference comparison in wi-fi 2.4 GHz and 5 GHz bands, in *2017 International Conference on Information and Digital Technologies (IDT)* (IEEE, 2017), pp. 106–112
13. V. İlçi, E. Gülal, R.M. Alkan, Performance comparison of 2.4 and 5 GHz WiFi signals and proposing a new method for mobile indoor positioning. Wirel. Pers. Commun. **110**, 1493–1511 (2020)
14. G. Naik, J. Liu, J.-M.J. Park, Coexistence of wireless technologies in the 5 GHz bands: A survey of existing solutions and a roadmap for future research. IEEE Commun. Surv. Tutor. **20**(3), 1777–1798 (2018)
15. L. Tian, S. Santi, A. Seferagić, J. Lan, J. Famaey, Wi-Fi HaLow for the internet of things: an up-to-date survey on IEEE 802.11 ah research. J. Netw. Comput. Appl. **182**, 103036 (2021)
16. M. Alam, N. Ahmed, R. Matam, F.A. Barbhuiya, Analyzing the suitability of IEEE 802.11 ah for next generation internet of things: A comparative study. Ad Hoc Netw. **156**, 103437 (2024)
17. S. Sur, I. Pefkianakis, X. Zhang, K.-H. Kim, WiFi-assisted 60 GHz wireless networks, in *Proceedings of the 23rd Annual International Conference on Mobile Computing and Networking* (2017), pp. 28–41
18. R.C. Daniels, J.N. Murdock, T.S. Rappaport, R.W. Heath, 60 GHz wireless: up close and personal. IEEE Microw. Mag. **11**(7), 44–50 (2010)
19. Y. Ghasempour, C.R.C.M. Da Silva, C. Cordeiro, E.W. Knightly, IEEE 802.11 ay: next-generation 60 GHz communication for 100 Gb/s Wi-Fi. IEEE Commun. Mag. **55**(12), 186–192 (2017)
20. C.J. Hansen, WiGiG: multi-gigabit wireless communications in the 60 GHz band. IEEE Wirel. Commun. **18**(6), 6–7 (2011)
21. E.J. Oughton, W. Lehr, K. Katsaros, I. Selinis, D. Bubley, J. Kusuma, Revisiting wireless internet connectivity: 5G vs Wi-Fi 6. Telecommun. Policy **45**(5), 102127 (2021)
22. E.J. Oughton, G. Geraci, M. Polese, M. Ghosh, W. Webb, D. Bubley, The future of wireless broadband in the peak smartphone era: 6G, Wi-Fi 7, and Wi-Fi 8. Authorea Preprints (2024)
23. S.S. Murad, R. Badeel, B.B. Abdal, T. Rahman, T. Al-Quraishi, Introduction to Wi-Fi 7: a review of history, applications, challenges, economical impact and research development. Mesop. J. Comput. Sci. **2024**, 110–121 (2024)
24. Purple AI. The history of Wi-Fi. Accessed on 27 Feb 2025
25. P. Brenner, IEEE 802.11 a technical overview (1997)
26. IEEE. IEEE 802.11a standard. Accessed on 27 Feb 2025
27. IEEE. IEEE 802.11b standard. Accessed on 27 Feb 2025
28. B. O'hara, A.l. Petrick, *IEEE 802.11 Handbook: A Designer's Companion* (IEEE Standards Association, 2005)
29. K. Medepalli, P. Gopalakrishnan, D. Famolari, T. Kodama, Voice capacity of IEEE 802.11 b, 802.11 a and 802.11 g wireless LANs, in *IEEE Global Telecommunications Conference, 2004. GLOBECOM'04.*, vol. 3 (IEEE, 2004), pp. 1549–1553
30. A. Kotanen, M. Hannikainen, H. Leppakoski, T.D. Hamalainen, Positioning with IEEE 802.11 b wireless LAN, in *14th IEEE Proceedings on Personal, Indoor and Mobile Radio Communications, 2003. PIMRC 2003.*, vol. 3 (IEEE, 2003), pp. 2218–2222

31. H. Boland, H. Mousavi, Security issues of the IEEE 802.11 b wireless LAN, in *Canadian Conference on Electrical and Computer Engineering 2004 (IEEE Cat. No. 04CH37513)*, vol. 1 (IEEE, 2004), pp. 333–336
32. A. Athanasopoulos, E. Topalis, C. Antonopoulos, S.A.K.S. Koubias, Evaluation analysis of the performance of IEEE 802.11 b and IEEE 802.11 g standards, in *International Conference on Networking, International Conference on Systems and International Conference on Mobile Communications and Learning Technologies (ICNICONSMCL'06)* (IEEE, 2006), pp. 141–141
33. S. Choi, J. del Prado Pavon, 802.11 g CP: a Solution for IEEE 802.11 g and 802.11 b Inter-Working, in *The 57th IEEE Semiannual Vehicular Technology Conference, 2003. VTC 2003-Spring*, vol. 1 (IEEE, 2003), pp. 690–694
34. R. Khanduri, S. Rattan, A. Uniyal, Understanding the features of IEEE 802.11 g in high data rate wireless LANs. Int. J. Comput. Appl. **64**(8), 1–5 (2013)
35. Apple. Apple Reinvents the Phone with iPhone. Accessed on 27 Feb 2025
36. D.T. Chelmins, H.D. Bguyen, L.R. Foore, Assessment of the 802.11 g wireless protocol for lunar surface communications. Technical report, NASA (2009)
37. E. Perahia, IEEE 802.11 n development: history, process, and technology. IEEE Commun. Mag. **46**(7), 48–55 (2008)
38. T. Paul, T. Ogunfunmi, Wireless LAN comes of age: understanding the IEEE 802.11 n amendment. IEEE Circuits Syst. Mag. **8**(1), 28–54 (2008)
39. Y. Xiao, IEEE 802.11 n: enhancements for higher throughput in wireless LANs. IEEE Wirel. Commun. **12**(6), 82–91 (2005)
40. E.H. Ong, J. Kneckt, O. Alanen, Z. Chang, T. Huovinen, T. Nihtilä, IEEE 802.11 ac: Enhancements for very high throughput WLANs, in *2011 IEEE 22nd International Symposium on Personal, Indoor and Mobile Radio Communications* (IEEE, 2011), pp. 849–853
41. S. Narayan, C. Jayawardena, J. Wang, W. Ma, G. Geetu, Performance test of IEEE 802.11 ac wireless devices, in *2015 International Conference on Computer Communication and Informatics (ICCCI)* (IEEE, 2015), pp. 1–6
42. M. Yang, B. Li, Z. Yan, MAC Technology of IEEE 802.11 ax: progress and tutorial. Mob. Netw. Appl. **26**, 1122–1136 (2021)
43. D.-J. Deng, K.-C. Chen, R.-S. Cheng, IEEE 802.11 ax: next generation wireless local area networks. In *10Th International Conference on Heterogeneous Networking for Quality, Reliability, Security and Robustness* (IEEE, 2014), pp. 77–82
44. A.F. Rochim, B. Harijadi, Y.P. Purbanugraha, S. Fuad, K.A. Nugroho, Performance comparison of wireless protocol IEEE 802.11 ax vs 802.11 ac, in *2020 International Conference on Smart Technology and Applications (ICoSTA)* (IEEE, 2020), pp. 1–5
45. J.S. Sánchez-Mahecha, S. Céspedes, J. Bustos-Jiménez, QoS evaluation of the future high-efficiency IEEE 802.11 ax WLAN standard, in *2018 IEEE Colombian Conference on Communications and Computing (COLCOM)* (IEEE, 2018), pp. 1–6
46. H. Yang, D.-J. Deng, K.-C. Chen, On energy saving in IEEE 802.11 ax. IEEE Access **6**, 47546–47556 (2018)
47. M.S. Afaqui, E. Garcia-Villegas, E. Lopez-Aguilera, IEEE 802.11 ax: challenges and requirements for future high efficiency WiFi. IEEE Wirel. Commun. **24**(3), 130–137 (2016)
48. IEEE Standards Association. The evolution of Wi-Fi technology and standards. Accessed on 17 Feb 2025
49. M. Alabi, The evolution of Wi-Fi Standards: From Wi-Fi 6 to Wi-Fi 7. *ResearchGate* (2023)
50. A. Garcia-Rodriguez, D. López-Pérez, L. Galati-Giordano, G. Geraci, IEEE 802.11 be: Wi-Fi 7 strikes back. IEEE Commun. Mag. **59**(4), 102–108 (2021)
51. S. Chauhan, A. Sharma, S. Pandey, K.N. Rao, P. Kumar, IEEE 802.11 be: a review on Wi-Fi 7 use cases, in *2021 9th International Conference on Reliability, Infocom Technologies and Optimization (Trends and Future Directions) (ICRITO)* (IEEE, 2021), pp. 1–7
52. MathWorks. Overview of Wi-Fi 7 or IEEE 802.11be. Accessed on 27 Feb 2025
53. C. Chen, X. Chen, D. Das, D. Akhmetov, C. Cordeiro, Overview and performance evaluation of Wi-Fi 7. IEEE Commun. Stand. Mag. **6**(2), 12–18 (2022)
54. G. Cheng, Z. Qian, Y. Jiang, Z. Yang, T. Rajamanickam, Wi-Fi 7 product development and test methods, in *Wi-Fi 7: Principles, Technology, and Applications* (Springer, 2024), pp. 297–389
55. C. Deng, X. Fang, X. Han, X. Wang, L. Yan, R. He, Y. Long, Y. Guo, IEEE 802.11 be Wi-Fi 7: new challenges and opportunities. IEEE Commun. Surv. Tutor. **22**(4), 2136–2166 (2020)
56. E. Khorov, I. Levitsky, I.F. Akyildiz, Current status and directions of IEEE 802.11 be, the future Wi-Fi 7. IEEE Access **8**, 88664–88688 (2020)
57. E. Reshef, C. Cordeiro, Future directions for Wi-Fi 8 and beyond. IEEE Commun. Mag. **60**(10), 50–55 (2022)
58. L. Galati-Giordano, G. Geraci, M. Carrascosa, B. Bellalta, What will Wi-Fi 8 be? A primer on IEEE 802.11 bn ultra high reliability. IEEE Commun. Mag. **62**(8), 126–132 (2024)
59. E. Oughton, G. Geraci, M. Polese, V. Shah, Prospective evaluation of next generation wireless broadband technologies: 6G versus Wi-Fi 7/8. Available at SSRN 4528119 (2023)
60. Y.-H. Liu, C.-P. Ku, T.-D. Chiueh, Design and implementation of a 5G NR transmitter with Wi-Fi coexistence by beamforming and power control. IEEE Open J. Commun. Soc. (2024)
61. S. Bhattarai, J.-M.J. Park, B. Gao, K. Bian, W. Lehr, An overview of dynamic spectrum sharing: ongoing initiatives, challenges, and a roadmap for future research. IEEE Trans. Cogn. Commun. Netw. **2**(2), 110–128 (2016)
62. S. Sagari, S. Baysting, D. Saha, I. Seskar, W. Trappe, D. Raychaudhuri, Coordinated dynamic spectrum management of LTE-U and Wi-Fi networks, in *2015 IEEE International Symposium on Dynamic Spectrum Access Networks (DySPAN)* (IEEE, 2015), pp. 209–220
63. H.-A. Safavi-Naeini, S. Roy, S. Ashrafi, Spectrum sharing of radar and Wi-Fi networks: the sensing/throughput tradeoff. IEEE Trans. Cogn. Commun. Netw. **1**(4), 372–382 (2015)
64. F. Hessar, S. Roy, Spectrum sharing between a surveillance radar and secondary Wi-Fi networks. IEEE Trans. Aerosp. Electron. Syst. **52**(3), 1434–1448 (2016)
65. A. Khan, P. Nand, B. Bhushan, E.M. Alazzawi, A. Jamil, A review of spectrum sharing in 6G for IoT environment, in *2024 2nd International Conference on Artificial Intelligence, Blockchain, and Internet of Things (AIBThings)* (IEEE, 2024), pp. 1–6
66. W. Fan, J. Han, Y. Su, X. Liu, F. Wu, B. Tang, Y. Liu, Joint task offloading and service caching for multi-access edge computing in WiFi-cellular heterogeneous networks. IEEE Trans. Wirel. Commun. **21**(11), 9653–9667 (2022)
67. S.K. Datta, C. Bonnet, MEC and IoT based automatic agent reconfiguration in industry 4.0, in *2018 IEEE International Conference on Advanced Networks and Telecommunications Systems (ANTS)* (IEEE, 2018), pp. 1–5
68. A. Jeknić, E. Kočan, Multi-link operation for performance improvement in Wi-Fi 7 networks, in *2024 28th International Conference on Information Technology (IT)* (IEEE, 2024), pp. 1–4

69. G. Cena, M. Rosani, S. Scanzio, Packet steering mechanisms for MLO in Wi-Fi 7, in *2024 IEEE 29th International Conference on Emerging Technologies and Factory Automation (ETFA)* (IEEE, 2024), pp. 1–4
70. Ruckus Networks. Wi-Fi 7 Multi-Link Operations (MLO). Accessed on 27 Feb 2025
71. AI & ML in next-gen networks: Boosting efficiency. Accessed on 04 Mar 2025
72. J.D. Rosenthal, M.S. Reynolds, Wideband OFDM Backscatter with Limited-Bandwidth Antennas for WiFi-6 and future high-data-rate backscatter systems, in *2024 9th International Conference on Smart and Sustainable Technologies (SpliTech)* (IEEE, 2024), pp. 1–6
73. C. Rey-Moreno, Z. Roro, W.D. Tucker, M.J. Siya, N.J. Bidwell, J. Simo-Reigadas, Experiences, challenges and lessons from rolling out a rural WiFi mesh network, in *Proceedings of the 3rd ACM Symposium on Computing for Development* (2013), pp. 1–10
74. P. Dutta, S. Jaiswal, R. Rastogi, Villagenet: a low-cost, IEEE 802.11-based mesh network for connecting rural areas. Bell Labs Tech. J. **12**(2), 119–131 (2007)
75. M. Quispe, J. Olivares, J. Samaniego, R. Morán, Technical and economic analysis of TVWS and 5.8 GHz WiFi systems for rural areas, in *2022 IEEE XXIX International Conference on Electronics, Electrical Engineering and Computing (INTERCON)* (IEEE, 2022), pp. 1–4
76. W. Zhang, J. Yang, G. Zhang, L. Yang, C.K. Yeo, TV white space and its applications in future wireless networks and communications: a survey. IET Commun. **12**(20), 2521–2532 (2018)
77. D. Johnson, N. Zlobinsky, A. Lysko, M. Lamola, S. Hadzic, R. Maliwatu, M. Densmore, Head to head battle of TV White Space and WiFi for connecting developing regions, in *e-Infrastructure and e-Services for Developing Countries: 8th International Conference, AFRICOMM 2016, Ouagadougou, Burkina Faso, December 6-7, 2016, Proceedings 8* (Springer, 2018), pp. 186–195
78. M.D.S. Akhtar, G. Mathur, O. Kravchenko, M. Rakhra, Connecting the unconnected: bridging the digital divide with affordable satellite through enabled smartphones, in *E3S Web of Conferences*, vol. 453 (EDP Sciences, 2023), p. 01054
79. S. Subba Rao, Bridging digital divide: efforts in India. Telematics Inform. **22**(4), 361–375 (2005)
80. HMVR Herath. Starlink: a solution to the digital connectivity divide in education in the global south. arXiv preprint arXiv:2110.09225 (2021)
81. E.J. Oughton, Policy implications of broadband access technologies. J. Telecommun. Policy **47**(1), 123–135 (2023)
82. C.A. Middleton, A framework for investigating the value of public wireless networks. TPRC (2007)
83. J. van den Velden, B.M. Sadowski, Creating public value with municipal Wi-Fi networks: a bottom-up methodology. Digit. Policy, Regul. Gov. **25**(2), 77–103 (2023)
84. Wi-Fi Alliance. The state of connectivity: Wi-Fi Momentum in 2024. Accessed on 11 Mar 2025
85. EIN Presswire. Wi-Fi 6 Global Market 2024 to Reach $76.32 Billion by 2028 at Rate of 34.2%. Accessed on 11 Mar 2025
86. RCR Wireless News. Global Wi-Fi 7 adoption. Accessed on 11 Mar 2025
87. Fortune Business Insights. Wi-Fi as a service market size, share & covid-19 impact analysis. Accessed on 11 Mar 2025
88. Market Digits. Public Wi-Fi market analysis and forecast. Accessed on 11 Mar 2025
89. TechCentral. Kenya's brck: All the details. Accessed on 27 Feb 2025
90. tele.net. OFC Track: RailTel's initiatives to expand its fibre network. Accessed on 11 Mar 2025
91. E. Puka, P. Herrmann, Mitigating dead spots in cellular networks with the hybrid communication protocol camfloop, in *2021 IEEE International Intelligent Transportation Systems Conference (ITSC)* (IEEE, 2021), pp. 1041–1048
92. D. Anderson, K.A. Shruthi, D. Crawford, R.W. Stewart, Evolving spectrum sharing methods, standards and trials: TVWS, CBRS, multefire and more. *Spectrum Sharing: The Next Frontier in Wireless Networks* (2020), pp. 59–74
93. Wireless Broadband Alliance, *Rural Wi-Fi connectivity challenges, use cases and case studies*. Technical report, WBA (2022)
94. N. Liu, D. Plets, K. Vanhecke, L. Martens, W. Joseph, Wireless indoor network planning for advanced exposure and installation cost minimization. EURASIP J. Wirel. Commun. Netw. **1–14**, 2015 (2015)
95. Electronic Design. Understanding wireless range calculations. Accessed on 11 Mar 2025
96. C. Chanpinit, S. Tanchonpong, N. Pichaiya, P. Champrasert, Wireless access point installation planning with techno-economic comparison, in *2022 37th International Technical Conference on Circuits/Systems, Computers and Communications (ITC-CSCC)* (IEEE, 2022), pp. 1–4
97. A. Hameed, A.N. Mian, J. Qadir, Low-cost sustainable wireless internet service for rural areas. Wirel. Netw. **24**, 1439–1450 (2018)
98. M.A. De Souza, H.P. Kuribayashi, P.A. Saraiva, F.D.S. Farias, N.L. Vijaykumar, C.R.L. Francês, J.C.W.A. Costa, A techno-economic framework for installing broadband networks in rural and remote areas. IEEE Access **9**, 58421–58447 (2021)
99. U. Gupta, Y.G. Kim, S. Lee, J. Tse, H.-H.S. Lee, G.-Y. Wei, D. Brooks, C.-J. Wu, Chasing carbon: the elusive environmental footprint of computing, in *2021 IEEE International Symposium on High-Performance Computer Architecture (HPCA)* (IEEE, 2021), pp. 854–867
100. Yu. Xiao, Y. Cui, P. Savolainen, M. Siekkinen, A. Wang, L. Yang, A. Ylä-Jääski, S. Tarkoma, Modeling energy consumption of data transmission over Wi-Fi. IEEE Trans. Mob. Comput. **13**(8), 1760–1773 (2013)
101. M.T. Kübler, Carbon footprint of an internet service provider: Exploring the carbon reduction potentials of a residential router applying ecodesign. Master's thesis, KTH Royal Institute of Technology (2022)

Future Networks

7

So far, we have looked at several commercially existing technologies, their architecture, business cases, and costing models, and we have realized that no single technology is the solution to bridging the digital divide. A combination of these solutions would drive connectivity across the globe. Looking ahead, the focus shifts to future networks and their potential to revolutionize connectivity and services that are powered by advancements in AI, edge computing, and 6G, and their role in bridging the digital divide. These solutions promise enhanced performance, scalability, and efficiency; additionally, to lower the energy consumption and carbon emissions. Future networks are poised to enable transformative applications, foster innovation and economic growth, and boost connectivity in regions with poor or no coverage.

What are **future networks**? Future networks are the communication technologies designed for rapidly increasing needs for hyper-connectivity that are faster, more secure, and more resilient, and focus on building a digitally inclusive society. They also aim to help reach the United Nations (UN)'s sustainable goals, especially net-zero emissions, climate change, and innovation goals [1]. Future networks play a key role in technological advancements by laying out the foundation, providing necessary document references, standardization, and a timeline for research (up to 10 years) [2]. These technologies are expected to enable a fully digital world with connected humans, machines, things, and places, supporting realistic virtual experiences. Accordingly, this chapter focuses on understanding what future networks are, which technologies are driving these capabilities, and how they could potentially help achieve the target of building a digitally inclusive ecosystem for all.

7.1 Key Drivers for Future Networks

Future networks are shaped by several key drivers, with seven main factors playing a pivotal role [2]:

- **An Automated World**: With the advancement in technologies, more and more objects and physical things are being connected with humans and automated, which would help to improve efficiency, computation, network management, interaction among them, and data management [3]. Digital twins help replicate real-world scenarios digitally on a small scale to test their performance and design solutions to tackle the challenge. IoT is helping monitor devices and human data to assess their health in real-time and take appropriate steps [4]. This need for integrated sensing would help solve many challenges the world faces in improving people's quality of life, such as VR, haptic applications, personal communication, and camera measurements.
- **Massive Connected Machines and AI**: Over the years, billions of machines are expected to be connected via the Internet, and will become more intelligent using AI and supported technologies [5, 6]. The network platform would be improved by the "everything connected" concept, and AI-AI communication to help improve the performance efficiency, QoS, predictive maintenance, improved resource utilization, and network reliability.
- **Transmission of Sensory Feelings via Internet**: In today's world, there is a need to transmit beyond data and voice over the Internet, such as touch, smell, visuals, and other technologies that engage human sensory organs [7]. The application areas would include haptics, holograms, AR/VR, remote healthcare, remote operations of machinery, brain-computer communication, contact lens support AR, exoskeletons, and universal translator earbuds [8].
- **Coverage Across the Globe and Interplanetary Communication**: With the increasing use of the connected everything concept, there is a need to provide 100% coverage for bridging the digital divide, seamless handoff, and reliable communi-

S. Koratagere Anantha Kumar, *Wireless Broadband Access to Bridge the Digital Divide*, Textbooks in Telecommunication Engineering, https://doi.org/10.1007/978-3-032-09261-8_7

cation networks [9]. There is a need to provide improved and reliable communication in indoor scenarios and all modes of transportation. The ongoing research focuses on the integration of terrestrial networks with NTN to provide these facilities to provide seamless connectivity across the globe.

- **Pervasive Edge Computing**: The telecommunication networks are moving toward the virtualization of components and functionalities, which would help in increased resource utilization, QoS, scalability, remote monitoring and maintenance of the network, and network performance [10]. Cloud computing helps process data at the edge rather than in a central system. Network virtualization eases the network scaling process and improves the network metrics in peak and off-peak hours.
- **Trustworthy Infrastructure**: The key threat when everything is connected to everything else is the security and resilience offered by the network [11]. Each technology and device has different security policies that it needs to adhere to, and there is a potential for data being hacked while passing through various security levels. Once the data is hacked, the hacker can remotely seize control of critical infrastructure and sensitive data. This is a critical threat and leads to the breakdown of essential services. Protection of personal data has become a need of the hour in all countries across the globe, especially Europe, which has followed it very strictly for many years now [11, 12]. Also, trustworthiness should be a measurable quantity to assess performance over the long run.
- **Cognitive Network**: In this, the network is expected to require zero or minimal human intervention during the deployment and operations [13]. In a cognitive network, the network can upgrade itself by observing the needs of the demands to be catered to, their usage behavior, traffic requirements, the intent of the user, and channel QoS. The cognitive networks are expected to be enabled using AI and learning from experiences to customize their services as the user's needs [14]. The key network functions where this feature is very critical and suitable include optimization of the radio interface, automation of network management, and orchestration.

Future network design considerations should include the digital divide and also ensure an equitable digital society for all. Without deliberate considerations and focused actions, these technologies could potentially widen the disparities among those who are connected and those who aren't, leaving the underserved communities behind. Technologies that constitute the future networks should take pricing and accessibility into account, as these factors will influence both their ability to scale and their utilization. The pricing strategies should be inclusive to ensure that everyone can afford the services. The next factor is technology integration across the different layers, as well as their enabling technologies, which would help in network reaching the underserved communities. The future network should also consider energy efficiency, sustainability, and carbon emissions to align with climate change goals.

7.2 Technologies Driving Future Networks

Many cutting-edge technologies are influencing the development of future networks to provide incredibly fast, reliable, and intelligent communication and revolutionize telecommunication and related sectors that are interrelated. An outline of the major technologies influencing the development of future networks and their role in bridging the digital divide may be seen below.

7.2.1 6G Research

The vision of 6G involves the target to provide seamless connectivity where the digital world merges with real-world senses. The focus of 6G is to create a human-centric, connectivity-for-all, sustainable, and efficient society to boost people's digital literacy and connect everyone [15]. Connecting the unconnected and meaningful connectivity are the key focus areas of 6G. The next important question would be, what should 6G focus on? What are its features?

6G would be a newer generation of technologies that would support the integration of different forms of communication, such as terrestrial and NTN, ultra-dense networks with seamless roaming across different technologies, and virtualization. The key features that should be incorporated into the design of the 6G architecture include cloud support, network design simplicity, flexibility, scalability, energy efficiency, resilience, sustainability, and robustness. The open-source aspects and dynamic network requirements need to be considered for 6G architecture to provide wide-area to personal cell communications

Table 7.1 KPIs capabilities in 6G networks

KPI	6G
Peak data rate	1 Tbps
Experience data rate	1 Gbps
Spectral efficiency	60 b/s/Hz
Experienced spectral efficiency	3 b/s/Hz
Area traffic capacity	1 Gb/s/m^{s}
Maximum bandwidth	100 GHz
Connection density	10^7 device/km^2
Energy efficiency	1 Tb/J
Latency	10 μs
Jitter	1 μs
Mobility	1000 Kmph

having 1 Tbps peak data rates, ultra-low latency ($< 1ms$), improved indoor coverage, cyber-physical security, and reliability in the range of 10^{-9}. The key capabilities of 6G are described in Table 7.1 [16].

6G is expected to include all 5G spectrum bands, low bands to the lower part of the THz communication band, and the visible light region, which helps focus on both coverage and capacity. 6G KPIs would vary across frequency bands; for example, low-spectrum bands offer wide-area coverage, whereas THz bands support ultra-high data rates. The research in 6G would focus on the complex design of equipment supporting those capabilities, integration to newer waveforms, modulation schemes, spectrum sensing strategies, hardware impairments, and protocols for different communication layers. Figure 7.1 shows IMT vision 2030 and beyond for 6G, which includes IMT vision 2020 goals such as eMBB, mMTC, and URLLC, and additionally includes design principles applicable to all usage scenarios such as connecting the unconnected, ubiquitous connectivity, AI, and communication integration, integrated sensing and communication (ISAC), sustainability, resilience, and security [17–19].

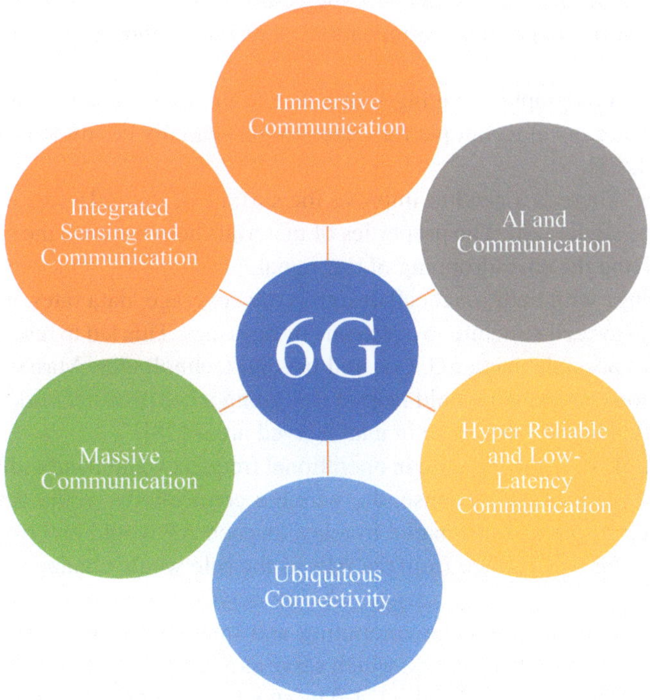

Fig. 7.1 6G key technology enablers

ISAC is a new concept that involves radar-like sensing capabilities in mobile networks, which would help in combining sensing and communication using shared hardware and spectrum, rather than having independent operational requirements. This functionality can be achieved by using advanced processing capabilities and algorithms; however, there is a need to assess the potential business model to understand the potential ROI. Similarly, **AI and communication** is gaining popularity to enhance network performance. The industry is exploring the full potential of next generation mobile networks (NGMN) using AI and ML technologies to build a network that can dynamically adapt to the users' channel conditions, optimize resource allocation, usage, and cost, enable revenue incentives, lower energy consumption, and transform vertical businesses. Finally, the final factor **ubiquitous connectivity**, has been explored in older technologies but has not been achieved. Hence, this is part of 6G and includes features such as seamless connectivity, in-building and outdoor connectivity, smooth handovers, uninterrupted connectivity both in urban and rural areas, and integration with satellites with no disruption in the QoS.

6G use cases include extreme capacity xhaul, device-to-device connectivity, enhanced hotspots, connectivity in remote areas, smart rail connectivity, multi-sensory extended reality, industrial automation, and autonomous mobility [16]. The research also needs to focus on sustainable business models and business cases, since high upfront 5G infrastructure costs pose a significant barrier to 6G adoption for many businesses.

7.2.1.1 Key Enablers for 6G

The key enablers for infrastructure supporting 6G KPIs vary depending on the intended use cases to offer seamless connectivity in underserved regions, and are discussed below:

- **Massive Beamforming and mMIMO**: The mMIMO technology offers enhanced benefits for antenna spectral efficiency, energy efficiency, data throughput, coverage at the cell edge, signal strength, and user QoE along with QoS [20]. With a higher number of antennas, spatial multiplexing can support more users than a single antenna; for example, using massive MIMO in 1 MHz of spectrum in 100 MHz would offer the same capabilities as 100 MHz of spectrum in the 100 GHz band [15].
- **Network Slicing**: 6G extends network slicing to overcome 5G challenges such as static slicing and SLA maintenance for slice tenants [21]. In 6G, network slicing is expected to support dynamic slicing with dynamic resource segmentation with guaranteed SLAs with the potential to support dynamic resource requirements and a myriad of applications for each tenant [22]. This would help improve spectral efficiency, resource utilization, network resilience, security, interoperability, efficient network planning, network congestion, optimization of resource allocation, zero-touch network, and uniform standardization protocols [23].
- **Holographic Radio**: To facilitate holographic imaging-level, ultra-high density, and pixelated ultra-high resolution spatial multiplexing, holographic radio is a novel technique for generating a spatially continuous electromagnetic aperture using the principle of interference [24, 25].
- **Intelligent Reflective Surfaces**: There is a need to improve the surface of the reflective materials to amplify the signals and prevent eavesdropping of the signal [26]. The properties of materials help improve the signal's absorption and enhance the signal quality while minimizing the eavesdropping of the signal.
- **User-Centric Cells**: The challenge with traditional cell design is the coverage, data rates, and edge cell performance [27]. Hence, there is a need to redesign the cell structure to overcome these issues. This led to research ideas toward personal cells and user-centric cells, which are possible using 5G, 6G, and beyond technologies. Massive MIMO supports user-centric cells with distributed network controls, which would help in designing a cell-free network where the user is supported by nearby APs and is useful for extending the coverage in underserved areas [28].
- **Integrated Access and Backhaul**: With the increase in operational frequency and bandwidth, there is a need for integrated access and backhaul. As the network becomes dense, the wireless connectivity would act both as access and backhaul technology [29]. This integration is different compared to relay, as each AP would process information packets and have variable loads. There is a chance for interference from the APs as the role would change as per the demand [30].

With 6G expected to bring breakthroughs in edge computing and intelligence, it is even more important than ever to safeguard end users from privacy risks and cyber-attacks, which serve as a critical weak spot in the network [31]. 6G is a work in progress and is different by research at multiple levels [17]. The standards for 6G are expected to be released by 2030, and around the same time, commercial rollout of 6G will begin worldwide. 6G explores ultra-high capacity, which is AI-driven to build a network that is affordable, inclusive, and adaptive in bridging the digital divide.

7.2.2 Antenna Technologies and MIMO

Advancements in antenna technologies are crucial in improving the QoS, coverage, and number of connections per RAN. These advancements boost the network performance and range, which in turn enhances the coverage, cost-effectiveness, energy consumption, and better QoS using fewer antennas in underserved communities. Future networks are set to be revolutionized by several key developments in fields such as extra large massive MIMO (XL-mMIMO), RIS, cell-free massive MIMO (CFmMIMO), beamforming, metamaterials and metasurfaces, hybrid beamforming, and ISAC.

The mMIMO deployments have motivated engineers to come up with innovative antenna solutions for integrating with large structures such as airports, shopping malls, and stadiums. This type of deployment is terms as **XL-mMIMO** technology, which involves the usage of a large number of antennas to improve the coverage, capacity, number of simultaneous users, signal strength, and QoS; it also helps in reducing interference and supports the vision of ubiquitous connectivity, especially in indoor areas [32]. Figure 7.2 shows the block diagram for XL-mMIMO. There are three main ways to design and develop XL-mMIMO:

- an antenna array of extra-large dimensions,
- large intelligent system that is implemented with a discrete array of antennas, and
- distributed antenna systems that are cooperating to achieve a common goal.

CFmMIMO uses a large number of distributed antennas to serve large users without cell boundaries as traditional cellular systems. CFmMIMO is a user-centric approach as shown in Fig. 7.3 [27], which results in a higher QoS, QoE, and spectral efficiency. This solution supports power optimization, efficient interference management, a user-centric approach, robust connectivity, eliminates cell-edge performance issues, and enhances coverage. This novel idea has its limitations and challenges that need to be addressed, such as scalability, capabilities of intelligent reflective surfaces (IRS) to implement advanced beamforming capabilities, timing synchronization, channel conditions, data compression, power consumption, backhaul capacity constraints, and spectrum allocation [33].

Next, spectrum and energy efficiency have improved with the introduction of **hybrid beamforming**, which is a combination of analog and digital beamforming techniques [34]. Analog beamforming uses phase shifters and is cost- and power-efficient, while digital beamforming offers precise control and creation of multiple beams; however, it is not efficient in terms of cost, power consumption, and complexity. Hybrid beamforming balances these factors by performing initial RF processing in the analog domain and then further refining it in the digital domain. Hybrid beamforming uses high-dimensional analog phase shifters and power amplifiers, and lower-dimensional digital signal processing (DSP) units [35]. The QoS offered by the network enabled with hybrid beamforming and mmWaves would exponentially increase as hardware efficiency, computation efficiency, and available spectrum efficiency improve [36]. This technology offers operators highly lucrative benefits, including greater cost and power efficiency, enhanced flexibility, and improved scalability

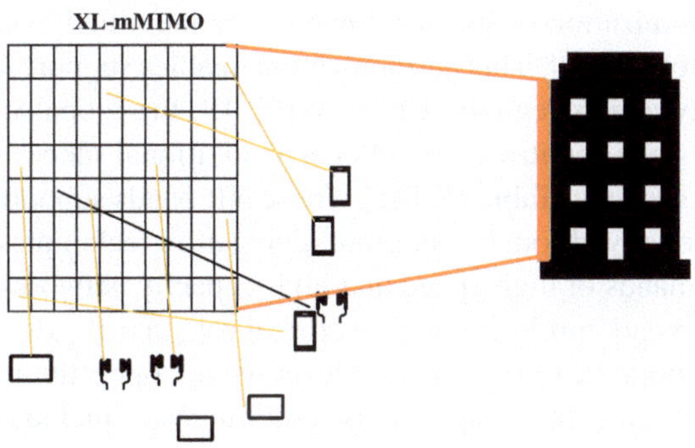

Fig. 7.2 XL-mMIMO block diagram

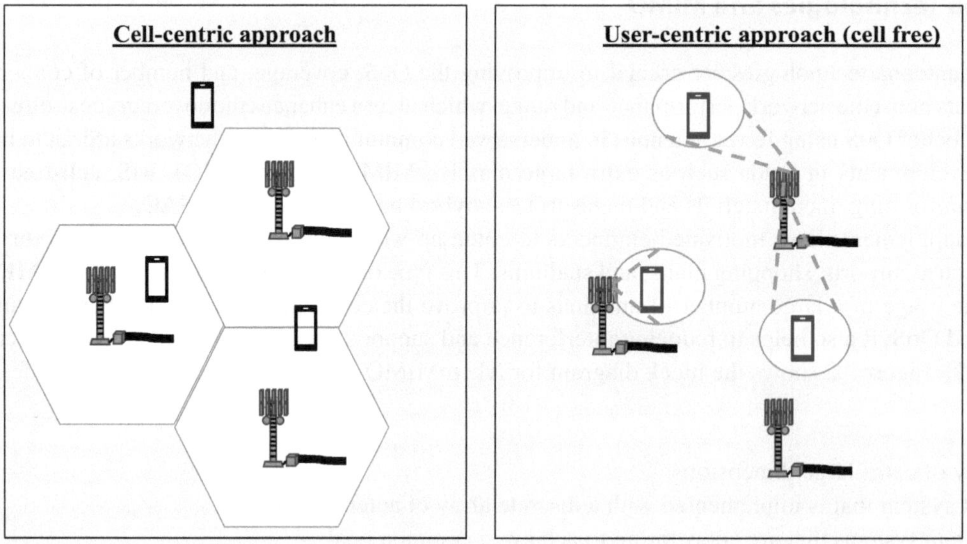

Fig. 7.3 Cellfree massive MIMO

7.2.3 Millimeter Waves - mmWaves

There is a scarcity of spectrum bandwidth because the demand for higher data rates and traffic generated is constantly rising. Sub-6 GHz bands are crowded, and there is a need to offload the transmission to different spectrum bands to improve user QoS and QoE [37]. Researchers and operators are looking for solutions in bands over 6 GHz, for example, millimeter waves (mmWaves) (30–300 GHz), which have features such as high data rates and low wavelengths (30–1 mm), and offer signal propagation range from a few 100 m to a maximum of 1 km with data rates up to 10 Gbps. The modified mmWave communication in rural areas helps in achieving a FWA up to 14 km in radius with gigabit-class speeds [38]. The advancement in technology helps in the miniaturization of antenna sizes, optimizes the power and performance, and integrates with large antenna arrays in compact devices. Hence, mmWave constitutes the small cells in 5G and beyond technologies.

Since the coverage area of mmWaves is relatively small, their deployment necessitates a dense network of BSs. This significantly increases costs due to the heightened requirements for hardware, wiring, backhaul connections, meticulous planning, installation, and ongoing maintenance. Another challenge during mmWave signal propagation is LOS between the transmitter and receiver; otherwise, the signal is degraded by clutter and gets blocked or absorbed by the surroundings. The mmWave signals are sensitive to atmospheric conditions such as gases, moisture, rain, and humidity; thereby reducing the maximum propagation distance as the signal is either absorbed or degraded [39].

At WRC-19, the need for global harmonization of spectrum bands was identified, particularly for mmWave frequencies, to ensure the global scalability of technologies [40]. This harmonization enables support for diverse applications and industry verticals, fostering seamless connectivity across regions. Under WRC-19, the globally harmonized mmWave bands span 24.25–27.5 and 37–43.5 GHz. However, some countries have allocated additional spectrum for mmWave applications alongside these globally harmonized bands, as shown in Table 7.2 [41]. These NR bands support various bandwidth sizes, including 50, 100, 200, 400, and 500 MHz. Additionally, through spectrum aggregation techniques, the total available bandwidth can be significantly increased to meet the demands of high-speed and high-capacity networks.

For advanced technologies, 5G and beyond, mmWave plays a critical role in offloading data traffic from macro cells [42]. This helps alleviate network congestion, improve QoS, and fully leverage the opportunities of hyper-connectivity in densely populated areas. Some of the potential revenue-generating use cases for mmWave include FWA in rural, suburban, urban, and dense-urban areas, surveillance, video broadcasting and streaming, crowded area capacity, last-mile backhaul complement, AR/VR, smart factories, industry 4.0 IoT sensors, and street macro applications such as traffic management, footfall studies, and mobility sensors.

Table 7.2 mmWave frequency bands defined in 3GPP Release 17

NR band	Bands
n257	26.5–29.5 GHz (USA, Korea, Japan)
n258	26.5–29.5 GHz (global)
n259	39.5–43.5 GHz (Europe)
n260	37.0–40.0 GHz (USA)
n261	27.5–28.35 GHz (USA)
n262	47.2–48.2 GHz (USA)
n263	57.0–71.0 GHz

7.2.4 Li-Fi

Li-Fi is a technology that transmits data through an light emitting diode (LED) bulb by varying the intensity of brightness (on/off) faster than human eyes can follow. It is a revolutionary future network technology that offers high-speed, secure, and reliable wireless communication using visible light, ultraviolet, and infrared spectrum bands. The data capacity limitations in the RF spectrum bands, especially in Wi-Fi and other wireless technologies, motivated the researchers to explore Li-Fi. The key working principle for Li-Fi depends on the modulation of light intensity to transmit data and is resistant to electromagnetic interference.

Li-Fi is a niche technology with significant potential to bridge the digital divide in ultra-dense areas. Functionally similar to Wi-Fi, Li-Fi transmits data using light waves instead of radio waves. It offers advantages such as seamless handover, enhanced security due to its inability to penetrate walls, and a limited range that reduces hacking potential. Moreover, Li-Fi can function even without a direct LOS between the transmitter and receiver. Its operational similarity to Wi-Fi in terms of protocols has led to Li-Fi becoming part of the IEEE 802 working group (WG), emphasizing its potential as a complementary communication technology. In July 2023, IEEE 802.11bb published its standards for light-based networking, providing a vendor-neutral framework to support and standardize the growing Li-Fi market. These standards aim to facilitate interoperability and accelerate the adoption of Li-Fi technology across various applications.

Figure 7.4 illustrates the Li-Fi architecture, depicting a room equipped with an LED bulb that not only provides illumination but also enables data transmission and reception. This setup allows seamless connectivity for devices within the room. The working principle of Li-Fi involves a sequence of steps as follows: First, the LED module transmits data by modulating the light intensity at extremely high speeds, imperceptible to the human eye. Next, the modulated light waves propagate within a confined area, such as a room. Following this, the intended device, equipped with a photodetector, receives the light signals and converts them into electrical signals. Finally, the electrical signals are decoded back into their original form in the data processing step, enabling seamless communication between devices. In an advanced configuration, a fiber optic cable is connected to the LED bulb, linking it to a data source or network backbone. Some of the potential functionalities and their applications offered by Li-Fi consist of high-speed data transmission, spectral efficiency, secure and private connection, and integration with existing technologies. Li-Fi is still in its early stage and not yet mature to bridge the wide-scale digital divide. Once the technology matures and the necessary infrastructure is widely available, it should be reevaluated.

7.2.5 THz

The microwave backhaul has been installed in rural areas but lacks sufficient bandwidth to support NGMN for providing higher data rates and capacity with low latency for rural scenarios. Alternative solutions using THz spectrum bands are being explored for achieving high-capacity wireless backhaul capacity [43]. For THz communication, the carrier frequencies typically fall within the range of 100–1000 GHz. This frequency range lies between the microwave and infrared regions of the electromagnetic spectrum and offers unique advantages for next-generation wireless communication systems. For rural communities and other underserved regions where laying down a fiber network is difficult, THz backhaul is considered a cost-effective solution in bridging the digital divide [44].

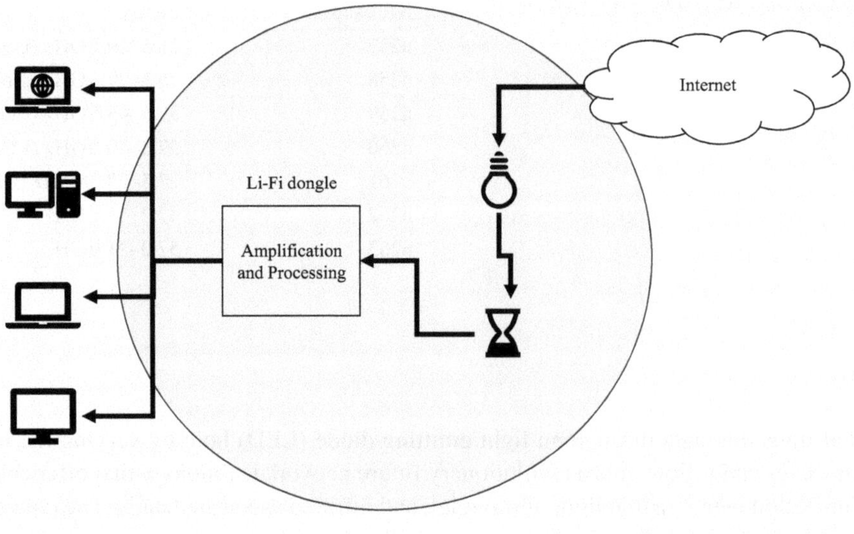

Fig. 7.4 Li-Fi architecture

THz communication has been explored for many years and was initially explored for cosmic studies and astronomy applications. However, the lack of suitable sources and detectors limited their scope and practical uses of THz technology. With the advent of time, researchers developed photoconductive antennas, time-domain spectroscopy, quantum cascade lasers (QCLs), and THz-time-domain spectroscopy (TDS), which would help in generation, detection, imaging, and spectroscopy for THz, respectively. THz systems have improved performance and capabilities thanks to enabling innovations such as graphene-based THz devices and metamaterials. THz research is poised to support future 6G networks.

The THz communication offers large bandwidths, higher capacity, greater directivity compared to microwaves, and strong diffraction, making it suitable for both LOS and NLOS communication [45]. THz frequencies enable significantly higher data rates up to terabits per second, and it suitable for applications in 6G, AR/VR, and high-speed data centers. THz frequency bands offer shorter wavelengths; therefore, they are valuable for imaging, sensing, and localization. Similarly, their high directivity makes them suitable for indoor, dense deployments and dense urban applications as the last-mile, and for remote rural areas as a backhaul.

THz is powered by several technologies supporting their features, such as advanced graphene materials, beamforming, mMIMO, RIS, integrated photonics, and ML [46]. The extremely high data rates offered by THz communication are a key enabler for advanced applications such as tactile Internet, holographic telepresence, and autonomous systems.

THz communication systems have several shortcomings that need to be addressed to improve visibility in wider market use cases, such as high propagation losses, penetration losses, shorter range, and network design complexities. The absence of global standards for operations, interfaces, spectrum bands, and policies further impedes their widespread deployment.

7.2.6 Reconfigurable Intelligent Surfaces

RIS is a transformative technology for wireless communication that can dynamically control EM waves to enhance the signal strength, minimize interference, improve QoS, and boost overall network performance [47]. RIS along with smart radio environments (SREs) are capable of over-the-air signal programmability in a cost- and energy-efficient manner by seamlessly integrating to optimize their analog and digital components for the communication and localization sensing [47]. RIS is expected to play a crucial role in bridging the digital divide by enhancing the coverage of the network, especially in low-signal, obstructed environments, or challenging terrain, by reducing the dependencies on a large number of antennas and BSs.

RIS is an artificial planar surface consisting of low-powered or passive metamaterials to manipulate incoming EM waves. The entire signal manipulation by RIS is performed in the analog domain to enhance the overall energy efficiency of the network [47]. RIS technology has been actively explored by various stakeholders, including academia, industry, policymakers, standards organizations, and the research and engineering communities [48]. RIS faces a critical industry challenge: it cannot selectively amplify signals within a specific band, and instead amplifies all signals in that band, which prevents it from distinguishing targeted users from malicious attackers. RIS could become a promising solution for advanced wireless communication systems, when it can offer selective amplification to enhance network efficiency, minimize interference, and ensure optimized resource utilization. Hence, to overcome this issue, intelligent environments serve the purpose of identifying the intended user, improving data security, and preventing the eavesdropping of data. 5G, 6G, and beyond technologies consider RIS as a potential enabler in improving the QoS and QoE for applications where it would play a crucial role in developing their business case, for beamforming, beam steering, enhanced coverage, interference management, and improved energy efficiency [49].

7.2.7 Trusted Connectivity

The foundation of modern-day communication, business, and critical infrastructure is telecommunication networks [50]. In the digital era, networks that are secure, trustworthy, and resilient are essential. The services must be functional and safe against a variety of risks, such as phishing and cyberattacks, as well as natural catastrophes and man-made errors of judgment. The telecommunications sector, which depends on and supports multiple other critical infrastructures, must explore strategies and address challenges to ensure networks are secure, trustworthy, and resilient networks. When extending connectivity to unconnected populations, stakeholders must ensure that the network is reliable and fosters confidence and trust, particularly among marginalized communities. This would encourage them to adopt digital connectivity and related technology as well as services.

7.2.7.1 Resilience

Network resilience is another critical factor driving innovation in 5G and 6G technologies. But what exactly is network resilience?

Any network would have unexpected breakdowns that would significantly impact their businesses, which requires them to select alternative backup options [51]. Therefore, network resiliency is a concept that describes how fast a network responds and resumes normal operations during an unexpected interruption to its services. Note that the higher the network resiliency for a network, the shorter its downtime, and vice versa. Strategies for network resilience ought to respond to any interruptions to the network, such as known (the engineers anticipate these challenges), unknown knowns (unpredictable, but expected, such as the life of the equipment), and unknown unknowns (totally unexpected, such as natural disasters and accidents). The study of network resilience is important to the network function, network structure design, and scientific significance [51].

A resilient network design should be able to survive communication failure (a situation where a network is unable to deliver communication services). Since the telecommunication industry is a critical infrastructure, steps should be in place during such scenarios. Some of the common causes of communication failure include cable cuts, hardware malfunctions, faulty software errors, power outages, accidents, human error, natural disasters, malicious physical attacks, war, and electronic attacks. Communication networks are subject to power outages and frequent upgrades that make them susceptible to security breaches, network failures, integration failures, vulnerability to external threats, and hacking of open networks.

Resilience for communication networks is defined as the ability to meet the customers during emergency scenarios [52]. This definition was further modified to suit the advanced telecommunication technologies "that resilience pertains to the capability of a network architecture to provide and maintain acceptable service levels for clients, regardless of whether nodes or links fail" [51, 53]. The process of ensuring network resilience involves several key steps—scheduled checks and maintenance, extra resources in case of sudden breakdown, isolation of failed networks, policies and frameworks for crisis management, and natural disaster preparedness [53].

The challenge in building a resilient network is the cost and complexity associated with building and maintaining it. Resilient networks are indispensable in sectors where uninterrupted connectivity is vital, such as healthcare, financial services, military operations, emergency services, and businesses. In underserved areas, the challenge lies in keeping the services running in remote areas or disaster-prone regions where the infrastructure is minimal, and the cost to restore the network is exorbitant. Hence, building a resilient network in underserved communities is critical during the lifecycle of the network.

7.2.7.2 Trustworthy

Today's citizens' security and well-being rely on critical infrastructure such as telecommunication, finance, energy distribution, and transportation [54]. Trustworthiness in networks encompasses the reliability, integrity, correctness, and authenticity of the network and its components. It ensures that the network behaves as expected and that data is accurate and unaltered during transmission and reception. In modern communication networks, enhancing one aspect of trustworthiness must not come at the expense of another. The network should remain trustworthy even when other components fail, when there is network disruption, human error, operational issues, and attacks by hostile parties [54]. Trustworthiness allows the network to become resilient and increase the usage of these technologies for other critical infrastructure and support integration with other technologies such as AI, IoT, and quantum computing.

Networks can use a selection of technologies and protocols to guarantee the reliability of the data. Some of the key techniques include secure authentication and authorization using blockchains, multi-factor authentication (MFA), and zero-trust architecture, data encryption using quantum encryption, network monitoring and anomaly detection using AI, privacy-preserving protocols using federated learning and differential privacy, dynamic resource allocation using AI and real-time trust challenges, and identity management to ensure that the right person is accessing the intended data transmission.

Ensuring a network's trustworthiness requires attention throughout its lifecycle, from design through post-operation; it encompasses various trust relationships, including human-human, machine-human, and machine-machine collaborations. The enabling technologies constitute isolated computing, audit trails, authentication, SLA automation, privacy preservation, quantum-ready encryption, reputation, and trust schemes. Modern networks face several pressing challenges, particularly in ensuring trust and security, such as ransomware, phishing, distributed denial-of-service (DDoS) attacks, data privacy, the lack of global standardization, integrating outdated legacy systems, and reliability and security across large-scale deployments. Addressing *ethical and social* considerations also demands attention, particularly in preventing AI misuse and ensuring data storage and processing comply with regulations such as general data protection regulation (GDPR). Addressing these challenges is vital to creating trustworthy, secure, and scalable networks. Though these factors increase the TCO, these are essential in components that help in gaining consumer trust and loyalty toward the broadband access solutions in the long term.

7.2.8 AI, GenAI, and ML

Over the years, the telecom sector has experienced significant change, which has been hastened by the incorporation of technologies such as AI, ML, and generative AI (GenAI). These technologies have helped the system become efficient in terms of operations, performance, energy consumption, resource allocation, customer experience, maintenance, and repairs. They have paved the path for future advancements. But what are these technologies and their role in the telecommunications industry?

- **AI** refers to the simulation of human intelligence in machines that can perform a range of functions, including problem-solving, decision-making, complex computation, mimicking human senses, learning and adaptation, and natural language processing [55]. With continuous learning capabilities, AI powered networks can predict maintenance, forecast breakdowns, and automate customer support.
- **ML** refers to a subset of AI in which a machine learns the system behavior and operations through data. It involves the development and study of statistical algorithms from the data that can learn the unseen data without explicit instructions [55, 56]. ML is widely used for applications related to network optimization, resource allocation, fault detection, customer behavior analysis, and data usage analysis. This would help in understanding the pattern and forecasting future values.
- **GenAI** refers to another subset of AI that focuses on creating new content such as text, radio, images, videos, software codes, music, and games, based on the pattern in the data. In the telecommunications industry, GenAI is used in generating synthetic data for network simulations, automating customer interactions, writing code, and developing innovative services [57]. In addition to increasing the productivity of a company, this technology helps them make better decisions, improve their analytics, and develop into more dynamic and agile organizations.

AI and related technologies play a crucial role in managing and optimizing the complexity in the network with the advent of 5G, IoT, edge computing, and advanced secure networks. AI algorithms predict network traffic patterns and allocate resources based on the predicted demand to manage the network and avoid congestion. AI helps in optimizing the resource allocation for network slicing, edge computing, beamforming, and mMIMO. They maximize the signal strength and minimize the interference, which in turn improves the network QoS while ensuring that the consumers are satisfied by assisting in the creation of individualized services for the customers and the provision of seamless experiences. Furthermore, AI and other associated technologies contribute to identifying anomalies, banning suspect calls, surveilling calls and communications to prevent unfortunate events, and the provision of powerful fraud protection. AI is also used to optimize bandwidth allocation and energy consumption, and offer stakeholders business insights to make informed decisions.

These technologies also play a vital role in connecting the unconnected and enhancing the digital ecosystem in underserved regions. For example, AI/ML algorithm would help in efficient optimization of network planning in terms of terrain, population density, spectrum availability, demand forecast, willingness-to-pay, and tower placement location of technologies to be deployed in these regions. ML would also help in performing predictive maintenance of the network to reduce the network downtime and lower its cost. The widespread application of GenAI in helping bridge the digital divide is by addressing the usage gap, mainly language barriers, and digital literacy. GenAI can generate and translate content in the local language; furthermore, it creates personalized, interactive training content to uplift the underserved communities and increase their awareness of digital connectivity and support voice interactions using digital services.

Additionally, AI is being explored as a solution for the digital divide in areas, such as providing accessibility and inclusivity for individuals with disability, developing personalized educational content for underserved regions, providing access to a chatbot with information relevant to local communities, offering language translation tools for enhancing the digital literacy of the people, and helping them to participate in digital economy [58].

7.3 Summary

This chapter explored future networks and their role in bridging the digital divide to enhance the global usage and access gaps. These technologies could potentially be designed with meaningful connectivity in hindsight to ensure that the digital divide is not widening further. The key drivers and technologies driving future networks should help in building a digitally inclusive society. While the trajectory and maturity of the future networks remain uncertain, their development should be guided by building equitable digital access for all.

Exercises

7.1 Evaluate the feasibility of deploying 6G technologies in rural areas. Discuss:

(a) The potential use cases of 6G in improving agriculture, healthcare, and education in rural regions.
(b) The challenges in integrating 6G networks with existing infrastructure in underserved areas.

7.2 Analyze the application of mmWave technology to provide high-speed internet in rural areas. Discuss:

(a) The advantages of using mmWave for bridging the connectivity gap in rural regions.
(b) The limitations and technical challenges in deploying mmWave networks over long distances in sparsely populated areas.

7.3 Li-Fi is a promising alternative to traditional wireless communication. Evaluate:

(a) The practicality of deploying Li-Fi systems in rural schools and healthcare clinics to enhance connectivity.
(b) The potential benefits of Li-Fi in terms of energy efficiency and reliability in resource-limited regions.

7.4 Massive IoT has the potential to transform rural agriculture. Discuss:

(a) Use cases of mIoT in improving crop monitoring, soil analysis, and irrigation systems in rural farms.
(b) The economic and technical barriers to implementing IoT at scale in smallholder farms.

7.5 RIS technology is being explored as a way to improve network coverage. Analyze:

(a) How RIS can improve connectivity in rural areas with challenging terrain and poor coverage.
(b) The technical and financial feasibility of deploying RIS in low-density regions.

7.6 Quantum secure networks offer high levels of data protection. Evaluate:

(a) The role of quantum secure networks in safeguarding financial transactions in rural areas.
(b) The challenges of deploying quantum secure networks in low-resource environments with limited technical expertise.

7.7 AI and ML are transforming network operations. Discuss:

(a) The potential of AI/ML in optimizing resource allocation and maintenance of rural communication networks.
(b) Examples of AI-driven solutions that can enhance the reliability of rural internet access.

7.8 THz technology is expected to support ultra-fast data rates. Evaluate:

(a) The applicability of THz communication in enabling high-speed connectivity for rural research and innovation centers.
(b) The limitations of THz technology in long-range communication and its suitability for rural use cases.

7.9 Network resilience is critical for ensuring reliable communication in rural areas. Discuss:

(a) Strategies to build resilient networks that withstand extreme weather events in rural regions.
(b) The importance of backup and redundancy systems in providing seamless connectivity during natural disasters.

7.10 Energy efficiency is a key focus for next-generation networks. Evaluate:

(a) The potential energy savings from adopting 6G networks in rural areas.
(b) The role of renewable energy sources like solar power in supporting energy-efficient 6G deployment in rural regions.

7.11 Examine the mathematical model for understanding the link budget calculations for future networks, such as 6G, THz, mmWaves, and Li-Fi, in terms of range, attenuation, and QoS.

References

1. A. Yastrebova, R. Kirichek, Y. Koucheryavy, A. Borodin, A. Koucheryavy, Future networks 2030: architecture & requirements, in *10th International Congress on Ultra Modern Telecommunications and Control Systems and Workshops (ICUMT)* (IEEE, 2018), pp. 1–8
2. CTOTechTrends2020. Future network trends. Accessed on 29 Oct 2024
3. A.M. Hadjiantonis, G. Ellinas, Towards autonomic wireless networking for the future internet, in *2010 Fourth IEEE International Conference on Self-Adaptive and Self-Organizing Systems Workshop* (2010), pp. 63–66
4. Y. Al-Hazmi, A. Willner, O.O. Özpehlivan, D. Nehls, S. Covaci, T. Magedanz, An automated health monitoring solution for future internet infrastructure marketplaces, in *26th International Teletraffic Congress (ITC)* (2014), pp. 1–6
5. K. Karathanasopoulou, A.-C. Maroudis, G. Dimitrakopoulos, E. Panagiotopoulos, J. Violos, Multi-level cognitive, risk-aware reconfiguration of the level of autonomy in highly automated vehicles, in *IECON 2022 - 48th Annual Conference of the IEEE Industrial Electronics Society* (2022), pp. 1–6
6. A.A. Barakabitze, N. Barman, A. Ahmad, S. Zadtootaghaj, L. Sun, M.G. Martini, L. Atzori, QoE management of multimedia streaming services in future networks: a tutorial and survey. IEEE Commun. Surv. Tutor. **22**(1), 526–565 (2019)
7. J. Serrano-Puche, Affect and the expression of emotions on the internet: an overview of current research. *Second International Handbook of Internet Research* (2020), pp. 529–547
8. L. Jalal, M. Anedda, V. Popescu, M. Murroni, Internet of things for enabling multi sensorial TV in smart home, in *2018 IEEE Broadcast Symposium (BTS)* (IEEE, 2018), pp. 1–5
9. R. Borralho, A. Mohamed, A.U. Quddus, P. Vieira, R. Tafazolli, A survey on coverage enhancement in cellular networks: challenges and solutions for future deployments. IEEE Commun. Surv. Tutor. **23**(2), 1302–1341 (2021)
10. Z. Ning, P. Dong, X. Wang, S. Wang, X. Hu, S. Guo, T. Qiu, B. Hu, R.Y.K. Kwok, Distributed and dynamic service placement in pervasive edge computing networks. IEEE Trans. Parallel Distrib. Syst. **32**(6), 1277–1292 (2020)
11. F. Severino, A. Canciani, C. Felicioli, V. Gervasi, A. Pelosi, A. Natali, W. Salvatore, S. Russo, Trustworthy AI for infrastructure monitoring: a blockchain-based approach. Procedia Struct. Integr. **62**, 276–284 (2024)

12. B. Van Caenegem, T. Skordas, Community research activities in secure and trustworthy ICT infrastructures. Telecommun. Syst. **35**, 89–97 (2007)
13. R.W. Thomas, D.H. Friend, L.A. DaSilva, A.B. MacKenzie, Cognitive networks. *Cognitive Radio, Software Defined Radio, and Adaptive Wireless Systems* (2007), pp. 17–41
14. D. Wang, B. Song, D. Chen, D. Xiaojiang, Intelligent cognitive radio in 5G: AI-based hierarchical cognitive cellular networks. IEEE Wirel. Commun. **26**(3), 54–61 (2019)
15. H. Saarnisaari, S. Dixit, M.-S. Alouini, A. Chaoub, M. Giordani, A. Kliks, M. Matinmikko-Blue, N. Zhang, A. Agrawal, M. Andersson, et al., A 6G white paper on connectivity for remote areas. arXiv preprint arXiv:2004.14699 (2020)
16. N. Rajatheva et al., White paper on broadband connectivity in 6G–6G research visions. Rajatheva, Ed. Oulu, Finland, University of Oulu (2020)
17. ITU. IMT 2030 and beyond. Accessed on 29 Oct 2024
18. R. Giuliano, From 5G-Advanced to 6G in 2030: New Services, 3GPP advances and enabling technologies. *IEEE Access* (2024)
19. M. Dimitris, S. Malik, Driving 5G-Advanced to 6G. Accessed on 29 Oct 2024
20. F.A. Pereira de Figueiredo, An overview of massive MIMO for 5G and 6G. IEEE Lat. Am. Trans. **20**(6), 931–940 (2022)
21. W. Wu, C. Zhou, M. Li, H. Wu, H. Zhou, N. Zhang, X.S. Shen, W. Zhuang, AI-native network slicing for 6G networks. IEEE Wirel. Commun. **29**(1), 96–103 (2022)
22. A. Thantharate, C. Beard, ADAPTIVE6G: adaptive resource management for network slicing architectures in current 5G and future 6G systems. J. Netw. Syst. Manage. **31**(1), 9 (2023)
23. H. Chergui, L. Blanco, L.A. Garrido, K. Ramantas, S. Kukliński, A. Ksentini, C. Verikoukis, Zero-touch AI-driven distributed management for energy-efficient 6G massive network slicing. IEEE Network **35**(6), 43–49 (2021)
24. A. Clemm, M.T. Vega, H.K. Ravuri, T. Wauters, F. De Turck, Toward truly immersive holographic-type communication: challenges and solutions. IEEE Commun. Mag. **58**(1), 93–99 (2020)
25. H. Zhang, H. Zhang, B. Di, M. Di Renzo, Z. Han, H.V. Poor, L. Song, Holographic integrated sensing and communication. IEEE J. Sel. Areas Commun. **40**(7), 2114–2130 (2022)
26. W. Qingqing, S. Zhang, B. Zheng, C. You, R. Zhang, Intelligent reflecting surface-aided wireless communications: A tutorial. IEEE Trans. Commun. **69**(5), 3313–3351 (2021)
27. H.A. Ammar, R. Adve, S. Shahbazpanahi, G. Boudreau, K.V. Srinivas, User-centric cell-free massive MIMO networks: a survey of opportunities, challenges and solutions. IEEE Commun. Surv. Tutor. **24**(1), 611–652 (2021)
28. Ö.T. Demir, E. Björnson, L. Sanguinetti, et al., Foundations of user-centric cell-free massive MIMO. Found. Trends® Signal Process. **14**(3-4), 162–472 (2021)
29. C. Madapatha, B. Makki, C. Fang, O. Teyeb, E. Dahlman, M.-S. Alouini, T. Svensson, On integrated access and backhaul networks: current status and potentials. IEEE Open J. Commun. Soc. **1**, 1374–1389 (2020)
30. A.H. Jazi, S.M. Razavizadeh, T. Svensson, Integrated access and backhaul in cell-free massive MIMO systems. IEEE Access (2023)
31. B. Mao, J. Liu, W. Yingying, N. Kato, Security and privacy on 6G network edge: a survey. IEEE Commun. Surv. Tutor. **25**(2), 1095–1127 (2023)
32. E. De Carvalho, A. Ali, A. Amiri, M. Angjelichinoski, R.W. Heath, Non-stationarities in extra-large-scale massive MIMO. IEEE Wirel. Commun. **27**(4), 74–80 (2020)
33. S. Elhoushy, M. Ibrahim, W. Hamouda, Cell-free massive MIMO: a survey. IEEE Commun. Surv. Tutor. **24**(1), 492–523 (2021)
34. A.F. Molisch, V.V. Ratnam, S. Han, Z. Li, S.L.H. Nguyen, L. Li, K. Haneda, Hybrid beamforming for massive MIMO: a survey. IEEE Commun. Mag. **55**(9), 134–141 (2017)
35. I. Ahmed, H. Khammari, A. Shahid, A. Musa, K.S. Kim, E. De Poorter, I. Moerman, A survey on hybrid beamforming techniques in 5G: architecture and system model perspectives. IEEE Commun. Surv. Tutor. **20**(4), 3060–3097 (2018)
36. J. Zhang, Yu. Xianghao, K.B. Letaief, Hybrid beamforming for 5G and beyond millimeter-wave systems: a holistic view. IEEE Open J. Commun. Soc. **1**, 77–91 (2020)
37. A.S. Gillis, millimeter wave (mmwave). Accessed on 29 Oct 2024
38. Ericsson. Ericsson, nbn Co and Qualcomm achieve 5G mmWave breakthrough extending range to 14 km with gigabit speeds, March 2025. Accessed on 23 Jun 2025
39. Ericsson. The unique capabilities of 5g mmwave. Accessed on 29 Oct 2024
40. M.J. Marcus, ITU WRC-19 spectrum policy results. IEEE Wirel. Commun. **26**(6), 4–5 (2019)
41. Telegraphic Japan. A look at the current status of 5G and millimeter waves at home and abroad. Accessed on 29 Oct 2024
42. W. Hong, Z.H. Jiang, C. Yu, D. Hou, H. Wang, C. Guo, Y. Hu, L. Kuai, Y. Yu, Z. Jiang, et al., The role of millimeter-wave technologies in 5G/6G wireless communications. IEEE J. Microw. **1**(1), 101–122 (2021)
43. M.A. Imran, M. Zennaro, O.R. Popoola, L. Chiaraviglio, H. Zhang, P. Manzoni, J. van de Beek, R. Stewart, M.A. Cox, L.L. Mendes, et al., Exploring the boundaries of connected systems: communications for hard-to-reach areas and extreme conditions. Proc. IEEE (2024)
44. National Science Foundation. NSF Award Search: Award #2238132 – RINGS: Collaborative Research: AI-Native Wireless Systems with Distributed Intelligence and Open Protocols, 2022. Accessed on 23 Jun 2025
45. X. Guofu, M. Skorobogatiy, Wired THz communications. J. Infrared, Millim., Terahertz Waves **43**(9), 728–778 (2022)
46. Z. Chen, X. Ma, B. Zhang, Y. Zhang, Z. Niu, N. Kuang, W. Chen, L. Li, S. Li, A survey on terahertz communications. China Commun. **16**(2), 1–35 (2019)
47. A. Ghaneizadeh, P. Gavriilidis, M. Joodaki, G.C. Alexandropoulos, Metasurface energy harvesters: State-of-the-art designs and their potential for energy sustainable reconfigurable intelligent surfaces. IEEE Access **12**, 160464–160494 (2024)
48. E. Basar, M. Di Renzo, J. De Rosny, M. Debbah, M.-S. Alouini, R. Zhang, Wireless communications through reconfigurable intelligent surfaces. IEEE Access **7**, 116753–116773 (2019)
49. Y. Liu, X. Liu, M. Xidong, T. Hou, X. Jiaqi, M. Di Renzo, N. Al-Dhahir, Reconfigurable intelligent surfaces: principles and opportunities. IEEE Commun. Surv. Tutor. **23**(3), 1546–1577 (2021)
50. D. Tipper, Resilient network design: challenges and future directions. Telecommun. Syst. **56**, 5–16 (2014)
51. X. Qi, G. Mei, Network resilience: Definitions, approaches, and applications. J. King Saud Univ.-Comput. Inf. Sci. 101882 (2023)
52. R.J. Ellison, D.A. Fisher, R.C. Linger, H.F. Lipson, T. Longstaff, Survivable network systems: an emerging discipline. CiNii Labs (1997)

53. V. Sourlas, O. Ascigil, I. Psaras, G. Pavlou, Enhancing information resilience in disruptive information-centric networks. IEEE Trans. Netw. Serv. Manage. **15**(2), 746–760 (2018)
54. F.B. Schneider, S.M. Bellovin, A.S. Inouye, Critical infrastructures you can trust: where telecommunications fits. Technical report, Cornell University (1998)
55. J.S. Gero, F. Sudweeks, *Artificial Intelligence in Design 96* (Springer, 2012)
56. E. Alpaydin, *Machine Learning* (MIT Press, 2021)
57. GenerativeAI. Your ai guide. Accessed on 29 Oct 2024
58. V. Božić, Artifical intelligence as the reason and the solution of digital divide. Lang. Educ. Technol. **3**(2) (2023)

Part III
Integrating Digital Strategies

Part I discussed broadband access, the causes of the digital divide, and potential technologies and strategies to bridge this gap. Meanwhile, Part II looked deeper into the potential technological solutions that would help in bridging the digital divide. The technologies could be part of ground, air, space, or an integration of the layers. It also highlighted the economic and environmental framework that helps in examining the accessibility, availability, and affordability of the technologies. Additionally, the future networks highlighted the need for higher QoS along with reliability. The key takeaway from here would be to guide future initiatives and foster collaboration to look for innovative solutions for bridging the digital divide.

Part III focuses on demand-side challenges to bridge the digital divide, and involves a focus on problems beyond the technology deployment standpoint. This helps in appreciating other equally significant factors that lead to the digital divide: affordability, priorities, electricity, digital literacy, limited access to advanced technologies, market dynamics, market policies and regulations, access to affordable devices, and marginalized communities. Several countries across the globe have initiatives and collaborative approaches to address these challenges. These pilot trials and case studies help understand how to tackle the digital divide.

These initiatives with technological assessment would help in formulating policies, regulations, and standards at the local, national, regional, and international levels. Effective frameworks should ensure that the future is digitally inclusive for all and offers meaningful connectivity. Furthermore, Part III highlights that addressing the challenge in question is a collaborative effort among the stakeholders, and each of them has a specific role to play. Finally, this part concludes by exploring future works that act as a guideline in the quest for building a digitally inclusive society for all. These questions would involve the community, researchers, policymakers, industry leaders, regulators, governments, and other stakeholders to collaborate to work towards building a digitally empowered society.

Global Initiatives: Connecting the Unconnected 8

Connecting the unconnected is a global challenge that requires efforts at all levels, not just technological interventions, which is the supply-side solution. To build an inclusive digital economy, it is crucial to address the underlying demand-side barriers (digital literacy, gender divide, culture, and priorities), to encourage reforms in connecting all. This chapter presents an overview of meaningful connectivity and why it plays an important role in bridging the divide. The emphasis then shifts to the entire innovation lifecycle, from R&D to the commercialization of a technology, goods, or service, with the goal of closing the gap. The chapter concludes by spotlighting a diverse range of the global projects and studies led by stakeholders with the goal of closing the digital divide and advancing meaningful connectivity.

8.1 Meaningful Connectivity

Many communities living mostly in remote and rural areas, as well as urban areas, struggle to gain access to digital services. As per ITU [1]:

- 2.5 billion people are yet to be connected to the Internet, who would require "universal coverage".
- Several million people are living within the coverage of modern telecommunication coverage but don't have meaningful connectivity.

What is meaningful connectivity? It is a level of connectivity that allows all users to have a safe, secure, satisfying, enriching, and productive online experience at an affordable cost [1]. As per ITU, anyone who has had access to the Internet from any location or device within the last 3 months is considered connected. This definition refers to the quality standard for Internet access and assesses the value offered to the users in terms of communication, opportunities, and learning outcomes. Meaningful connectivity is a prime focus for several organizations, including ITU and UNESCO; and many countries [2, 3]. The focus is now to bridge the digital divide among the people by focusing on meaningful connectivity and increasing the digital adoption among the users irrespective of gender, age, income, location, and language [4].

There is a need to build a digitally inclusive society that focuses on equal access to ICT for all, promotes digital literacy, and encourages participation of all in democratic processes and digital governance [2]. The key challenges with meaningful connectivity exist across the globe, even in urban areas where the users have access to advanced telecommunication infrastructure but are unable to fully benefit from advanced broadband infrastructure, and include [5]:

- **Poverty**: As highlighted earlier, affordability remains one of the primary challenges in achieving meaningful connectivity, particularly in low-income regions. Poverty is a complex situation where an individual lacks the financial resources and essentials needed for a minimum standard of living. This demand-side barrier persists even when users can afford broadband subscriptions; the compounded costs of Internet-enabled devices, power consumption, and essential accessories often place digital access beyond their financial reach. These cost-related constraints are among the most significant obstacles preventing marginalized communities from participating in the digital economy.
- **Low priority**: For economically challenged individuals, the priorities are the essential needs in terms of clean water, food, sanitation, healthcare, education, and employment opportunities; they first target to afford these services. Internet

S. Koratagere Anantha Kumar, *Wireless Broadband Access to Bridge the Digital Divide*, Textbooks in Telecommunication Engineering, https://doi.org/10.1007/978-3-032-09261-8_8

connectivity is not on their priority list. However, broadband access is a critical factor in providing these essential services, and there is a need to raise awareness among these individuals, especially among marginalized and indigenous groups.

- **Literacy and education**: There are approximately 7164 languages spoken throughout the globe; however, only a few hundred languages are supported by web pages on the Internet. Additionally, most of the digital content is available in English, having a share of 49.5%, followed by Spanish at 6%, German at 5.6%, Japanese at 5%, French at 4.4%, Russian at 3.9%, and Portuguese at 3.8% [6]. Meanwhile, the rest of the language content has less than 3% share of the web content. ICT skills are dependent on the literacy skills of the users; though there are text-to-speech and speech-to-text capabilities, not everything is supported by this. Internet access holds limited value for individuals who struggle with literacy and face challenges related to language barriers.
- **Lack of awareness**: Around 42% of the global population lives in remote and rural areas, and among these, there is a significant number of people, especially in marginalized and remote communities, who lack awareness of the benefits offered by digital connectivity. There is a need to increase their participation in securely accessing digital services [7].
- **Lack of digital skills**: Some individuals are educated but don't have the necessary skills to use digital services, for example, older adults [8]. There is a need to train these people to support their healthy aging population and build their confidence in using digital services in building a digitally inclusive society [9, 10]. Another group that falls in this category is women and kids, who would also require special training to confidently use digital services.
- **Exorbitant usage-based pricing models**: In many low-income and developing countries, the pricing plans of Internet-based services are significantly dependent on the data consumed and are priced exorbitantly high [11, 12]. This restricts the usage of the Internet by the users, as they can only afford to pay a limited amount. There is a need for an innovative cost structure to encourage underprivileged individuals to explore digital services and realize meaningful connectivity targets.
- **Miscellaneous factors**: They are social and cultural factors that prevent specific groups from accessing digital services for reasons such as technology not being built to address women's needs, discrimination in many aspects of social life, including employment, income, and education, technophobia, lack of access and training, and low digital skills confidence [13]. With increasing cybercrime against Internet users and digital arrests, individuals are skeptical about using the Internet, especially for digital and financial safety [14]. In some instances, people have lost their funds, are unable to access their profiles, have their national digital identity stolen, and beyond [15, 16]. Cybercrime poses a serious threat not only to individuals but also to large enterprises, and is one of the most pressing issues contributing to global financial losses amounting to billions of dollars annually [17].

Figure 8.1 shows the pillars of the digital divide that need to be addressed to provide meaningful connectivity for all. Digital utilization must be prioritized alongside digital infrastructure rollout to ensure the success of initiatives aimed at bridging the digital divide [18]. These two would go hand in hand in building a digitally inclusive society. Building robust and accessible digital infrastructure is the foundation for offering broadband services. It involves the deployment of broadband technologies such as wired and wireless technologies using different backhaul technologies to provide reliable Internet services in urban, rural, and remote regions. Furthermore, there is a need to provide public computer kiosks, affordable smartphones, and device access to underprivileged communities to encourage the digital utilization of the digital infrastructure. This would involve collaboration on a large scale among the different stakeholders–government, businesses, community, international organizations, and academia–to optimize the deployment strategies to meet the demand using the limited funds. There is also a need to create the content in the local language using AI, and to educate individuals in the confident use of end user devices to promote widespread adoption [19]. When industries that deliver essential services adopt digital tools to meet individuals' basic needs, it encourages users to explore the benefits of connectivity and fosters greater engagement with digital platforms.

Therefore, investing in the infrastructure solely without building a user base that is capable of utilizing those services would lead to resource underutilization and an unsustainable business model [20]. Hence, digital infrastructure and digital utilization should advance together to maintain the supply and demand that would build a sustainable business model capable of contributing toward UN SDG targets [21].

8.2 Research to Commercialization of Digital Divide Solutions

The pathway of technological solution from R&D to commercialization is an interplay of processes, including the development of technology and its related policies, regulations, and standards. The process of converting an idea into a commercial technological solution is complex and involves several steps where different stakeholders play a key role at each level.

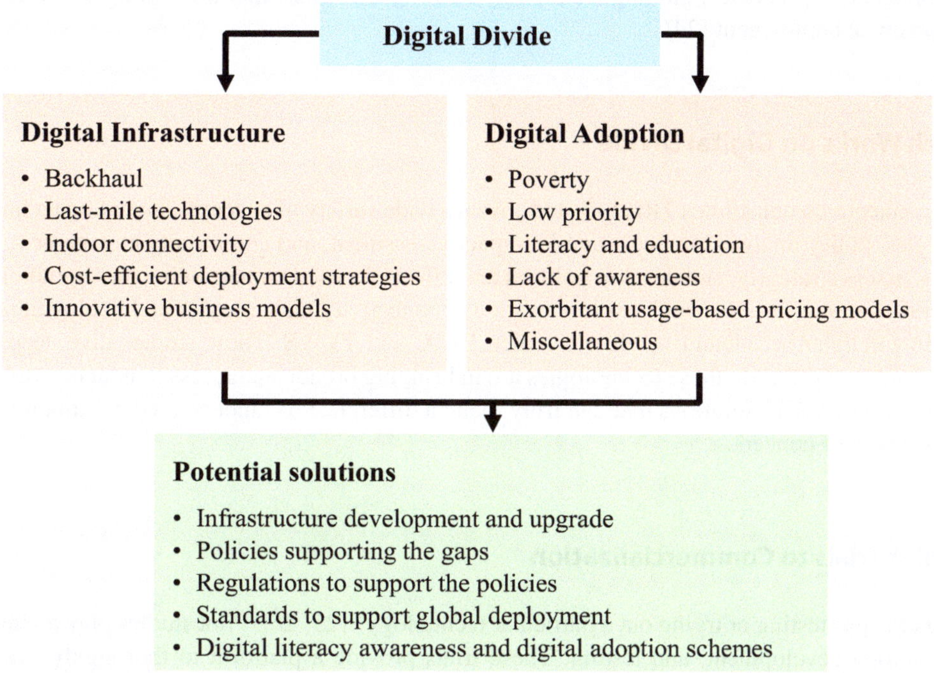

Fig. 8.1 Digital infrastructure and digital adoptions

The innovation lifecycle typically comprises the following stages: idea generation and preliminary research, applied R&D, prototyping and validation, field testing, regulatory approvals and certifications, pre-commercialization activities, market launch and commercialization, followed by post-market monitoring.

Technology readiness levels (TRLs) represents a widely adopted framework for evaluating the maturity of a technology by analyzing its conceptual foundation, technical requirements, and demonstrated capabilities [22]. This framework operates on a scale ranging from 1 to 9, where level 1 indicates the initial stage of development, and level 9 signifies a fully mature technology, as depicted in Fig. 8.2 [23]. The advantages of using TRLs include fostering a shared understanding of the technology's status,

Phase	TRL	Hardware	Software
Research	1	Basic principles	
	2	Concept and application formulations	
	3	Concept validation and proof of concepts	
Development	4	Experimental pilot	
	5	Demonstration pilot	
	6	Industrial pilot	
Deployment	7	First implementation	Industrialization detailed scope
	8	A few record of implementation	Release version
	9	Extensive implementation	

Fig. 8.2 Technology readiness level

enhancing risk management processes, guiding technology funding decisions, and facilitating the transition of technology from research to practical deployment [24].

8.2.1 Research Works on Digital Divide

Research on rural connectivity constitutes TRL 1–3, and covers a wide variety of concepts, including technological evaluation, deployment strategies, policy analysis, socioeconomic impact assessment, and community engagement. These studies seek to identify barriers, assess strategies, and develop policies and efforts focused on narrowing the digital divide and enhancing connectivity in rural communities. Research explores various connectivity technologies and deployment models tailored to rural areas, including cellular broadband, satellite Internet, FWA, and TVWS. These studies dive deep into exploring how practical, efficient, and cost-effective these technologies are in bringing broadband access to underserved communities. They aim to shine a light on innovative solutions that can truly make a difference for underserved communities. This helps in the development of the proof of concept.

8.2.2 Rural Pilot Trials to Commercialization

A pilot study refers to "pre-testing or trying out a particular technology" [25, 26]. Pilot studies play a vital role in the journey of invention, technology development, and testing. These trials provide a platform to thoroughly evaluate the feasibility and functionality of a technology in real-world scenarios. By simulating practical environments, pilot studies help uncover critical insights into the technology's capabilities, system behavior, performance metrics, associated risks, and costs involved. Additionally, they address and clarify misconceptions related to the intended use case, ensuring that the deployment aligns with both user expectations and operational requirements. This iterative process ultimately helps refine the technology, paving the way for more effective, scalable, and efficient implementations.

Pilot studies are instrumental in assessing the likelihood of success for large-scale deployments while evaluating their social, human, and environmental impacts [26, 27]. By testing technologies or systems in controlled, smaller-scale environments, these studies provide insights into potential challenges and opportunities that may arise during broader implementation. They help gauge how the deployment could influence social structures, improve human well-being, and align with environmental sustainability goals. This comprehensive understanding enables stakeholders to make informed decisions, optimize strategies, and mitigate risks before proceeding with full-scale deployment [28].

Rural connectivity pilot trials are essential for evaluating the feasibility of the technology in terms of technical analysis, economic analysis, business models, market uptake, pricing strategies, demand-supply balance, policy, decision-making, legal requirements, spectrum licensing, and regulation formulation. These trials help the stakeholders make an informed decision on the performance, effectiveness, and suitability of technologies to bridge the digital divide. These trials help identify challenges, validate technologies, and examine deployment strategies for extending broadband connectivity in regions with poor or no coverage and connecting underserved communities.

Pilot trials assist the stakeholders in evaluating the network performance based on the feedback from the end users, such as their needs, usage patterns, satisfaction with using the services, and QoE. It highlights hurdles with infrastructure deployment in underserved regions, including legislative limitations, poor backhaul connectivity, lack of electricity, and topographical barriers. Pilot trials also provide the results on the viability and economics of implementing digital infrastructure, such as satellite ground stations, fiber-optic cables, and tower deployments, which are evaluated through field testing.

Pilot trials help educate constitutional and regulatory choices for rural connectivity, such as spectrum allotment, licensing requirements, permitting processes, and financing sources. Regulatory sandbox efforts might help pilot trials by exempting certain regulatory restrictions to stimulate innovation and experimentation in digital divide solutions. Overall, pilot trials play a vital role in bridging the digital divide, empowering rural communities, reducing the gender gap, and creating employment opportunities in remote areas. They offer critical lessons to achieve universal connectivity that help refine technologies, improve deployment strategies, and inform policy development.

A diverse group of stakeholders is involved in pilot test trials, who contribute specific expertise, resources, perspectives, and analytics; these constitute TRL 4–6. [29]. For example, government regulatory bodies, TSPs, technology vendors and equipment manufacturers, academia and research organizations, NGOs, investors and development agencies, policy think tanks and advocacy groups, standardization bodies, media, and public relations [30].

The next question is: who participates in pilot trials and testbed deployments?

The following play a significant role in pilot trials:

- **Government and regulatory bodies** provide funding and incentives for private organizations to deploy and test the technology of interest. The regulatory bodies monitor compliance with legal safety standards and formulate policies, regulations, and spectrum allocation to support pilot trials. For example, the spectrum sandbox helps in testing in a controlled environment, examining the demand assessment, experimenting with wireless technologies, and setting the price for spectrum bands.
- **TSPs** deploy the network infrastructure for the connectivity trials to evaluate service reliability, performance, and scalability.
- While the **academia and research organizations** deploy the network to conduct studies and provide insights on the technical, social, and economic aspects. These trials would help in developing newer test trials to suit rural environments.
- The parties would receive funding from **investors and development agencies** for deployment, who would study the pilot trials for assessing the solution's financial feasibility and long-term prospects.
- For pilot trials, there is a high dependency on **technology vendors and equipment manufacturers**, who are tasked with designing and building the critical components such as BSs, antennas, user devices, cables, and other necessary hardware, to meet the networks's KPIs and ensure their functionality and reliability.
- Meanwhile, **standardization bodies** play an essential role by developing and ensuring that the equipment complies with global standards and maintains interoperability for enabling the global scalability and seamless integration of the technology.
- The involvement of **local communities and end users** is vital for pilot trials as they participate as the testers of the services and provide valuable feedback on usability, accessibility, mobility, and impact. This collaborative approach ensures that the solutions developed are not only technically sound but also socially relevant and impactful.
- Pilot trials receive support from **NGOs** in the form of financial aid and initiatives aimed at skill development for residents, thereby enhancing the effectiveness and long-term impact of pilot trials.
- **Policy think tanks and advocacy groups** translate the results of pilot trials into actionable and scalable policies. They work with wider groups to formulate rules, regulations, and legal norms to achieve their targets, such as sustainability, inclusivity, affordability, and long-term social impact.
- Finally, the **media and public relations teams** play a pivotal role in raising awareness about pilot trials and highlighting the success of such outcomes. They shape public perceptions and manage stakeholder communications.

Each of these stakeholders plays a vital role in ensuring the success of test trials, from planning and execution to analysis and scaling up of the solutions. Once the technology reaches TRL 7, the prototype is tested in the operating environment to understand the achievable KPIs of the pilot deployment. This helps in enhancing the design to achieve the target KPIs in the realistic operating conditions. Once this criterion is met, that is, when the technology has proven to work in its final form and under expected working conditions, then the system design is complete and certified for production for commercial deployment, and this constitutes the TRL 8. Finally, a technology reaches TRL 9 when it is actually deployed and is functioning successfully in real-world conditions at a scale. The technology is validated for market readiness and operational use globally.

8.2.3 Corporate Social Responsibilities

CSR initiatives to address the digital divide include improving access to technology, developing digital literacy, building community empowerment, and pushing for policy reform. Corporations may help establish more fair and inclusive societies by investing in digital inclusion activities, ensuring that everyone has the chance to participate and flourish in the digital economy [31]. They leverage the resources, expertise, and investments and influence the other stakeholders to promote digital connectivity, which is inclusive and offers access to services for all. There are several ways in which firms can contribute toward bridging the digital divide, as shown in Fig. 8.3.

With CSR initiatives, firms can donate or subsidize the distribution of end user devices to underserved communities, schools, libraries, and other non-profit organizations to help increase access to technology and improve digital literacy among the end users in these areas [32, 33]. They could also provide funds for projects whose objective is to bridge the digital divide or

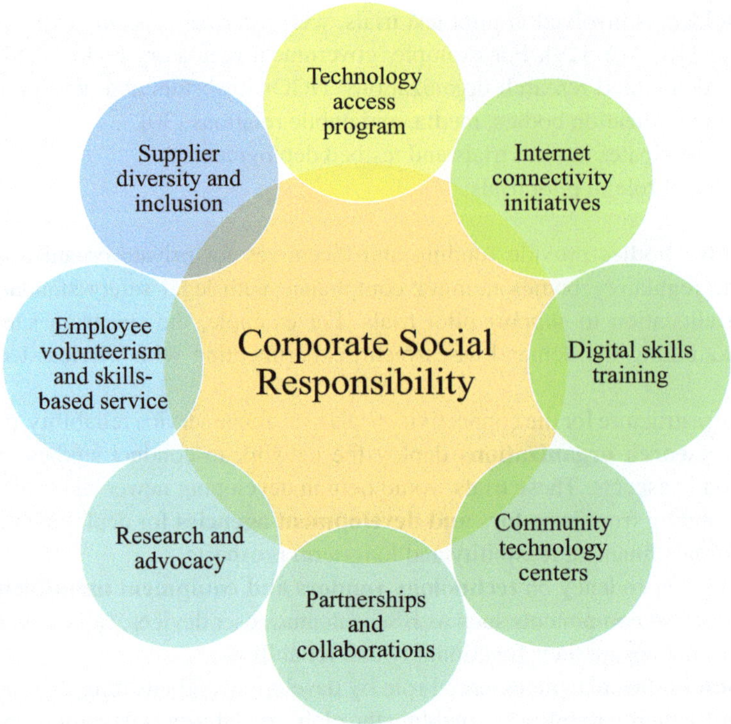

Fig. 8.3 CSR initiatives

even provide a free digital path for low-income households in rural and remote areas, also termed as digital pedestrian [34, 35]. They could also develop and sponsor digital skills training programs and establish community technology centers equipped with computers, the Internet, and technology resources to empower the locals. They can also provide technical assistance and foster local entrepreneurship by supporting individuals and businesses in delivering digital technologies and services.

8.3 Initiatives to Bridge the Divide

As economic progress fuels the adoption of digital technologies, the unequal benefits further widen this gap, especially in underserved regions. To address this, there is a need for a deeper exploration of business models and the social impact of digital divide initiatives via research and pilot trials, including PPP, cost-effective solutions, and targeted programs to foster equitable digital inclusion across nations. This section focuses on pilot trials and initiatives aimed at bridging the digital divide. It examines the deployments carried out as part of these efforts and highlights the key takeaways derived from the associated research and practical experiences in different continents.

8.3.1 Africa

Several studies have shown that around 1 billion people who are living in Africa are either living in underserved regions or unconnected from the Internet [1]. Therefore, several global initiatives focus on building a digital infrastructure that is digitally inclusive for all, and empowering these communities with digital skills would lay a solid foundation for long-term development and inclusive growth across the continent.

One of the large-scale initiatives by the World Bank is *Digital Economy for Africa (DE4A)*, which is a transformative program launched to ensure that every individual, business, and government in **Africa** is digitally enabled by 2030 [36]. The initiative aims to align with the African Union's "*Digital Transformation Strategy for Africa*" and aims to harness the potential of digital technologies to drive:

- economic growth
- reduce poverty
- improve service delivery across the continent.

Digital infrastructure, digital public platforms, digital financial services, digital businesses, and digital skills are among the subjects that are the focus of the program. (Note: While the aims and targets are common among different global initiatives. Therefore, this section would focus on providing an in-depth explanation for the African scenario.) Now, let's categorize the initiatives toward the main focus areas:

- **Bridging the Gender Digital Divide**
  One of the biggest roadblocks to achieving inclusive digital transformation in Africa is the gender digital divide [37]. Some of the common gender divide barriers include sociocultural norms restricting women's online activity, restricted technology access, and costly services. Gender equality-inducing policy interventions are critical in solving this. Investing in digital literacy and providing women and girls with special training can enable them to develop essential skills. The public and private sectors need to collaborate to build an inclusive digital economy and empower the marginalized communities as well. Awareness campaigns can reach social norms and encourage digital inclusivity even further. These strategies can complement the SDGs, enhance economic growth, and reduce gender inequality.
  In Sub-Saharan African countries, the gender digital divide prevents women from participating in economic activities. The key initiatives target providing subsidized devices, and offering digital literacy trainings and STEM education for girls to bridge this gender gap [38]. The women's enterprise fund and the Kenyan government's initiative to bridge the gender digital divide focus on providing financial support, training, and digital tools to enhance their economic opportunities [39, 40]. Additionally, the Kenya *National Digital Master Plan* aims to expand women's involvement in financial services (mobile money transfers and WhatsApp) and enable women market traders [41, 42]. Meanwhile, gendered digital skills programs for women in Nigeria focus on context-specific training, partnerships, and monitoring frameworks to improve digital literacy and economic opportunities [43].
- **Community-led Internet**
  As highlighted by the World Economic Forum (WEF), the community-led initiatives are playing a critical role in bridging the digital divide in Africa [44]. These organizations and initiatives work at the grassroots level to provide connectivity to people in remote and rural areas, and especially to the underserved communities. Some notable key initiatives include *Zenzeleni* in South Africa and *Mawingu* in Kenya [45], where local communities have utilized the capabilities of Wi-Fi in connecting several large- and small-scale communities at affordable cost and offering services to the end users. For end users in regions that are in conflict-affected areas, such as North Uganda, the community-led initiative has helped in getting access to education, job opportunities, and personal development. In Uganda, *BOSCO* uses solar-powered systems to connect with satellites to offer Internet connectivity via portal devices funded by donors. Similar to this, the *iNethi* platform in Cape Town uses mesh routers, open-source software, and cellular BSs to provide the Ocean View township with free educational materials, file-sharing services, and chat platforms. These projects face significant obstacles from incumbent operators, regulatory constraints, limited public awareness, a complex licensing process, and high CAPEX.
- **Policy Reforms and Public-Private Partnerships**
  Policy reforms play a key role in bridging the digital literacy and affordability gaps, especially where users live within a coverage area of a network but remain unconnected, as being connected is not their priority. For addressing this gap, PPP are driving innovation solutions, such as KT Corporation in Rwanda is expanding the 4G coverage to 97% of the population, Microsoft's Airband, and Google's Loon; and international organizations, such as ITU, GSMA, and the Carnegie Endowment for International Peace, conduct research studies and provide critical insights and guiding principles for policymakers to improve their strategies in bridging the digital divide [46].
  Research indicates that a 10% increase in broadband penetration can boost the GDP for countries in Africa by 2.5% [47]. Furthermore, to encourage the growth in GDP, Mozambique's "*Digital Acceleration Project*," is supported by the World Bank initiatives to improve the digital literacy, which helped in driving the adoption from 15% in 2015 to 32% in 2021 [48].
- **Research Bodies, Investors and Development Agencies**
  The *Digital Transformation with Africa (DTA)* initiative, launched in December 2022, is a $350 million initiative with an additional $450 million in mobilized funds aimed at reducing Africa's digital divide and anchoring the program around the three pillars: Digital Economy and Infrastructure, Human Capital Development, and Digital Enabling Environment [49]. The objective of this funding is to improve connectivity, foster innovation, and support digital inclusion.

Under this initiative, several key infrastructure projects and collaborations are supported, and it includes the *Lobito Trans-Africa Corridor*, which enhances connectivity in Angola and the Democratic Republic of Congo (DRC) with $100 million in funding from the U.S. Export-Import Bank. Additionally, USAID's "*Dinheiro Digital é Melhor*" project in Angola invests $5 million to expand digital finance ecosystems. Vice President Kamala Harris has encouraged private sector and philanthropic support to advance Africa's digital inclusion efforts under the *DTA* initiative.

Another large project is the *AfricaConnect* program, which is a transformative project aimed at improving digital connectivity and fostering collaboration among the stakeholders for building a sustainable digital society [42]. This project is a partnership between the EU and African *National Research and Education Networks (NRENs)* and focuses on building a pan-African Internet infrastructure to boost access to digital services and opportunities arising from it. The project is divided into phases, and focuses on improving connectivity for research and educational institutions, bridging the digital gap, investing in skills, and training African researchers, educators, and IT professionals to manage local digital infrastructure, and fostering international collaborations across different regions in Africa. By empowering African research and education sectors, the *AfricaConnect* program is a vital step toward building a digitally inclusive continent, driving socioeconomic development, and advancing Africa's integration into the global knowledge economy using community schools as centers for community kiosks for empowering residents with digital literacy and skills.

Another initiative, the *Digital Village Initiative (DVI)*, is an ambitious program aimed at leveraging digital technology to transform rural and underserved areas in Africa into hubs of innovation, connectivity, and economic opportunity [50]. It is a large collaborative project across different sectors and consists of multiple stakeholders, including governments, private sector players, and development organizations that focus on social, economic, and environmental conditions in underserved regions. The initiative seeks to address the digital divide and promote inclusive growth by integrating digital solutions and literacy, as well as driving digital platforms and services into rural development. The initiative involves the deployment of broadband networks using Wi-Fi hotspots and mobile broadband connectivity in remote areas, along with the setting up of digital devices to access the Internet services. Pilot trials have been launched across several countries in Africa and have helped improve the digital literacy, education, and adoption rates in these countries.

8.3.2 Asia

Asia has the world's largest offline population, with substantial underserved regions that lack appropriate Internet connectivity. Approximately half of the world's disconnected people live in Asia, underscoring ongoing issues in infrastructure, price, and digital literacy. India accounts for around 684 million individuals who do not have access to the Internet in 2025, accounting for roughly 48% of the population. Similarly, China's offline population is 336 million, or 23.6% of the total population. Despite advancements in urban areas, rural and remote communities in these nations have connectivity challenges. Several initiatives and government-led programs aim to narrow this digital gap. However, major work remains to provide fair and affordable access across the area.

Another huge budget initiative is *Universal Service Obligation Funds (USOF)*, a policy and funding mechanism designed to ensure that underserved and rural areas receive access to basic and advanced telecommunications services, which is financed through contributions from TSPs and is managed by the government or regulatory body [51]. These initiatives are often carried out through PPP, linking funding from the government and private sector expertise. The *USOF* manages initiatives such as *BharatNet* in **India** ($5.5 billion), which aims to connect over 250,000 village panchayats with high-speed broadband using optical fiber [52]; and *Palapa Ring* ($1.5 billion), a nationwide broadband backbone in **Indonesia** [53]. The USOF method addresses the financial and logistical constraints of delivering telecom services in less economically viable places, hence promoting digital inclusion and socioeconomic progress. Several other countries have utilized different funds to boost their digital infrastructure, with different priorities; for example:

- Indonesia and Sri Lanka: Google's Loon project provided 4G LTE connectivity through high-altitude balloons.
- Bangladesh: ICT for education, equipped over 35,000 rural schools with digital classrooms.
- Southeast Asia: The *ASEAN* Smart Cities Network enhanced urban connectivity and digital services.
- Pakistan: The *USOF* expanded broadband access in remote regions such as Baluchistan.
- China: The Digital Villages Initiative boosted rural incomes through 5G networks and e-commerce.
- Myanmar: The *SMART* Myanmar Initiative promoted digital literacy and entrepreneurship.
- Japan: The Local 5G Networks program improved precision agriculture, healthcare, and disaster preparedness.
- South Korea: The *Digital New Deal* invested in 5G, AI, and digital literacy post-COVID-19.
- Mongolia: The *Nomadic Internet Connectivity* initiative provided satellite-based Internet for remote communities.

Let's explore a few Asian countries and their focus areas, initiatives, and lessons in building a digitally inclusive society.

- **India**
  India has around 600 million people yet to be connected to the Internet and millions more who can't afford to become connected [54]. The *Pradhan Mantri Gramin Digital Saksharta Abhiyan (PMGDISHA)* and *common service centers (CSC)* have been pivotal in enhancing rural connectivity in India by focusing on enhancing the digital skills of rural citizens and providing access to government services, e-commerce, and financial services. Along with this, the Indian government has launched initiatives such as *E-sign, PayGov, Digital Locker, E-TTAL, Jan Dhan Yojana, the National Digital Literacy Mission,* and *E-Bhasha* [55], to bridge the digital divide between the urban and rural areas. These services also offer secure solutions for facilitating financial transactions and storing documents digitally. These initiatives help in building a digitally inclusive society and meet the government's targets to include everyone in the country's development journey [56].
  BharatNet is complemented by *PM-WANI* to enable affordable public Wi-Fi in remote and rural areas [57]. There are several PPP and CSR initiatives in India; for example, reaching 1500 villages throughout India, Microsoft's *Airband Initiative*, in collaboration with Indian ISP AirJaldi, seeks to give rural communities access to affordable internet [58].
- **China**
  China's forward-looking vision for the future development of broadband connectivity involves cooperation among the stakeholders and aims to narrow the digital divide across the country as well as worldwide. Global Internet development hinges on collaboration with less-developed economies and the strengthening of digital skills to ensure inclusive deployment. China has significantly invested in building and expanding digital infrastructure that offers fiber-optic broadband, 5G, and other technologies to address the digital divide. China has also prioritized building digital literacy skills to ensure integration of rural economies into the digital landscape and utilized the *USOF* program to subsidize Internet access in remote regions to ensure affordable services for the end users [59]. China's commitment to digital literacy and digital inclusion is reflected in initiatives for bridging the digital divide between youth and the old; for example, *Huawei's "Seeds for the Future,"* which trains youth, professionals, and elders in emerging technologies and builds their confidence in using the technology [60, 61].
- **Singapore**
  In terms of the telecom sector, Singapore is an established market. But this sophisticated market is also confronted with the problem of the digital divide. Singapore has implemented several initiatives to bridge the digital divide, focusing on inclusive access, digital literacy, and technological adoption [62]. Some of the initiatives include *National Digital Literacy Programme (NDLP)* focusing on building the digital skills of students for learning and future employment prospects, and promoting tech-savvy education; *Digital for Life Movement* includes initiatives for digital literacy to support marginalized groups and low-income families, and *Home Access Programme* provides affordable connectivity and subsidized rate end user devices for low-income families to be an inclusive digital world. They also focus on building the digital skills of older adults with unique initiatives, such as the *Seniors Go Digital Programme* [63].
- **Saudi Arabia**
  Saudi Arabia has made significant strides in addressing the digital divide, digital literacy, digital gender divide, and digital inclusion through various initiatives aimed at deploying and expanding the infrastructure and adoption to foster economic and social growth [64]. *Saudi Vision 2030* is a broad national plan aiming to diversify the economy and promote digital transformation for all segments of society, particularly in underserved regions [65]. Some of the key initiatives supporting these targets are the National Digital Transformation Program, the "*Saad*" Initiative, Saudi Ministry of Communications and Information Technology (MCIT) runs initiatives aimed at improving digital literacy across all age groups, the government's "*Yesser*" initiative for e-governance-related use cases, and support for women's empowerment in technology programs. These projects are part of Saudi Arabia's wider initiatives to establish an inclusive digital environment where all citizens, no matter where they are located or their social background, have access to enjoy the benefits of the digital transformation [66].
- **UAE**
  UAE has taken various efforts to narrow the digital gap and ensure equal access to digital services [67]. *UAE Vision 2031* articulates a strategic objective to make the UAE an exemplary digital economy by increasing access to technology in rural and underserved communities. The initiatives of smart government improve access to public services through digital platforms, facilitating remote communities' access to necessary services. The UAE national program for AI drives technology uptake among industries, institutions of learning, and vulnerable populations to facilitate digital inclusion. The UAE has been heavily investing in the telecommunication infrastructure with goals of increasing the technology adoption both in urban and rural areas under the "*Etisalat's 5G Network*" initiative [68]. In the education field, the education initiative of digital transformation in education guarantees rural students laptops and internet access, enhancing learning equality [69, 70]. The women in tech program also enables women to engage in the digital and tech industries, eliminating gender gaps in digital access.

8.3.3 Europe

Though the overall Internet adoption rate, throughput, and QoS are significantly higher in Europe, several studies have highlighted the need to bridge the digital divide in these places [71, 72]. Turkey, Bulgaria, Romania, Cyprus, and Poland have the worst coverage, data rates, latency, and QoS across Europe, while the Nordic countries offer the highest QoS. EU faces challenges in bridging the digital divide and is focusing on approaches to enhance the infrastructure, digital skills, affordability, and regulatory frameworks [73]. Through the *Connecting Europe Broadband Fund (CEBF)* program, the EU is improving high-speed broadband access in underserved areas by deploying infrastructure that is capable of supporting 1 Gbps connectivity and full 5G coverage in underserved areas by 2025 [74].

For bridging the digital divide in other regions, the *Digital Decade Policy* program was introduced to emphasize infrastructure development, digital skills, and online public services while ensuring inclusivity through digital rights and principles [75]. Research on Internet usage in the EU highlights the digital disparities and categorizes the users into five types [72]:

- Non-users
- Sporadic users
- Instrumental users
- Entertainment users
- Advanced users.

The studies have shown that around 60% of users fall in the first two categories, and the main factors for low digital engagement were age and ease of Internet access. There are several initiatives launched to bridge the digital divide, including the *Digital Education Action Plan* and the *Digital Single Market Strategy* [76, 77]. To boost the digital literacy among the users, cities such as Ghent, Bordeaux, and Barcelona lead projects such as *Digital Divide Observatory* and localized AI solutions, offering digital tools, training, and community support [77].

Several European mayors are emphasizing the need for sustainable digital inclusion through sustainable funding and PPP. Some of the key initiatives in bridging the digital divide and their adoption in Europe include:

- **France**
  Digital France 2025 focuses on expanding high-speed broadband access, particularly in rural areas, with a goal of universal fast Internet by 2025. This program also highlights the need for digital skills development and public-private collaborations to build an inclusive digital ecosystem for a long-term vision.
- **Germany**
  Digital Strategy 2025 prioritizes digital education by integrating information science, programming, and algorithms into school curricula. These investments in digital infrastructure and collaboration with the private sector aim to bridge the digital gap across different age groups and regions.
- **Netherlands**
  Dutch Digitalization Strategy and *Connectivity Action Plan* focus on high-speed Internet access, digital skills development, and ensuring that all sectors benefit from digital advancements while safeguarding privacy and trust.
- **Nordic countries**
  Nordic countries also prioritize bridging the digital divide by expanding broadband access and promoting digital literacy, particularly in rural areas. Finland focuses on digital literacy in education, Sweden invests in rural connectivity, while Denmark and Norway emphasize accessibility through PPP.
- **United Kingdom**
  Digital Development Strategy 2024–2030 aims to cut the worldwide digital divide by half and aims to bring around 800 million people who are currently unconnected online by 2030 [78]. The plan promotes digital inclusion through infrastructure investments, broadband expansion, and policies that support marginalized communities.
  Get Online Greater Manchester provides free or discounted connectivity and digital skills training to social housing residents [79]. Additionally, initiatives such as *Digital Helping Hands* aim to tackle digital exclusion by offering tech support to vulnerable groups.

Bulgaria, **Poland**, and **Romania**, similarly, are focused on increasing broadband availability, especially in rural zones, and enhancing digital literacy [80]. Another major initiative, the "*National Strategy for Digital Agenda*," seeks to address the digital divide through investments in infrastructure and education, as well as PPP to increase broadband penetration.

In **Cyprus**, initiatives are more aligned with integrating digital tools into public services and ensuring digital skills training for marginalized communities [81]. Cyprus also emphasizes improving broadband access and promoting digital literacy through community programs.

8.3.4 North and South America

Although the USA and Canada have attained over 90% Internet penetration, still digital divide still exists for remote and rural regions, which is more dominant in indigenous areas in both North and South American continents. More than 100 million individuals in North and South America have never used the Internet, according to studies by ITU and the World Bank [82, 83]. The primary challenges across these continents are affordability, expansive geography, ensuring connectivity for indigenous communities, and low digital literacy. Satellite programs throughout the Americas play an essential role in crossing the digital divide, especially in rural and underserved areas. Starlink (SpaceX) takes the lead with its low-latency LEO satellites, offering fast Internet to rural communities throughout the U.S., Canada, Mexico, Brazil, and Chile, covering Amazonian and Indigenous communities. Amazon's Project Kuiper, currently under development, plans to launch thousands of satellites to provide low-cost connectivity throughout North and South America.

Several initiatives in these continents are working toward bridging the digital divide. Here, we would look at a few main initiatives:

- **United States**
  Bipartisan Infrastructure Law invests $65 billion in broadband expansion, with the *BEAD* program aiming for universal broadband by 2030, with an aim to provide connectivity to 100% of households in the USA [84]. Programs such as the *Affordable Connectivity Program (ACP)* and *Digital Navigator* initiatives support affordability and digital literacy [85]. PPP, such as those with schools and libraries, further enhance digital inclusion and support the growing focus on technology inclusion through education and workforce development programs. Under the CSR scheme in underserved urban areas, businesses provide infrastructure for learning and innovation. Programs in indigenous community areas address connectivity gaps, with additional funding allocated to Native American communities.
- **Canada**
  Universal Broadband Fund ($3.2 billion) targets 98% connectivity by 2026 [86]. Through innovation, affordability initiatives, and cross-sector collaborations, Canada strives to close urban-rural and socioeconomic divides. Some of the popular initiatives in Canada include "*Connecting Families*" offers affordable Internet; "*ISED Canada*" focuses on building resilient infrastructure; and the "*Digital Literacy Exchange Program*" promotes digital inclusion [87].
- **Brazil**
  Programa Nacional de Banda Larga (PNBL) and the *Digital Inclusion Program* expand broadband, focusing on the Amazon region [88]. "*Connected Education*" provides schools with Internet, and satellite technologies enhance rural access [89] are helping with the digital transformation for a sustainable society in Brazil. These programs aim at digital literacy training, especially for women, youth, and SME. Brazil has been utilizing satellite technologies to improve coverage in remote areas, which has also improved access to online learning and financial services. Brazil's increasing e-commerce sector is a beneficiary of these digital divide programs, enhancing economic inclusion. Government collaboration with international institutions also increases financing and support.
- **Mexico**
  Internet para Todos extends free broadband to rural areas [90], and the *Red Compartida* Project boosts competition, reducing Internet costs and improving coverage [91] are two of the main digital divide initiatives in Mexico. It is aimed at providing free and low-cost broadband to rural and poor urban areas. Mexico's approach emphasizes public venues such as schools, hospitals, and community centers as points of connectivity to increase digital inclusivity and build literacy skills. Investments in expanding 4G and fiber-optic infrastructure are key to bridging the digital divide and would encourage digital literacy, targeting women, youth, and disadvantaged groups.
- **Argentina**
  The key focus for the Argentinian government is offering affordable devices, empowering women, children, and older adults, expanding existing infrastructure, and supporting digital literacy training. This would help to reduce the inequalities, foster innovation, and build a digitally empowered society. Some of the key initiatives aimed at minimizing the digital divide include *Plan Conectar*, which expands fiber-optic infrastructure; *Conectar Igualdad*, which provides students with laptops [92]; and investments in satellite technology to improve rural connectivity [93].

- **Colombia**
  The Colombian government wants to develop supportive policies for digital transformation in schools, underserved areas, towns, and villages. They are also supporting digital skill development for women and youth to foster entrepreneurship and economic participation by these groups. The key initiatives include "*Vive Digital*," which focuses on expanding rural broadband and promoting digital literacy initiatives [94]. Public Wi-Fi zones and telecom partnerships ensure affordable access, fostering economic inclusion [95].
- **Uruguay**
  In Uruguay, some of the large-scale initiatives to bridge the digital divide span across "*Plan Ceibal*," which offers free laptops and school connectivity [96], broadband expansion, and digital literacy programs that support marginalized communities and rural areas [97]. For improving digital adoption, public Wi-Fi hotspots provide support to local communities and marginalized sectors. The government collaborates with TSPs to boost and expand the network coverage as well as QoS.

8.3.5 Oceania

Oceania consists of several Island countries with varying sizes, and it is significantly challenging to deploy telecommunication infrastructure in these regions that consist of several islands and expansive geography with low population density. The focus of the digital divide initiatives in these regions is improving Internet access in remote and underserved regions, mainly Pacific Island nations. Let's examine a few key countries in Oceania, for example:

- **Australia**
  The Australian Government's *Better Connectivity Plan for Regional and Rural Australia* is an all-encompassing strategy to improve digital connectivity in rural and regional towns [98]. The plan is to invest over $1.1 billion in projects that are meant to enhance mobile and broadband services, boosting economic growth and the standard of living in the regions. This is part of the government's over $2.2 billion investment in regional communities. The project primarily focuses on providing connectivity in regions where services are not economically viable, targeting remote and rural areas, rural businesses, First Nations communities, and disaster-prone regions [99]. The aim of the project is also to improve network resilience and public safety. It aims to facilitate the targets through the broadcasting resilience program, telecommunications disaster resilience innovation program, *Mobile Black Spot Program (MBSP)*, *Mobile Network Hardening Program*, and *Regional Roads Australia's Mobile Program* [98]. *MBSP* is a dedicated fund to improve mobile network coverage in areas with a digital divide and includes collaboration with MNOs to maximize the infrastructure resource usage. Support for First Nations communities includes a budget of around $32.5 million to enhance digital inclusion among them as part of the broader "*Closing the Digital Gap*" initiative [100]. By 2026, they would want to achieve UN SDG Target 17, which refers to equal levels of digital inclusion for all. These sustainability projects face significant logistical challenges, extreme weather conditions, and technical complexities during infrastructure deployment; the government, however, is actively working to overcome these barriers.
- **New Zealand**
  Though New Zealand is technologically advanced for the telecommunications sector, it has over 20% of the population who are digitally unconnected for several reasons, such as cost, digital literacy gaps, or trust issues, particularly in rural areas and among certain demographics [101]. For bridging this gap, the government has adopted the "*Digital Strategy for Aotearoa*" promotes inclusion through which it helps to provide digital services and tools to access them in underserved communities and marginalized sectors, especially for people with disabilities. Approximately 42% of digital inclusion initiatives operate nationwide, while others are region-specific.
  National initiatives, such as the *Digital Inclusion Alliance Aotearoa*, coordinate smaller programs, promoting synergy between efforts across regions. These programs tackle barriers like affordability, access, and digital literacy comprehensively [101]. The *Ultra-Fast Broadband Initiative* program aims to improve rural connectivity, providing Internet access to regions where digital exclusion is prevalent. The key strategic themes for bridging the digital divide in New Zealand include trust (Mahi Tika), inclusion (Mahi Tahi), and growth (Mahi Ake); using these, the government wants to build a thriving digital economy that supports sustainable development.
- **Pacific Nations**
  Except for Australia and New Zealand, the rest of the Pacific Island Nations are utilizing the digital divide initiatives [102, 103], such as:

- *One Laptop Per Child (OLPC)* program to provide affordable laptops and satellite-based Internet to remote areas
- *Rural Internet Connectivity System (PACRICS)* to bridge digital connectivity in underserved communities
- World Bank-supported *Pacific Regional Connectivity Project* to lay submarine cables and satellite ground systems.
- Pacific Islands Forum aims to improve Internet access, digital literacy, and ICT infrastructure, digital financial services, and e-governance to enhance inclusion and access to public services.

8.4 Summary

The digital divide remains a pressing challenge globally, disproportionately impacting rural, remote, and underserved populations. Efforts to bridge this gap require a combination of policy reforms, technological innovations, pilot trials, and targeted initiatives by governments, private organizations, and development agencies. Across continents, varying approaches have shown success in improving connectivity, whether through satellite technologies, PPP, community-led Internet solutions, or gender-focused initiatives. While urban areas experience rapid digital growth, rural regions still require strategic interventions, such as affordable Internet services, digital literacy programs, and infrastructure development, to ensure equitable access to digital opportunities. Collaborative, multi-stakeholder approaches are critical to addressing these disparities, fostering economic growth, and ensuring digital inclusion for all.

Exercises

8.1 Examine the success of community-led internet projects in Africa by analyzing:

(a) The key objectives and strategies employed in such initiatives to bridge the digital divide.
(b) The socioeconomic improvements observed in rural communities that benefited from these projects.

8.2 Explore efforts to reduce the gender digital divide in African regions. Discuss:

(a) Specific programs or initiatives aimed at improving access for women and girls.
(b) The broader societal changes resulting from these gender-focused interventions.

8.3 Analyze the impact of policy reforms in India on digital inclusion by discussing:

(a) Key policies implemented to enhance rural connectivity and their outcomes.
(b) The role of public-private partnerships in addressing the digital divide in rural India.

8.4 Evaluate the use of satellite internet technology in connecting remote Pacific Island nations. Discuss:

(a) How satellite technology has addressed the unique connectivity challenges of island regions.
(b) The impact on education, healthcare, and disaster management in these communities.

8.5 Investigate the role of CSR initiatives in bridging the digital divide in Europe. Analyze:

(a) Key CSR projects implemented by corporations to improve digital access.
(b) How these initiatives have supported marginalized communities in Europe.

8.6 Explore the development of community networks in Latin American countries. Discuss:

(a) How these networks have empowered local communities to build and maintain their own connectivity.
(b) The long-term social and economic benefits observed from these initiatives.

8.7 Analyze the digital inclusion strategies of Nordic countries by discussing:

(a) The policies and technologies adopted to ensure universal internet access.
(b) How these strategies can serve as a model for other regions aiming to bridge the digital divide.

8.8 Evaluate the effectiveness of public Wi-Fi initiatives in addressing digital inequality in the United States. Discuss:

(a) How public Wi-Fi programs have improved internet access in underserved urban and rural areas.
(b) The challenges encountered in sustaining and expanding these initiatives.

8.9 Examine the role of policy reforms and investments in improving connectivity in Israel. Discuss:

(a) Key government initiatives to expand digital infrastructure in rural and urban areas.
(b) The role of international partnerships in accelerating digital transformation in Israel.

8.10 Analyze global research efforts in understanding and addressing the digital divide. Discuss:

(a) The key findings from research on the impact of digital exclusion on rural communities.
(b) How these studies have influenced policies and projects aimed at digital inclusion.
c) Choose four international research projects and compare them to have a deeper understanding of the initiatives.

References

1. International Telecommunication Union (ITU), Universal and meaningful connectivity (UMC). Accessed 17 Feb 2025
2. P.C. Santos, M.B. de Espíndola, M.E.B. de Almeida, Connected education innovation program: an analysis in light of the meaningful connectivity concept. Revista Educação e Cultura Contemporânea **21**, 11780–11780 (2024)
3. UN ESCAP et al., Towards meaningful connectivity: insights from Asia-Pacific case studies, in *Economic and Social Commission for Asia and the Pacific* (2021)
4. R. Radhakrishnan, A. María Rodríguez Pulgarín, T. Woodhouse, *8. Enabling Women's Digital Participation: The Case for Meaningful Connectivity* (Amsterdam University Press, Amsterdam, 2023), pp. 113–126
5. Ceneva, BharatNet: India's ambitious project to bridge the digital divide. Accessed 17 Mar 2025
6. Statista, Most common languages used for web content 2025. Accessed 17 Mar 2025
7. S. Knight, Delivering the digital region: leveraging digital connectivity to deliver regional digital growth. Aust. Planner **52**(1), 4–15 (2015)
8. A. Sepúlveda, The digital transformation of education: connecting schools, empowering learners. TIC EDUCAÇÃO, 249 (2020)
9. N. Mulati, M.N. Aung, M. Field, E.W. Nam, C.M.H. Ka, S. Moolphate, H. Lee, Y. Goto, N.H. Kweun, T. Suda, et al., Digital-based policy and health promotion policy in Japan, the Republic of Korea, Singapore, and Thailand: a scoping review of policy paths to healthy aging. Int. J. Environ. Res. Public Health **19**(24), 16995 (2022)
10. W.E. Loges, J.-Y. Jung, Exploring the digital divide: Internet connectedness and age. Commun. Res. **28**(4), 536–562 (2001)
11. C. Ali, *Farm Fresh Broadband: The Politics of Rural Connectivity* (MIT Press, 2021)
12. I. Sinha, Cost transparency: the net's real threat to prices and brands. Harv. Bus. Rev. **78**(2), 43–43 (2000)
13. J. Cooper, K.D. Weaver, *Gender and Computers: Understanding the Digital Divide* (Psychology Press, 2003)
14. E. Jane, Gendered cyberhate: A new digital divide?, in *Theorizing Digital Divides* (Routledge, 2017), pp. 186–198
15. C. Ingham, Digital arrest logs on police websites in Pennsylvania. Master's thesis, University of Pennsylvania, 2022
16. C. Stryker, *Hacking the Future: Privacy, Identity, and Anonymity on the Web* (ABRAMS, 2012)
17. O. Kovalchuk, M. Shynkaryk, M. Masonkova, Econometric models for estimating the financial effect of cybercrimes, in *2021 11th International Conference on Advanced Computer Information Technologies (ACIT)* (IEEE, 2021), pp. 381–384
18. H. Hara, R. Komatsu, N. Shiota, Architecture for digital technology utilization to accelerate digital innovation. Fujitsu Sci. Tech. J. **54**(3), 39–46 (2018)
19. A. Vasiljevs, R. Vīksna, N. Vacheva, A. Lagzdiņš, Advancing digital language equality in Europe: a market study and open-source solutions for multilingual websites, in *Proceedings of the 25th Annual Conference of the European Association for Machine Translation (Volume 1)* (2024), pp. 600–609
20. L. Ran, K. Wang, Q. Zhang, The impact of digital transformation on enterprise capacity utilization under the background of "dual circulation". J. Knowl. Econ., 1–26 (2024)
21. A. Nallasivam, K.N. Anantha Subramanya Iyer, K. Desai, S. Kautish, S. Ghoshal, et al., Universal access to internet and sustainability: Achieving the UN's Sustainable Development Goals, in *Digital Technologies to Implement the UN Sustainable Development Goals* (Springer, 2024), pp. 235–255
22. J.C. Mankins et al., Technology readiness levels. White Paper **6**(1995), 1995 (1995)
23. P. Blanc, B. Ducastel, J. Cazin, M. Al Blooshi, S.S. Al Dhaheri, M.H. Al Marzooqi, E. Maneux, P. Ciret, M. Sow, J.-C.-C. Massabuau, First-time implementation of innovative in situ biotechnology on an offshore platform in Arabian Gulf for continuous water quality monitoring and early leak detection, in *Abu Dhabi International Petroleum Exhibition and Conference* (SPE, 2017), p. D011S023R001

24. A. Olechowski, S.D. Eppinger, N. Joglekar, Technology readiness levels at 40: a study of state-of-the-art use, challenges, and opportunities, in *2015 Portland International Conference on Management of Engineering and Technology (PICMET)* (IEEE, 2015), pp. 2084–2094
25. D.F. Polit, C.T. Beck, *Essentials of Nursing Research: Appraising Evidence for Nursing Practice* (Lippincott Williams & Wilkins, 2010)
26. R. Owen, N. Goldberg, Responsible innovation: a pilot study with the UK engineering and physical sciences research council. Risk Anal.: Int. J. **30**(11), 1699–1707 (2010)
27. O. Hazzi, I. Maldaon, A pilot study: vital methodological issues. Bus.: Theory Pract. **16**(1), 53–62 (2015)
28. R.L. Glass, Pilot studies: what, why and how. J. Syst. Softw. **36**(1), 85–97 (1997)
29. W. Jiang, B. Han, M.A. Habibi, H.D. Schotten, The road towards 6G: a comprehensive survey. IEEE Open J. Commun. Soc. **2**, 334–366 (2021)
30. T.J. Rowley, The power of and in stakeholder networks, in *Stakeholder Management* (Emerald Publishing Limited, 2017), pp. 101–122
31. L. Lobschat, B. Mueller, F. Eggers, L. Brandimarte, S. Diefenbach, M. Kroschke, J. Wirtz, Corporate digital responsibility. J. Bus. Res. **122**, 875–888 (2021)
32. T. Busch, Capabilities in, capabilities out: overcoming digital divides by promoting corporate citizenship and fair ICT. Ethics Inf. Technol. **13**(4), 339–353 (2011)
33. G. Windsor, How do drivers of human capital systems and corporate social responsibility affect sustainability of grassroots level knowledge-based organisations which exercise digital divide elimination as their main social responsibility? PhD thesis, UNSW Sydney, 2015
34. C. Holst, D. Stelzle, L.M. Diep, F. Sukums, B. Ngowi, J. Noll, A.S. Winkler, Improving health knowledge through provision of free digital health education to rural communities in Iringa, Tanzania: nonrandomized intervention study. J. Med. Internet Res. **24**(7), e37666 (2022)
35. J. Noll, Policy and regulations for sustainable and inclusive connect. Accessed 29 Oct 2024
36. T.A. Sawadogo, M. Gueye, A.-E. Sarah Victoire Emmanuel Costafrolaz, P.V.-M. Nguyen, B. Bounabat, I. Hamdouch, M. Dieng, *Mali-Digital Economy Diagnostic* (Policy Commons, 2023)
37. B. Samir, K.G. Caroline, O. Raymond, Bridging the gender digital divide: Africa's imperatives. Accessed 29 Oct 2024
38. UN Women, Bridging the digital gender divide in sub-Saharan Africa: Strategies for inclusion (2024). Accessed 28 Mar 2025
39. Women Enterprise Fund (WEF), Kenya, Empowering women through financial support and digital tools (2024). Accessed 28 Mar 2025
40. Ministry of ICT, Innovation and Youth Affairs, Kenya. Kenya National Digital Master Plan 2022–2032 (2024). Accessed 28 Mar 2025
41. Safaricom, MPESA: Driving financial inclusion in Kenya (2024). Accessed 28 Mar 2025
42. AfricaConnect, Africa connect. Accessed 29 Oct 2024
43. UNDP Nigeria, Empowering women through digital skills programs in Nigeria: strategies for inclusive growth (2024). Accessed 28 Mar 2025
44. World Economic Forum, Bridging Africa's digital divide: the rise of community internet (2024). Accessed 28 Mar 2025
45. Mawingu, Mawingu networks. Accessed 29 Oct 2024
46. T. Blair, Institute for Global Change. Bridging Rwanda's digital divide: smart Rwanda Master Plan, in *TBI Partnerships Case Studies* (2023)
47. ITU, Economic contribution of broadband, digitization and ICT regulation. Accessed 29 Oct 2024
48. T. Casey, I. Pswarayi-Riddihough, Moving Mozambique toward a digital future. Accessed 29 Oct 2024
49. United States Government, Digital transformation with Africa (DTA) (2022). Accessed 29 Mar 2025
50. FAO, Digital villages initiative in Africa. Accessed 29 Oct 2024
51. J.C. Panzar, Funding universal service obligations, in *Handbook of Worldwide Postal Reform* (2008), pp. 98–109
52. National Portal of India, Universal service obligation fund. Accessed 29 Oct 2024
53. Digital Regulation Platform, Indonesia's Universal Service Obligation Fund. Accessed 29 Oct 2024
54. R. Annapurna, How India is using the internet. Accessed 29 Oct 2024
55. B.S.S. Kumari, P. Lavanya, P. Padmambika, Bridging the digital divide: addressing inequality and access disparities. Int. J. Indian Psychol. **12**(3) (2024)
56. S. Jaikaran-Doe, I. Hay, A. Fluck, D. Moltow, A Trinidad and Tobago case study of one response to the equity challenge in education within the digital era, in *Education and the UN Sustainable Development Goals: Praxis Within and Beyond the Classroom* (Springer, 2023), pp. 223–242
57. S. Uppuluri, *Bharatnet: The Way Forward* (Indian Institute of Management Bangalore, 2020)
58. A. Bajpai, *Rural wireless deployments in India* (2019), arXiv preprint arXiv:1909.08388
59. L. Zou, J.H. Shen, J. Zhang, C.-C. Lee What is the rationale behind China's infrastructure investment under the belt and road initiative. J. Econ. Surv. **36**(3), 605–633 (2022)
60. Y. Yao, H. Zhang, X. Liu, X. Liu, T. Chu, Y. Zeng, Bridging the digital divide between old and young people in China: challenges and opportunities. Lancet Healthy Longevity **2**(3), e125–e126 (2021)
61. S.Z. Yu, All under Huawei: China's new vision for a Tech Sinica, in *Huawei Goes Global: Volume I: Made in China for the World* (2020), pp. 87–114
62. I.Y.H. Ng, S.S. Lim, N. Pang, Making universal digital access universal: lessons from COVID-19 in Singapore. Univers. Access Inform. Soc. **22**(3), 1073–1083 (2023)
63. K.V. Bhanu Murthy, A. Kalsie, R. Shankar, Digital economy in a global perspective: Is there a digital divide? Transnational Corporations Rev. **13**(1), 1–15 (2021)
64. A. Saad, A. Fahim, Access digital divide in the Kingdom of Saudi Arabia (KSA): current state and measures for improvement. Int. J. Inform. Eng. Electron. Bus. **12**(3), 13 (2021)
65. R. Rahman, A. Qattan, Vision 2030 and sustainable development: state capacity to revitalize the healthcare system in Saudi Arabia. INQUIRY: J. Health Care Organ. Provision Financing **58**, 0046958020984682 (2021)
66. S. Muzafar, N.Z. Jhanjhi, Success stories of ICT implementation in Saudi Arabia, in *Employing Recent Technologies for Improved Digital Governance* (IGI Global, 2020), pp. 151–163
67. R.S. Sharma, I.A. Mokhtar, D.N. Ghista, A. Nazir, S.Z. Khan, Digital literacies as policy catalysts of social innovation and socioeconomic transformation: interpretive analysis from Singapore and the UAE. Sustain. Soc. Dev. **1**(1), 2167 (2023)
68. Etisalat Corporation, Satcoms. Accessed 29 Oct 2024
69. Government of UAE, Digital inclusion. Accessed 29 Oct 2024
70. UAE Ministry of Education, Digital transformation journey in education. Accessed 29 Oct 2024

71. A. Gomes, J.G. Dias, Digital divide in the European Union: a typology of EU citizens. Soc. Ind. Res., 1–24 (2024)
72. M. Feurich, J. Kourilova, M. Pelucha, E. Kasabov, Bridging the urban-rural digital divide: taxonomy of the best practice and critical reflection of the EU countries' approach. Eur. Plan. Stud. **32**(3), 483–505 (2024)
73. V. Bilozubenko, O. Yatchuk, E. Wolanin, T. Serediuk, M. Korneyev, Comparison of the digital economy development parameters in the EU countries in the context of bridging the digital divide. Probl. Perspect. Manag. **18**(2), 206 (2020)
74. European Commission, Connecting Europe Broadband Fund (CEBF): supporting high-speed broadband deployment in underserved areas. Accessed 29 Mar 2025
75. European Commission, Digital Decade Policy Programme 2030: advancing digital skills, infrastructure, and inclusion. Accessed 29 Mar 2025
76. N. Mar, Bridging the digital divide in the EU. Accessed 29 Oct 2024
77. B. Christine, Opinion bridging the digital divide: a call for European unity. Accessed 29 Oct 2024
78. UK Government, Digital development strategy 2024–2030. Accessed 29 Mar 2025
79. Greater Manchester Combined Authority, Get online greater Manchester. Accessed 29 Mar 2025
80. M.R. Szeles, M. Simionescu, Improving the school-to-work transition for young people by closing the digital divide: evidence from the EU regions. Int. J. Manpow. **43**(7), 1540–1555 (2022)
81. M. Pérez-Castro, M. Mohamed-Maslouhi, M. Montero-Alonso, The digital divide and its impact on the development of Mediterranean countries. Technol. Soc. **64**, 101452 (2021)
82. International Telecommunication Union, Facts and figures 2023. Technical report, International Telecommunication Union (2023). Accessed 27 July 2025
83. The World Bank, *Individuals Using the Internet (% of Population)* (2023), https://data.worldbank.org/indicator/IT.NET.USER.ZS. Accessed 27 July 2025
84. The White House, Bipartisan Infrastructure Law. Accessed 29 Mar 2025
85. Federal Communications Commission, Affordable Connectivity Program. Accessed 29 Mar 2025
86. Government of Canada, Universal Broadband Fund. Accessed 29 Mar 2025
87. Government of Canada, Digital Literacy Exchange Program. Accessed 29 Mar 2025
88. Government of Brazil, National Broadband Plan. Accessed on 29 Mar 2025
89. Telebras, Satellite program. Accessed 29 Mar 2025
90. Government of Mexico, Internet para Todos. Accessed 29 Mar 2025
91. Red Compartida Project, Red Compartida Project. Accessed 29 Mar 2025
92. Government of Argentina, Plan Conectar. Accessed 29 Mar 2025
93. ARSAT, Argentine Satellite Solutions. Accessed 29 Mar 2025
94. Government of Colombia, Vive digital. Accessed 29 Mar 2025
95. Government of Colombia, Public Wi-Fi zones. Accessed 29 Mar 2025
96. Plan Ceibal, Plan Ceibal. Accessed 29 Mar 2025
97. Government of Uruguay, Digital literacy programs. Accessed 29 Mar 2025
98. Australian Government, Better connectivity plan for regional and rural Australia. Accessed 29 Oct 2024
99. A. Marshall, C.-A. Wilson, A. Dale, Telecommunications and natural disasters in rural Australia: the role of digital capability in building disaster resilience. J. Rural. Stud. **100**, 102996 (2023)
100. S.S. Zheng, Remote first nations communities are among the most digitally excluded people in Australia: report. Accessed 29 Oct 2024
101. Digital Govt NZ, Digital inclusion and wellbeing in New Zealand. Accessed 29 Oct 2024
102. Ituani Centre, Ituani Centre: empowering rural communities in Vanuatu through ICT. *Ituani Community Reports* (2015)
103. Government of Papua New Guinea, National ICT policy, connect PNG initiative, and universal access scheme (2022). Accessed 24 Mar 2025

Empowering Digital Futures

9

Digital access is no longer a luxury but a fundamental enabler of participation in the modern economy, education, and governance. The development and deployment of technologies aimed at bridging the digital divide requires a holistic approach. It is shaped by the interplay between technology, policies, regulatory frameworks, standards, and community engagement. The effectiveness of technology depends not only on its design but also on inclusive policies, regulatory frameworks, and interoperability standards; and these elements play a critical role in reducing deployment costs and enhancing accessibility, especially in underserved regions. Thus, technology, policy, regulation, and standards must align cohesively to drive affordable and sustainable digital inclusion, reinforcing an inclusive digital environment and ecosystem for underserved communities.

Governance frameworks comprising inclusive policies, robust regulatory mechanisms, and interoperability standards play a vital role in bridging the digital divide by reducing deployment costs and enabling equitable digital access for all. This chapter explores how such frameworks foster accessibility through effective regulation and standardization. Through this lens, the reader would appreciate the need for a holistic approach where the stakeholders work together in providing affordable, accessible, and available digital access for all globally.

9.1 Policies and the Digital Divide

Policymakers are among the most important stakeholders in the effort to close the digital gap. They recommend policies that other stakeholders must adhere to. We need to understand the role policymakers play in bridging this gap.

9.1.1 What Is Policymaking?

Policymaking is a complex process involving a wide range of individuals and entities beyond politicians and civil servants actively participating in the development and shaping of the policy through dynamic networks that change with external conditions and political situations across the globe [1]. Some other key organizations and entities involved in policymaking include government agencies, quasi-governmental bodies, government think tanks, lobbyists, consultants, and academic researchers. Each of these organizations and entities plays a unique role in influencing, shaping, and implementing the policy. So, how do these stakeholders contribute to policymaking?

Policies are plans, courses of action, recommendations, or procedures that are intended to influence the decisions and provide guiding principles [2]. Policies are made by elected bodies that make fundamental decisions on how to achieve an intended target while adhering to regulations and standards set at different levels. Decision-making is a part of policymaking and has a dynamic relationship with it. The key purpose of policy development is to ensure that appropriate steps are taken toward:

- addressing social, economic, and environmental challenges, such as poverty, illiteracy, crime, climate change, public health crises, clean water, and an aging population,
- ensuring fairness and equity for all in accessing the resources and opportunities, such as digital connectivity, employment, voting rights, and business practices,
- providing clear guidelines for the decision-makers,

S. Koratagere Anantha Kumar, *Wireless Broadband Access to Bridge the Digital Divide*,
Textbooks in Telecommunication Engineering, https://doi.org/10.1007/978-3-032-09261-8_9

- managing risk and ensuring the safety of individuals, organizations, and systems,
- promoting stability and order in the country and across the globe to ensure smooth operations in governance, business, and public life,
- facilitating accountability for the behavior and performance of individuals and organizations, and
- encouraging innovation and progress across different sectors to make a positive difference in the lives of the people.

Each stakeholder defined policies on their own, but they convey a similar meaning. The general guiding principles for professionals who intend to contribute to public policy include [3]:

- **First Principle**: In a democratic society, the policymakers should not be biased toward any community or some segment of society but rather take a large view and value the interests of the whole society.
- **Second Principle**: The policymakers should focus on practical and achievable solutions rather than unrealistic and partial alternatives. They need to evaluate realistic options that align with the current societal values, existing institutions, and political feasibility.
- **Third Principle**: The effective policy inquiry should address the needs and concerns of the decision-makers; that is, it should provide insights that are useful and actionable for policymakers. Additionally, the policy research should present clear, evidence-based insights to those in power, even if the insights are challenging or uncomfortable. This would aid the policymakers in making an informed decision.
- **Fourth Principle**: The analysts and researchers work closely with the decision-makers to investigate specific policy questions and ultimately provide clear recommendations for action. Real-world policy decision-making is influenced significantly by external pressures, evolving conditions, and unforeseen complexities.

Achieving impactful policies requires persistence, adaptability, and continuous alignment with the policy question of interest to ensure the drive for positive change. The policy studies need to emphasize clear, concise, and relevant findings and the critical nature of honesty about research limitations. Often, it is hard for policymakers to know in advance what will work and what won't; because of this, there is an increasing dependency on building evidence and evaluation into the policymaking process for its success [4]. Therefore, there is a focus shifting toward an evidence-based policymaking process.

9.1.1.1 Formulation of Policies

Policymaking is both a structured, cyclical process and a dynamic, evidence-based approach. Policymakers tend to be approached by a large number of people who wish to influence their policies—citizens, institutions, and interest groups. Policy design is a rational, systematic approach designed to support decision-making in the public sector that involves generating information about the potential consequences of different policy choices to help policymakers develop an informed decision. The process should be transparent and empirically based on the existing knowledge, and results should be reproducible for accuracy. The policies are designed to encourage and support the growth of individuals and businesses that meet a set of realistic targets.

Figure 9.1 shows the steps involved in policy analysis that help in drafting the recommendations for the policymakers. The order of the steps could be changed depending on the policy analysis, and generally, there is feedback from one stage to another; most of the projects involve a subset of these steps [5]. At its foundation, policy design encourages clarity in objectives, identification of constraints, cost-benefit trade-offs, and long-term impact assessments. The process includes defining the problem, setting measurable objectives, identifying evaluation criteria, analyzing alternatives, and implementing and monitoring the chosen policy. Feedback mechanisms ensure that policies remain adaptable to changing conditions.

However, modern and innovative policymaking is increasing calls for integration with evidence-based research to address the needs [6, 7]. Evidence-based policymaking refers to the method of policy development that consults facts, assesses credible and relevant evidence, and builds scientific models to aid the decision-making process, rather than using theory and political views [4, 8]. To guarantee fair recommendations that are driven by the needs of the country or community rather than the desire to win over voters, all evidence-based policy studies should be objective. The stakeholders can create a thorough model that replicates real-world situations by cooperating. To improve science-policy interactions and, eventually, policy frameworks, it is essential to incorporate data on contemporary societal issues with the intricacies of scientific procedures [7].

Figure 9.2 shows the steps involved in evidence-based policy-making [9]. The steps involved here are slightly different from the generic policy analysis discussed earlier—recognize the problem, create demand for evidence, collate and assess the evidence, and deliver change and recommendations. A common barrier to effective, evidence-based decision-making

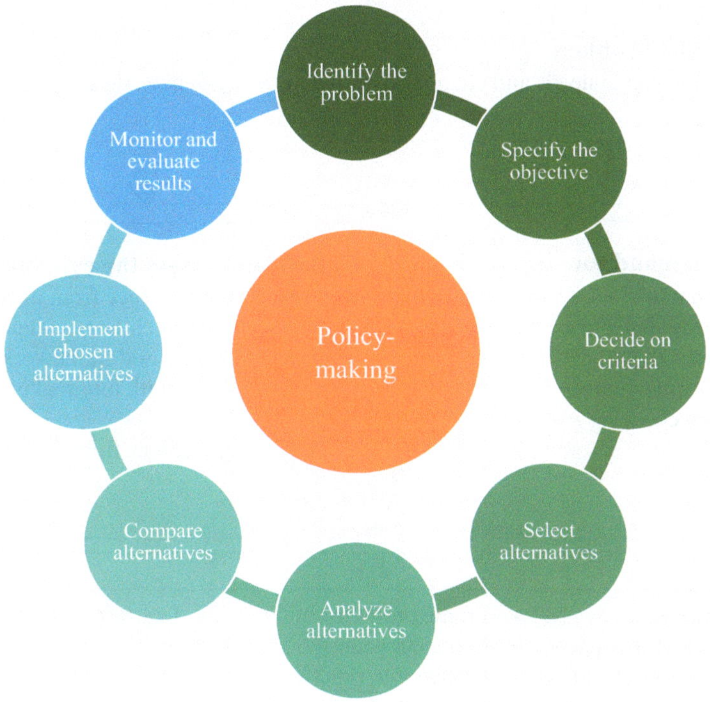

Fig. 9.1 Policies formulation cycle

is the inability of many processes and systems to integrate the available relevant evidence [9]. The data for the evidence-based studies is derived from trusted and reliable sources, including published literature (books or journals), manuals and reports (grey literature), personal experience, or information received through word of mouth from colleagues, peers, expert advisors, or local knowledge [9]. Evidence-based policy analysis is particularly valuable when the anticipated impact of the policy is significant. In such scenarios, reliable evidence helps understand and minimize uncertainty in both outcome and decision-making, especially when the investment scale or implementation timeline is significantly large [10].

Once the scene setting is done, there is a need to think from the decision maker's point of view to assess the economic, societal, and political benefits offered by a particular policy. Additionally, there is a need to consider financial benefits for the individuals. There is also a need to identify where the evidence can make a significant difference to take informed decision based on the success of previous policy implementation to encourage the participation of credible organizations, political support, and interventions. To ensure that the policy development is addressing the intended problem, there is a need to break down the assessment into small, achievable targets.

The success of the digital divide policies is measured through a combination of qualitative and quantitative metrics, as shown in Fig. 9.3 [11, 12]. The metrics are always helpful in comparing multiple potential policy reforms to the base case policy. Policymakers would use this data-driven, evidence-based strategy to make informed decisions about reducing the digital divide further and assess the suitability of policy reforms. The potential solution would depend on the location of interest, the demand-supply parameters, and the unique challenges of the particular area.

The metrics for assessing the impact of the policy would help to study the real and recognizable progress toward the goals of the digital divide and sustainability [11, 12]. In some cases, extra effort would be required to estimate the metrics, and in others, all the necessary data would be readily available. This is used across all sectors and around the globe to help policymakers, decision-makers, and investors understand the direction of the ongoing work. This assessment would study the impacts in terms of technological, economic, environmental, and social settings. Reports would also include potential future targets for each indicator.

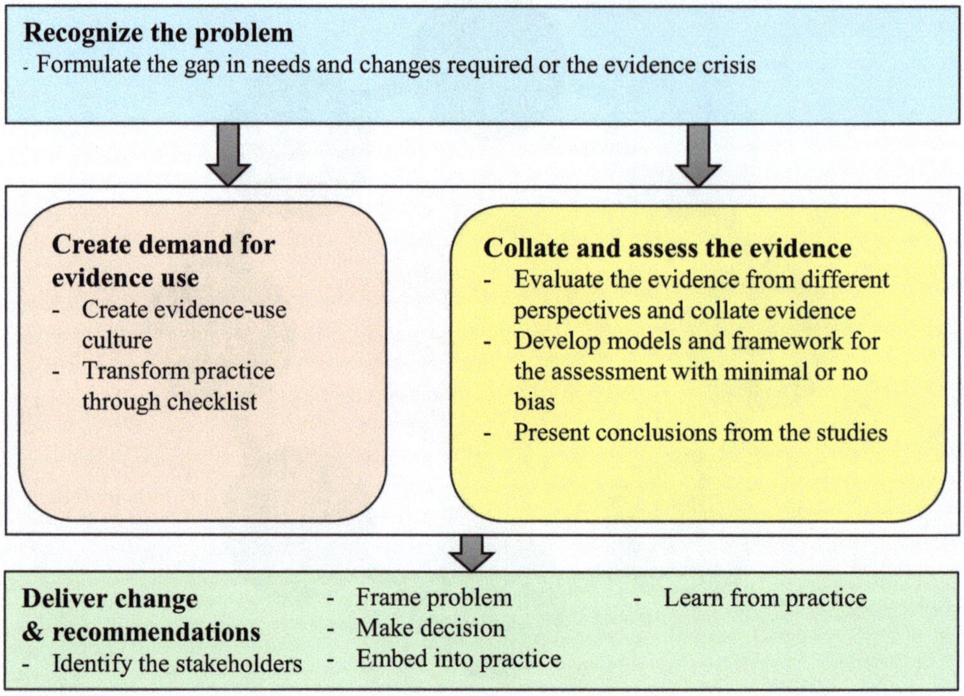

Fig. 9.2 Evidence-based policy-making

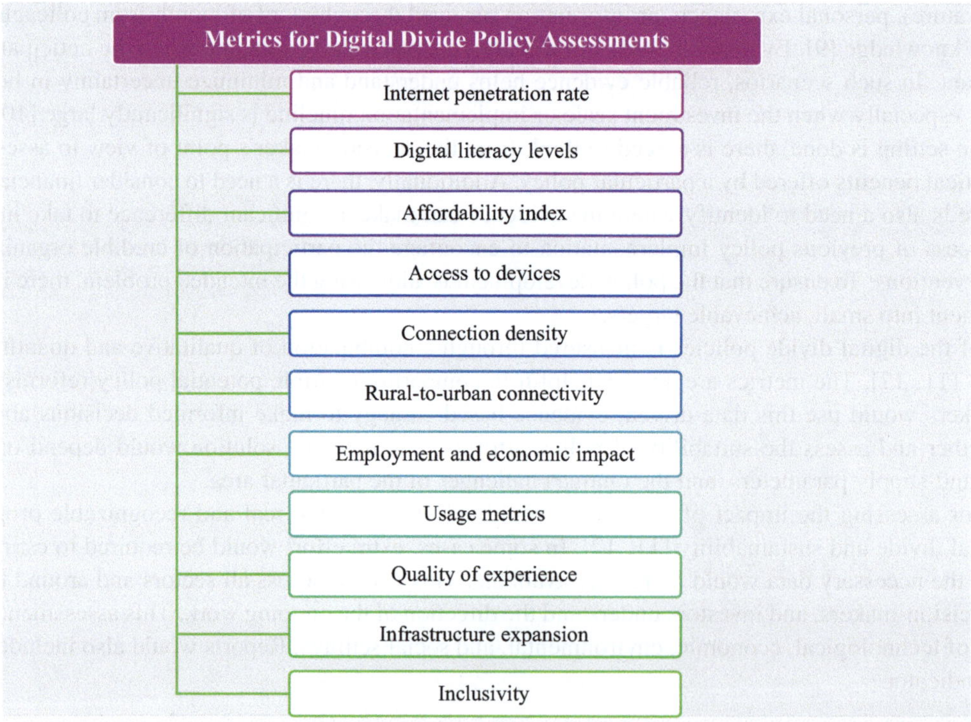

Fig. 9.3 Measurement of policy effectiveness

9.1.2 Policy for Bridging the Digital Divide

There is a need to explore the role policy and policymakers play in bridging the digital divide by developing and implementing long-term solutions to bridge the access gap across all age groups and strive for sustainable goals. This can be accomplished when rural policies are created with the user's needs—inclusive, affordable, secure device access; infrastructure availability; digital literacy; language hurdles; and the broader welfare of people, as well as the inclusion of marginalized communities. Some of the key policies suggested that are aimed at bridging the digital divide gap are [13]:

- **Infrastructure development**: Governments and organizations across the globe are working to expand broadband infrastructure in remote and rural areas at subsidized rates. For example, the *BharatNet* initiative in India and the *National Broadband Network (NBN)* initiative in Australia aim to enhance Internet speed access for all citizens as well as provide a Universal Service Guarantee to ensure that all their citizens have access to broadband services. One of the potential strategies that would lower the cost of Internet services and make it affordable for the masses would be the **building of community networks** [14]. Some key policies and reforms were initiated to support new entrants and foster competition from newer entrants, for example:

  - **Spectrum Allocation Reforms**: Many countries are now adopting transparent spectrum allocation processes that are fair for all the players to ensure that the new entrants are capable of accessing affordable resources without facing significant competition from the incumbents [15].
  - **Infrastructure Sharing**: There has been a growing demand to use open networks and permit different forms of infrastructure sharing at different levels across the globe to lower the overall costs in service provisioning [16].
  - **USOF**: Governments have expanded the USOF provision of connectivity in underserved regions by subsidizing newer initiatives and new entrants to bridge the digital connectivity in these regions [17]. EU has recognized the potential of ICT transforming the economic, social, and political lives of individuals. The key elements of universal service obligation (USO) in the telecommunication sector are availability, affordability, and accessibility [17].
  - **Consumer Protection and Transparency**: There is a need to protect consumer data and enlighten the users about digital safety habits [18]. The regulatory bodies in the USA, EU, UK, Japan, Australia, and several other countries have introduced measures to ensure transparency in service agreements and data collection policies. These policies are also beneficial for the consumers when the consumer data is being used to train AI without their knowledge or their data being sold for the targeted market research and firms [19]. This ensures that the firm is accountable for consumer data protection.

- **Affordable Access Program**: These policies aim to subsidize Internet access costs by providing affordable devices and service plans to users with low income [20]. Some of the policies developed by the policymakers to bridge the affordability gaps focus on subsidizing Internet plans that are either low or free for households with low incomes [21]. These plans include sufficient data for digital pedestrians who would require Internet of text services for job searching, education, and healthcare. Additionally, some policies also provide low-cost devices with Internet capabilities to encourage and enable users to connect to the Internet. Several countries across the globe are providing free public Wi-Fi hotspots in cities and rural areas to encourage digital inclusion. Additionally, the government policies also encourage rural deployment by community networks and operators by offering tax incentives for the service providers [22]. These programs also help in cultivating digital literacy skills and create awareness among the specifically targeted users about the benefits of being connected. These policies aim in building a sustainable ecosystem while fostering a self-sustainable business environment. Expanding and reforming subsidies is also vital to meet the demands of the market today. Policymakers would also need to determine where to find the funds to compensate for this additional subsidy that is anticipated [14].
- **Digital Literacy and Education**: The policies addressing usage gaps should ensure that everyone, regardless of age, background, gender, nationality, and circumstances, can access and utilize digital tools [23]. Some of the key initiatives based on the policies in this domain include digital literacy programs by governments, NGOs, and CSR initiatives that aim to teach people who can't afford to pay for digital literacy training and also people living in underserved areas. The target for the digital divide policies is to ensure that the next generation is equipped with digital literacy skills; this is possible by teaching students about and how to utilize information and communication technologies (ICT) from a young age. Educating individuals on how to use digital services safely and confidently is another goal of the policies.

The policies also emphasize the establishment and development of community learning centers that are Internet-connected and provide low- or no-cost computing facilities, and this type is popular throughout Europe and Africa [24, 25]. Furthermore, training teachers on how to use digital services to reach more students would assist students in improving their digital literacy skills [26]. Targeted policies for marginalized communities play an important role in digitally empowering people who are disproportionately affected by the digital divide.

These policies are being explored globally; however, there are several barriers faced in their implementation, which will be discussed in Sect. 9.4.

9.2 Regulations and the Digital Divide

9.2.1 What Is Regulation?

Regulations are rules or frameworks implemented via laws established by regulatory bodies to manage and control the behaviors of firms to ensure fairness, safety, and efficiency in terms of cost and service provided or goods sold [27]. It is impossible to study an important market worldwide without considering the impacts of constraints on the behavior of the firms due to one or more regulatory norms. A completely unregulated market is not viable and practical, and a regulated market has its own set of issues that must be addressed, which involve evaluating different regulatory instruments, regulatory processes, and the extent of regulation. These factors further need to consider the distribution of costs and benefits throughout the population arising from a particular set of regulatory activities. Furthermore, regulations should promote progress rather than impede economic expansion and the well-being of individuals and businesses.

Regulations are different compared to policies; regulations are a dedicated task for developing practical rules of the everyday market to protect the firms and consumers in a large marketplace [28, 29]. Policies are broad statements that outline the intent and focus on wider aspects. Meanwhile, regulations are specific, enforceable bylaws. The regulators need to ensure that the markets are stable for the investors, irrespective of the political changes across the globe [27, 30]. There are three main studies of regulation, whether theoretical or empirical [27], as follows:

- price and entry regulation in industries with a competitive market structure,
- price and entry in monopolistic industries and "qualitative" regulation that models market-failure problems linked indirectly to price, profits, and market structure, and
- environmental, health, occupational safety, and product-quality regulation.

9.2.1.1 Formulation of Regulations

Figure 9.4 describes the steps involved in the formulation of regulation [31]. The government agencies, organizations that influence regulations, or other regulating bodies identify a societal need or a gap that needs to be addressed by either reforming an existing regulation or creating a new one. Policies are high-level guiding principles that assist with compliance, promote operational efficiencies, and enhance the mission of addressing the gap. The regulatory bodies draft detailed rules and frameworks that would ensure the smooth implementation of policies to ensure the policy goals are achieved (note that not all policies get converted into regulations). To ensure that the regulations drafted are realistic and achievable, the regulatory bodies consult the stakeholders, such as industry representatives, R&D organizations, civil society, the public, and independent consultants. The stakeholders provide their responses for this consultation and review the views to refine the rules and framework.

Once the consultation responses are reviewed, there is a need to review the legal compliance with the existing law, and there is no conflict with the existing rules and regulations governing related matters before putting them forth in front of the government, and they would collaboratively approve and adopt the regulations. Once adopted, the regulation is implemented, and it would involve creating the necessary infrastructure, processes, policies, and compliance for the impacted firms and businesses. Regulatory bodies also monitor and enforce rules to ensure compliance with the regulations laid down. After a certain duration, the policies are reviewed to ensure that they are still valid and address the market needs.

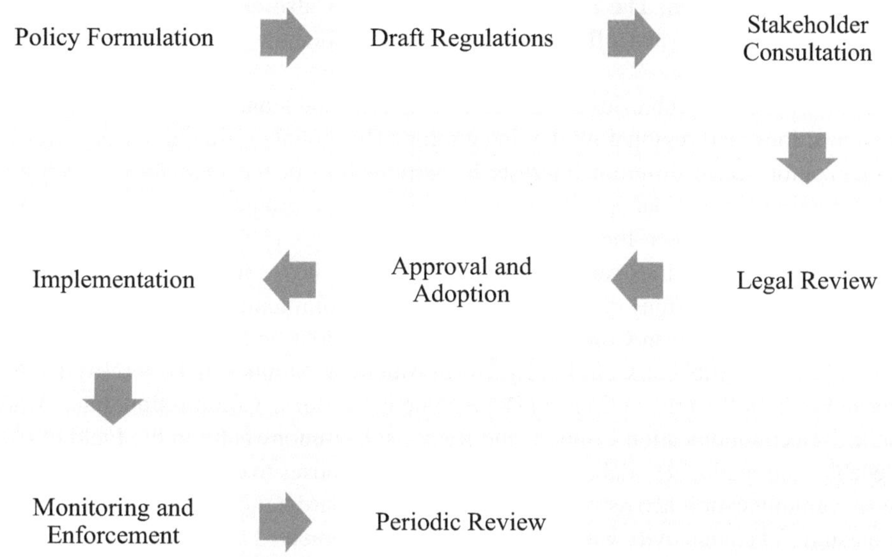

Fig. 9.4 Regulation formulation

9.2.2 Regulations for Bridging the Digital Divide

The telecom sector is a capital-intensive sector, and therefore, historically, telecommunication networks began as primarily monopolies in many countries, with a single entity controlling, deploying, operating, and providing services to the customers, especially with the rollout of landline services [32]. Although the industry has developed over time to accommodate technologies with advanced capabilities, it nevertheless exhibits monopolistic traits with regard to specific aspects, including infrastructure, connectivity to PoP, and spectrum allocation. In the telecommunication sector, regulations play a pivotal role in helping the telecommunications network become competitive, especially in the usage of spectrum, protection of incumbent users, market competition, pricing strategy, addressing market failures, and protecting the interests of both consumers and firms [33]. TSPs, equipment manufacturers, and other communication-related firms operate in competitive market structures, necessitating price and entry regulation to maintain fair competition. Additionally, the regulations ensure that the firms maintain minimum standards and quality for the products and services that they offer to the customers. These days, there is an additional focus on other aspects, including cybersecurity, service reliability, resilient networks, and consumer protection. The telecommunication regulations are thus a multi-faceted approach that is dependent on specific market structure and context.

Regulations in the telecom industry refer to a set of rules and guidelines that govern the operations and conduct of telecommunication companies, which are also applicable to any stakeholder who wishes to test and experiment with telecommunication network capabilities at ground, space, and air layers [34]. Several telecommunication regulatory bodies supervise and monitor telecommunication firms to ensure that firms abide by the requirements and rules. These regulatory bodies ensure that telecommunication firms provide and adhere to minimum QoS, privacy, standards, and practices [34].

Regulations are laid out at international, national, local, and regional levels to ensure smooth operations of businesses across the globe while supporting growth at each level [35]. There is a need to focus on regulations at different levels because the market is highly dynamic and interconnected, as discussed below:

- **Localized Level**: This addresses the challenges faced by the localized marketplace and community-specific issues, such as rural areas, infrastructure deployment, right-of-way issues, and local zoning for telecommunication towers. This ensures that digital infrastructure and its transformation are inclusive and accessible for all. These regulations are applicable to community networks and LAN deployments.

- **National Level**: This regulation at the country level sets rules for spectrum management, operational parameters, market competition, licensing, consumer protection, spectrum trading, spectrum sharing, infrastructure sharing, pricing models, legal procedures, and dispute resolution. The national-level regulations also ensure that operators deploy infrastructure and services in remote and rural areas and offer national roaming under the USO to promote digital inclusion in the telecommunication deployment.
- **Regional Level**: At this level, several countries collaborate to develop and standardize laws to guarantee roaming agreements, cross-border connections, and regional market integration. This collaboration happens with neighboring countries as well. Regional organizations have different interests in participating in the regulations, such as regional or global mandates to improve regulations, carry out specific regulatory functions, promote regulatory reforms, provide technical assistance, exchange information between the regulators and stakeholders in the regulatory process, and fund consulting studies, workshops, and other activities related to regulatory know-how [36]. Some selected regional organizations interested in telecommunication regulations include the East African Community (EAC), GCC, Southern African Development Community (SADC), and Andean Community, or economic blocs, such as European Union (EU), Asia-Pacific Economic Cooperation (APEC), Caribbean Community, North American Numbering Plan (NANP), African Development Bank (AfDB), European Bank for Reconstruction and Development, European Commission, Inter-American Development Bank, Mercosur, Pacific Telecommunication Council, and Regional Commonwealth in the Field of Communications.
- **Global Level**: This enables international collaboration across 180+ countries to ensure global interoperability of networks and services, seamless communication across broader equipment standardization, and the development of global broadband access to offer meaningful connectivity while connecting the unconnected [32]. Overseeing the domain name system (DNS), Internet Corporation for Assigned Names and Numbers (ICANN) makes sure that all Internet-connected devices communicate with each other seamlessly. A ITU-organized conference called WRC offers recommendations for worldwide spectrum harmonization to promote coordinated wireless technology use and minimize inter-network cross-border interference. WTO Trade Agreement is another well-known international institution that oversees and controls fair competition and nondiscriminatory market access for telecom access worldwide and guarantees seamless cross-border services. The World Bank is an international financial institution that is interesting to support regulatory reforms by providing funding, technical assistance, and policy advice to countries globally.

Telecommunication regulations face several challenges at any point in time. There is a need to update the regulations often to match the market demand and the changes in the telecommunication ecosystem [37]. As of 2025, there have been rapid advancements in technologies, market dynamics, and developmental goals across the globe, which are driving the need for reforms and newer regulations. Some of the key issues that need to be addressed over the next few years include regulating emerging technologies and future networks that were discussed in Chap. 7, efficient spectrum allocation of radio frequencies, coexistence of different technologies, bridging the digital divide, OTT services and net neutrality, network resilience, global harmonization, financial strain that hinders innovation and QoS, and affordability aspects [38]. Non-regulatory factors, such as trust among participating firms, also play a crucial role in the success and effectiveness of regulations [39]. The telecom firms must have a comprehensive understanding of these requirements and implement the necessary steps to comply with the regulations. Telecommunication regulations are quite complex, and are updated regularly to reflect new technological advancements. Therefore, telecom firms must keep abreast of the most recent regulations in order to avoid paying penalties, causing damage to their brand, and facing legal consequences [33]. This is quite a challenge for smaller telecom service providers (TSPs).

9.2.2.1 What Is Regulation in Bridging the Access Gap?

Telecommunication regulations play a crucial role in bridging the digital divide and ensuring that broadband services are available, reliable, and affordable for all. Some of the regulations that contribute toward minimizing this gap are access in rural areas, affordable services, encouraging different technologies and innovation that cater to rural and remote deployments, fostering digital inclusion policies, empowering marginalized communities, and strengthening consumer protection. These regulations involve different stakeholders collaborating to ensure the overall economic development and personal welfare of people living in underserved regions.

Tackling the digital divide requires innovative approaches across multi-dimensions to tackle issues including electricity access, education, affordability, pricing strategies, inclusive business models, service plans, infrastructure sharing, and related market competition. In addition to foundational infrastructure and affordability concerns, advanced digital equity includes training in digital skills, implementation of digital inclusion programs, development of digital competencies, accessible

Table 9.1 Roles of different organizations

Function	Responsible organization
Policy development	Government ministry or executive branch
Regulation	Separate regulatory authority
Network operations and service provisioning	Commercial and private service provider

content platforms, and protecting consumers from unfair pricing, data privacy violations, and subpar QoS. Providing free digital access for digital pedestrians to raise their awareness and learn digital skills, helps in fostering essential skills.

Effective policies and regulations developed from the study would help empower individuals, businesses, and the entire world to reap the benefits offered by digital connectivity and its transformation. Though the need for regulations varies with the conditions of the marketplace, however, all regulatory frameworks include certain critical elements that are common across all for an efficient framework, such as functionality aspects of regulatory authority, decision-making processes, accountability, consumer protection, dispute resolution, and enforcement powers

For a market-based supply model, Table 9.1 shows the role of different organizations and has the features compliant with WTO Regulation Reference Paper, where the regulator is separate from the telecommunications operator, and that can resolve interconnection disputes [36]. The roles of different organizations are

- **Government officials** recommend and set policies in the national interest that promote higher competition among the firms (incumbent service providers and new entrants as well) [36].
- **Regulatory bodies** are an independent body that implements the policies recommended by the government and working bodies in an objective and neutral manner to be impartial toward all firms, especially when it comes to market competition policy, interconnection, roaming, cross-border policies, and pricing models.
- **Commercial and Private Service Providers**: They drive the industry network expansion, service innovation, and technological advancements by investing in infrastructure, R&D, and supporting future networks to improve the QoS and expand connectivity. They also provide a range of services from voice to data to the subscribers and end users, catering to personal and business needs. The competition among the operators encourages affordable prices, improved QoS, and broader use cases to enhance digital transformation.

Bridging the digital divide would need efforts by all three organizations, contributing in their own way. For example, governments play a crucial role in developing and implementing national broadband strategies, allocating funds for rural broadband projects, and implementing digital literacy programs in rural areas [40, 41]. Similarly, the regulatory bodies establish rules for fair competition, ensure optimized usage of USOF for underserved communities, and promote infrastructure sharing to reduce deployment costs in rural areas while ensuring that it doesn't hinder urban deployment [25, 42]. Along the same lines, the role of telecommunication firms involves developing equipment and user-end devices, deploying infrastructure, and investing in R&D developing newer technologies to bridge the digital divide gap and expand connectivity [43, 44].

9.3 Standards and the Digital Divide

Standardization is the critical step to ensure the scalability, minimum QoS, and worldwide interoperability of the technology. An agreement among several agents to do certain vital activities by established roles is known as a standard [45–47].

9.3.1 What Are Standards?

Technological standardization refers to the process of development of technical standards and their global implementation to ensure consistency, reproducibility, and quality of products, systems, and services [48–50]. This is a crucial step to ensure interoperability and safety among different equipment, reduce the cost of developing and commercializing the technology or products, and foster innovation [51]. Standards allow different products and systems to interconnect with each other to form a larger ecosystem, irrespective of the manufacturing brand and model.

Standards play a vital role in ensuring technological convergence that is developed using a great number of intellectual properties. Global technology standards of wireless communication industries lay the foundation for a transparent, competitive ecosystem open for all and provide a continuous roadmap for evolution [52]. These standards help in the development to commercialization of the product, system, or service to the mass commercial market, contributing value to the larger ecosystem. The evolution of technology standards is shaped by macro- and micro-level forces, as it is driven by the ambitions and the actions of firms and industry associations [47]. There is a need for supply-side as well as demand-side standardization for overall technology standardization [47], as discussed below:

- For **supply-side** standardization, the producers, manufacturers, and developers who create the technological products and services establish best practices and tested solutions that are reproducible and achieve the desired results. This standard is based on a proven framework, principles, and rules that guide components to interact at different network layers and interconnect with a wider network ecosystem. This ensures that the data and logical controls flow seamlessly within the system.
- For **demand-side** standardization, there is a need to focus on consumers and their needs, which include uniformity, compatibility, ease of use, integration with the existing products, and interchangeability with different products. Standards should also help in developing vendor-independent devices so that consumers can avoid being locked into proprietary systems, offering flexibility and versatile choices. The standards should also emphasize how they can replace end users' devices with newer models in less time.

Therefore, global technology standardization should balance both the supply-side and demand-side standards. This is especially true for widespread, globally deployed technologies that adhere to a diverse landscape of regulations and policies across jurisdictions. Additionally, the balance factor consists of political, social, and economic factors on one side and supply-side and demand-side standards and manufacturing cost structures on the other [47]. The key benefits offered by technology standards include [52]:

- ensure system interoperability while enabling product differentiation,
- create newer markets for the technology products and expand the usage of existing technologies in newer markets,
- improve cost-efficiency by encouraging global standards and multi-vendor sourcing,
- reduce market risk in capital-intensive areas,
- improve quality and reliability, and contribute to the regulation development,
- enhance the technology continuously by collaborating with different organizations and standardizing bodies,
- understand the time-to-market products by combining standards with product development,
- opportunity to create and license valuable intellectual property and develop a solid framework to build a strong ecosystem for the technology, and
- enhance the overall ecosystem and produce the end user devices at a lower cost to improve affordability.

9.3.1.1 Formulation of Standards

Figure 9.5 shows how a standard is formulated in a generic standardization organization. The standardization process involves a series of steps aimed at developing, reviewing, and implementing technical standards to encourage wide-scale adoption of the technology. This is especially required for technologies that would be potential solutions for bridging the digital divide, as some parameters would be altered to meet the demand for the underserved regions. Some of the key steps involved in the development of standards [53, 54], include:

- **Identify needs and objectives** would help in addressing the gap that requires standards. This step would involve conducting market research, consulting stakeholders, and performing gap analysis and assessment for technical requirements, challenges, and opportunities in solving the problem. A standard committee is formed that assumes responsibility for the respective standard development, which would include the organization of study group (SG) to focus on a specific sector and its responsibilities.
- **Project approval process** includes drafting and submission of a standardization proposal that articulates the need for new standards or revision, defines the scope, specifies the objectives, outlines the technical requirements, analyzes stakeholders, implements guidelines, and performs benefits and impact analysis. Additionally, it also consists of details regarding the

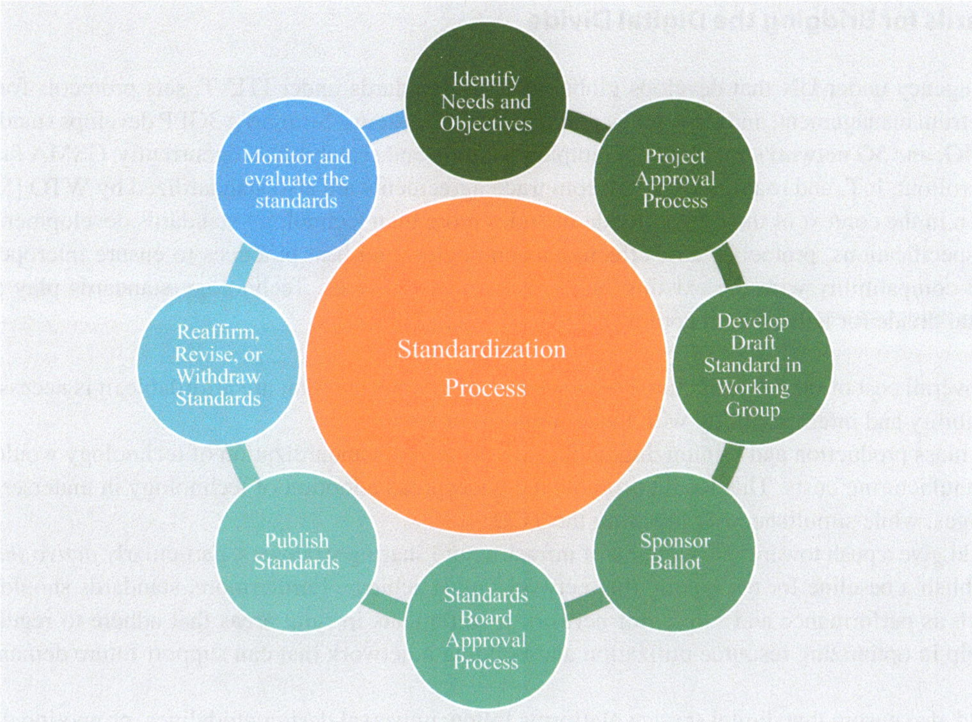

Fig. 9.5 Standardization stages

timeline, milestones, regulations, compliance, next steps, and action plan. Once the standard committee receives approval for standardizing bodies, they need to form WGs within the SG to perform the study.

- **Develop draft standard in WG**: A WG consists of individuals, researchers, representatives from companies, organizations, NGOs, non-profits, government agencies, equipment manufacturers, and other stakeholders who participate in the development of the standards. The representatives of a WG have an interest in developing standards and are managed by a WG chair. Once the draft is developed, it is required to ensure the formulated standard meets the standards checklist, copyright policy, metric policy, and several other policies.
- **Sponsor ballot**: Once the draft standard is reviewed, finalized, and approved by the WG, then it is submitted to the technical committee for review for consistency and alignment across related standards. Then the standards drafted would go through a balloting where the members who have ballot rights (including representatives from WG as well as outside the WG). They could vote in favor of or against the standard. The role of the balloting committee is to ensure a fair and rational process, where each representative casts votes and presents comments to develop a high-quality standard. All comments received during the ballot are considered. Each firm has a different system for proceeding with the ballot systems, but the results would determine whether a majority has been achieved that would support the standards or not.
- Once the **standards board approves the process**, then the standards drafts are published and are open for a certain duration for any individuals to submit comments and obtain a response from the WG. After this, the draft standard is reviewed, finalized, and approved by the WG, it is then submitted to the standards board committee, which would approve or disapprove based on whether the draft standards meet the review committee requirements. Once approved, it is published and publicly made available through official channels, websites, repositories, and publications.
- Finally, the standards are promoted to increase their adoption via channels like awareness campaigns, capacity-building workshops, webinars, training programs, and incentives to encourage compliance with standards to increase adoption. Since the standards are living documents, over time, they would be **reaffirmed, revised, or withdrawn** based on the monitoring and evaluation obtained for the standards.

9.3.2 Standards for Bridging the Digital Divide

ITU is a special agency under UN that develops global telecom standards under ITU-T, sets protocols for 5G, broadband deployment, spectrum management, and emergency communication systems. Similarly, 3GPP develops standards for interoperability for 3G, 4G, and 5G networks for global 5G implementation and scalability. Concurrently, GSMA facilitates uniform standards for 5G rollout, IoT, and roaming. The telecom trade agreements are also standardized by WTO [55].

Standardization in the context of the digital divide includes more than technology standards development; it involves the development of specifications, protocols, cost-effective technologies, and best practices to ensure interoperability, quality, affordability, and compatibility with the existing infrastructure and services. Technology standards play a crucial role in bridging the digital divide for underserved communities in the following ways:

- To reduce the overall cost of the infrastructure and provide user-end devices that are affordable, it is necessary to guarantee device compatibility and interoperability with the current technologies.
- By facilitating mass production and minimizing duplication of effort, standardization of technology would also contribute to reducing manufacturing costs. This would promote the widespread adoption of technology in underserved areas facing similar challenges, while simultaneously lowering the TCO.
- Standards would give a push toward the adoption of infrastructure sharing strategies, particularly *active sharing*, and *NHN*. Standards establish a baseline for the quality that networks must achieve. Furthermore, standards should outline specific guidelines, such as performance and safety, for network requirements in rural areas that adhere to regulations. It would additionally help in optimizing resource utilization and building a network that can support future demands, say, after 10 years.
- Standards could also ensure that digital service platforms follow universal design guidelines, promoting digital inclusivity for all. These standards would empower marginalized communities and enhance their digital skills by giving priority to support for local languages, offering affordable plans, and encouraging digital literacy initiatives.
- Standards should incorporate benchmarks to evaluate performance, reliability, and energy efficiency when deploying a combination of technologies for backhaul, middle-mile, and last-mile connectivity to ensure all possible scenarios could be utilized for bridging the digital divide. For this scenario, there is a need to ensure efficient spectrum resource allocations across different technologies based on the bands that the devices can support.
- Standards are required in establishing a uniform format for regulating the legal contracts that govern rural networks when national operators and local networks are operated by NHN providers or community engagement networks are connected. A standard legal contract would ensure transparency and uniformity in legal processes, minimize the time for negotiation and forming a collaboration, and resolve potential disputes. Additionally, standardized contracts may strengthen cooperation and confidence between the national operator and local network operator by facilitating the successful integration of their networks. This would create equitable and sustainable collaborations by establishing clear requirements and regulations.
- Standards ensure the facilitation of policies that seek to close the digital gap by offering a common framework that complies with laws established at various levels. It helps in recommending and implementing coherent policies for digital inclusion. Auditing and tracking the digital divide initiatives help in assessing their success rates, which would be made easier with this.

Hence, standardization efforts play a vital role in bridging the digital divide by setting a benchmark standard for different wireless broadband technologies.

9.4 Governance Failures in Enabling Digital Adoption

Despite the intended focus on developing an ecosystem of policies, regulations, and standards to bridge the digital divide globally, the realistic implementation of these falls short of intended outcomes. There is a need to focus on why the ecosystem is not able to achieve a global, equitable, and inclusive digital environment. For example, in real-life implementation of policies, regulations, and standards for SAGIN to bridge the digital divide is challenging and faces several obstacles along the way, as shown in Table 9.2. One of the foremost challenges in deploying and maintaining a telecommunication network in remote and rural areas is achieving interoperability due to a lack of compatibility with the existing infrastructure and also between the technologies.

Economic and socio-cultural factors are another major driving factor that is hindering digital inclusion efforts, as the priorities of the users are different, and being connected is not on their priority list due to several factors, as discussed earlier in Sect. 8.1 [40]. Additionally, there is a usage gap because of a lack of digital literacy skills in these regions, which is making it difficult for people to engage with digital services. The policies and regulations should focus on building a digitally inclusive society that also focuses on digital safety.

Next, regulatory and policy fragmentation and inconsistency across different levels are also major challenges. The lengthy process of obtaining approvals, spectrum licensing, along with coordination challenges among the stakeholders, and bureaucratic bottlenecks, further hinders the efficiency of the process. Addressing these problems would require focused efforts with a multi-dimensional approach that can ensure sustainable solutions in bridging the digital divide. For example, with the introduction of SAGIN, there is a shift of focus toward seamless integration of technologies; though it is still in the early stages of R&D. There is a need to develop policies, regulations, and standards that would fuel the widespread deployment of SAGIN.

9.4.1 Barriers in Policies

Barriers in policy formulation and execution; spanning from initial design to final implementation, can significantly hinder efforts to bridge the digital divide, often resulting in diminished impact or outright policy failure [56–60]:

- Policies are often designed to adopt a one-size-fits-all approach that are mostly fueling urban development, which is not suitable for addressing equitable digital adoption in underserved regions. This method doesn't differentiate between deployment locations, regions, gender, socioeconomic disparities, income, affordability, priorities, digital literacy, or user demand. The market scenarios are polar opposite in nature in the highly connected and unconnected regions.
- It is difficult to get the right balance in having a good competition policy against enabling operations to provide coverage to as many citizens as possible via universal coverage policies in some markets/countries. Failure in the implementation of these policies leads to further widening of the digital divide globally [61].
- During the stakeholder engagement, generally, there is low to no participation of stakeholders from the marginalized communities, and they are rarely consulted during policy formulation, though it is being developed for them. An active participation from this community would have a positive impact on it and would help in developing more suitable policies that would benefit them.
- In order to bridge the digital divide, many ministries from various sectors need to collaborate, including ICT, education, healthcare, transportation, road construction, town planning, and rural development. It is generally challenging to manage the cross-collaboration and create policies that do not conflict or overlap with one another, though.
- At the moment, there is a lack of a mechanism to monitor the effectiveness of a policy implemented. As discussed in Sect. 9.1.1.1, analysis of policy effectiveness helps in building impactful policies. However, in reality, there is an absence of a feedback loop or quantitative assessment which makes it harder to assess the digital inclusion outcomes and rate the effectiveness of the policies.

Table 9.2 Barriers in commercializing SAGIN to bridge the gap

Barrier	Challenge
High system complexity	Integrating space, aerial, and terrestrial layers requires sophisticated coordination and interoperability across domains
Deployment challenges	Real-world deployment–especially of UAVs and LEO satellites–faces logistical, regulatory, and environmental hurdles
Security vulnerabilities	SAGIN nodes are susceptible to cyberattacks like eavesdropping, spoofing, and denial-of-service
Spectrum management	Coordinating spectrum across layers and jurisdictions is complex and often lacks harmonized policies
Cost and ROI uncertainty	High capital expenditure with unclear short-term returns makes private investment risky
Standardization gaps	Lack of unified standards across layers impedes interoperability and vendor collaboration
Policy and regulatory lag	Existing telecom regulations are often not designed for cross-domain networks like SAGIN
AI and autonomy readiness	Real-time decision-making across dynamic environments requires advanced AI, which is still maturing

- Another problem with policies is that it is a short-term goal, and is dependent on the interest of the ruling party. The policies are subject to change on short notice and driven by the electoral cycle rather than a long-term digital inclusion goal.
- In order to comprehend the limits and constraints of choosing a certain strategy and formulating policy, it is necessary to integrate the results of policy studies, pilot trials, test trials, feasibility studies, and international case studies; however, this is quite challenging to achieve since not everything is documented properly, and available digitally. The development of a centralized resource center would help in minimizing resource and effort duplication globally.
- Typically, policies are announced with financial backing from governments and affiliated organizations, but might overlook the integration of a sustainable business model, which leads to the ultimate failure of the policy implemented. Several initiatives are focusing on bridging the digital divide; however, they lack a clear business model for initiative to sustain beyond the funding period.

Additionally, there are several policy failures which are widening the digital divide, such as the lack of enforcement of coverage obligations, poor spectrum assignment strategies that fail to align coverage and capacity with actual demand, lack of focus on developing digital skills, and mismanagement or underutilization of USOF in regions where they exist. Furthermore, the absence of effective infrastructure-sharing frameworks, inadequate protection of telecom infrastructure leading to vandalism, and limited efforts to promote ISPs and new market entrants further exacerbate the situation. Moreover, infrastructure gaps from the related sectors, such as unreliable electricity supply and poor road infrastructure, also hinder the expansion and sustainability of digital networks in underserved areas.

9.4.2 Barriers in Regulations

There is always scope for improving regulations, especially when addressing the digital divide. Some of the key regulatory barriers that hinder progress toward addressing the digital divide include rigid licensing frameworks for deployments in underserved regions, urban-centric spectrum regulatory frameworks, lack of incentives for rural and remote spectrum allocation, delayed approvals, lack of clear interoperability mandates, and slower development and adoption of regulations related to emerging technologies and solutions [62, 63]. Additionally, legal ambiguities, procedural inefficiencies, and weak enforcement mechanisms often act as significant barriers to progress in this area [64].

Regulations set by several governing bodies often inhibit new entrants from entering the market, aiming to preserve market stability, avoid disruptions that could negatively affect competition, and uphold quality standards through strict compliance with technical, security, and operational frameworks. However, such regulatory barriers drive the costs of the network deployment and services offered at higher costs to the consumers, which results in incumbent monopolies, and higher operational costs that are not supportive toward bridging the digital divide [65, 66]. Therefore, regulators also ensure there is fair competition, fairer spectrum allocation processes, unbundling of local loops, shared spectrum access, and other provisions to support newer entrants.

Moreover, the regulations related to consumer protection, privacy, data rights, and digital harms are yet to evolve to meet the current cybersecurity threats, which lowers digital adoption. Additionally, inadequate consumer protection against digital assaults and arrests, e-financial scams, identity theft, and the unauthorized sale of user data contribute to public mistrust and the perception that digital platforms are unsafe to use [67, 68]. Consumers and businesses face hurdles in reporting and resolving conflicts, while a parallel lack of trust among the operators in terms of interoperability, data security, KPIs, and sharing sensitive information results in poor implementation of end-to-end network sharing, especially when testing NHN strategy or spectrum sharing on large-scale deployments [69, 70].

9.4.3 Barriers in Standards

Similarly, addressing the barriers to using standards would help in developing solutions and technology that are suitable for deployment in underserved regions, which satisfy accessibility, availability, and affordability to minimize the digital divide [71]. One of the important barriers in the standardization process is the dominance of participation of stakeholders from developed countries over developing or underdeveloped countries, which exacerbates the digital divide [72]. This leads to a misalignment in the interest while developing the standards which is also fueled by a lack of interest and access to resources by the stakeholders in underserved communities to participate toward standard activities.

Another challenge with developing standards that address the digital divide is the time constraints, as standardization is typically a long cycle, and the technology would have evolved drastically, hindering the implementation of the technology for wide-scale deployments [73]. Also, standards are not focusing on the usage barriers, such as linguistics, digital literacy, per-capita income, affordability, operators' skills, and cultural relevance, which sometimes leads to the low adoption of the standards [74]. Another barrier to using standards to minimize the digital divide includes inconsistency of standards across different sectors and industry verticals, lack of quality control measures, and lack of binding with the standards in case of non-compliance, especially in terms of sustainability and accessibility, which results in weak implicit support for building an equitable digital ecosystem globally [67].

9.4.4 Future Direction

The barriers in policies, regulations, and standardization are not isolated roadblocks, but they form an interconnected network of governance limitations that address the digital divide globally. Mainly, the policy gaps arise from a lack of a clear picture of the local reality, while the regulations impede the entry of newer entrants and explore newer strategies in bridging the digital divide [75]. Standards, though focused on easing the interfaces and interoperability, are resulting in market monopoly situation as they lack the views of the users who are severely impacted by the digital divide [74].

To build an inclusive, digital, equitable ecosystem, the reforms should go beyond the technology and give equal importance to a holistic approach shaped by the interplay between technology, policies, regulatory frameworks, standards, and community engagement. There is a growing imperative to reimagine how governance is shaping and empowering the digital futures of underserved communities.

9.5 Summary

The rapidly evolving telecommunication ecosystem offers a wide range of solutions and opportunities to bridge the digital divide using wireless broadband technologies at affordable prices. Yet, this also presents us with unique challenges that demand a futuristic thinking approach to connect the unconnected. By harnessing the capabilities of newer and emerging technologies, crafting informed policies, implementing supportive regulations and standards, and adopting sustainable strategies, stakeholders would collaboratively move forward in building a digitally inclusive future for all.

Exercises

9.1 Analyze the importance of evidence-based policymaking in addressing the digital divide by discussing:

(a) How findings from global research efforts, like those mentioned in the previous chapter, can guide policy formulation.
(b) The role of pilot trials in testing and refining policies aimed at digital inclusion.

9.2 Investigate the role of regulations in enabling community-led internet projects. Discuss:

(a) Key regulations that have supported the development of community networks in rural areas globally.
(b) Challenges faced by community networks due to regulatory barriers and potential solutions.

9.3 Evaluate how standards in telecommunications influence satellite connectivity. Discuss:

(a) Standards that ensure interoperability and quality of service for satellite communication in remote regions, such as those in the Pacific Islands.
(b) The impact of these standards on improving education and healthcare services in isolated communities.

9.4 Examine the effectiveness of policies aimed at reducing the gender digital divide. Discuss:

(a) How targeted policies, like digital literacy programs for women, have been implemented in Africa.
(b) The challenges in measuring the long-term impact of such policies on social equity.

9.5 Explore recommendations for accelerating digital adoption in underserved areas. Discuss:

(a) How successful initiatives from Europe or Asia can be adapted to rural regions in Latin America.
(b) The role of global standards and partnerships in ensuring sustainable adoption of digital technologies.

9.6 Evaluate the role of public private partnership, in formulating telecommunication regulations. Discuss:

(a) How collaborations between governments and corporations have shaped policies in regions like India and Saudi Arabia.
(b) The effectiveness of these regulations in improving affordable and universal internet access.

9.7 Analyze the barriers to commercializing solutions for the digital divide. Discuss:

(a) Key challenges faced in scaling successful pilot trials globally.
(b) Suggestions for overcoming technological and financial obstacles in areas such as Oceania or Africa.

9.8 Discuss recommendations for evidence-based policy formulation in bridging the digital divide. Analyze:

(a) How lessons from Nordic countries' digital inclusion policies can guide global policymaking efforts.
(b) The importance of integrating research findings into policy frameworks to ensure effective outcomes.

9.9 Evaluate the role of global standards in ensuring universal connectivity. Discuss:

(a) How international organizations contribute to setting standards for digital technologies.
(b) The impact of these standards on enabling access to underserved populations in remote and rural areas.

9.10 Analyze how evidence-based regulations contribute to bridging the digital divide. Discuss:

(a) Real-world examples of regulation driving digital adoption, such as those implemented in the United States or Europe.
(b) Recommendations for improving the regulatory process to enhance its effectiveness in rural areas.

References

1. Fast Track Impact, *Why engage with policy?* Accessed 18 Mar 2025
2. Open University, *Introducing environmental decision making*. Accessed 18 Mar 2025
3. C.E. Lindblom, Who needs what social research for policymaking? Knowledge **7**(4), 345–366 (1986)
4. University of Oxford Blavatnik School of Government, Guide to evidence-based policymaking. Accessed 17 Mar 2025
5. W.E. Walker, Policy analysis: a systematic approach to supporting policymaking in the public sector. J. Multi-Criteria Decis. Anal. **9**(1–3), 11–27 (2000)
6. M. Fedorowicz, L.Y. Aron, *Improving Evidence-Based Policymaking: A Review* (Urban Institute, Washington, DC, USA, 2021)
7. W.F. Strydom, N. Funke, S. Nienaber, K. Nortje, M. Steyn, Evidence-based policymaking: a review. South Afric. J. Sci. **106**(5), 1–8 (2010)–
8. C.H. Oh, R.F. Rich, Explaining use of information in public policymaking. Knowl. Policy **9**(1), 3–35 (1996)
9. H. Downey, N. Boenisch, D. Browne, Y. Buckley, G. Cranston, L.V. Dicks, J. Hughes, C. Latrémouille, W.H. Morgan, M. Muir et al., Creating evidence-based policy and practice, in *Transforming Conservation: A Practical Guide to Evidence and Decision Making* (Open Book Publishers, Cambridge, 2022)
10. K. Bogenschneider, T. Corbett, *Evidence-Based Policymaking: Envisioning a New Era of Theory, Research, and Practice* (Routledge, 2021)
11. M.H. Snook, K. Garvin, G. Hayward, Sustainability programs: policies, metrics, and feedback, in *Sustainable Land Development and Restoration: Decision Consequence Analysis* (Butterworth-Heinemann, Burlington, 2010), pp. 13–27
12. R.K. Bullock, *Theory of Effectiveness Measurement* (Air Force Institute of Technology, 2006)
13. C.M. Paula, A.G. Fanfalone, G. Hitchings, M. Marshalian, M. Reisch, P. Veneri, V. Weber, *Bridging digital divides in G20 countries* (OECD, Technical report, 2021)
14. R. Seamans, Three policies to address the digital divide. Accessed 19 Mar 2025
15. M.E. Meena, J. Geng, Dynamic competition in telecommunications: a systematic literature review. SAGE Open **12**(2), 21582440221094610 (2022)
16. E.J. Oughton, Policy options for broadband infrastructure strategies: a simulation model for affordable universal broadband in Africa. Telematics Inform. **76**, 101908 (2023)

17. D. McMenemy, Internet access and bridging the digital divide: the crucial role of universal service obligations in telecom policy, in *International Conference on Information* (Springer, 2022), pp. 122–134
18. M.G. de Matos, I. Adjerid, Consumer consent and firm targeting after GDPR: the case of a large telecom provider. Manage. Sci. **68**(5), 3330–3378 (2022)
19. R. Mishra, D. Varshney, Consumer protection frameworks by enhancing market fairness, accountability and transparency (FAT) for ethical consumer decision-making: integrating circular economy principles and digital transformation in global consumer markets. Asian J. Educ. Soc. Stud. **50**(7), 10–9734 (2024)
20. C. Ali, *Farm Fresh Broadband: The Politics of Rural Connectivity* (MIT Press, 2021)
21. M. Noll, Internet and society: a review. J. Internet Stud. **12**, 34–56 (2018)
22. L. Bauchinger, A. Reichenberger, B. Goodwin-Hawkins, J. Kobal, M. Hrabar, T. Oedl-Wieser, Developing sustainable and flexible rural-urban connectivity through complementary mobility services. Sustainability **13**(3), 1280 (2021)
23. G.I. Sari, S. Winasis, I. Pratiwi, U.W. Nuryanto et al., Strengthening digital literacy in Indonesia: collaboration, innovation, and sustainability education. Soc. Sci. Human. Open **10**, 101100 (2024)
24. K. Allmann, G. Blank, Rethinking digital skills in the era of compulsory computing: methods, measurement, policy and theory. Inform. Commun. Soc. **24**(5), 633–648 (2021)
25. S. Okonkwo, Digital inclusion in Africa: bridging the divide (2025), Available at SSRN 5151540
26. H. Choudhary, N. Bansal, Addressing digital divide through digital literacy training programs: a systematic literature review. Digit. Educ. Rev. **41**, 224–248 (2022)
27. P.L. Joskow, R.G. Noll, Regulation in theory and practice: an overview. Stud. Public Regulat., 1–78 (1981)
28. P. Ranci, The early years of the energy regulator in Italy. Competition Regul. Netw. Ind. **20**(3), 250–257 (2019)
29. D. Weil, A. Fung, M. Graham, E. Fagotto, The effectiveness of regulatory disclosure policies. J. Policy Anal. Manag.: J. Assoc. Public Policy Anal. Manag. **25**(1), 155–181 (2006)
30. G. Majone, The European community between social policy and social regulation. J. Common Mkt. Stud. **31**, 153 (1993)
31. Regulations.gov. Learn about the regulatory process. Accessed 21 Mar 2025
32. C. Blackman, L. Srivastava, *Telecommunications Regulation Handbook* (World Bank and the International Telecommunication Union, Washington, DC, 2011)
33. Metavshn, Understanding compliance and regulations for telecoms. Accessed 19 Mar 2025
34. M. Maggetti, F. Di Mascio, A. Natalini, *Handbook of Regulatory Authorities* (Edward Elgar Publishing, 2022)
35. J.T.J. Penttinen, Standardization and regulation, in *The Telecommunications Handbook: Engineering Guidelines for Fixed, Mobile and Satellite Systems* (2015), pp. 23–48
36. H. Intven, *Telecommunications Regulation Handbook* (World Bank Publications-Books, 2000)
37. P. Colomo, Future-proof regulation against the test of time: the evolution of European telecommunications regulation. Oxf. J. Leg. Stud. **42**(4), 1170–1194 (2022)
38. TelcoSec News, Telecom regulators: navigating industry challenges and innovations. Accessed 21 Mar 2025
39. K. Verhoest, B. Redert, M. Maggetti, D. Levi-Faur, J. Jordana, Trust and regulation, in *Handbook on Trust in Public Governance* (Edward Elgar Publishing, 2025), pp. 360–380
40. Cenerva, BharatNet: India's ambitious project to bridge the digital divide. Accessed 17 Mar 2025
41. Federal Communications Commission (FCC), Affordable connectivity program (ACP). Accessed 21 Mar 2025
42. S.K.A. Kumar, E.J. Oughton, Infrastructure sharing strategies for wireless broadband. IEEE Commun. Mag. **61**(7), 46–52 (2023)
43. M. Zuhdi, E.T. Pereira, A. Teixeira, Trends in the telecom industry and opportunities for service providers, in *2011 13th International Conference on Transparent Optical Networks* (IEEE, 2011), pp. 1–4
44. J.Z. Cheng, J.Z. Tsyu, D.Y. Hsiao-Cheng, Boom and gloom in the global telecommunications industry. Technol. Soc. **25**(1), 65–81 (2003)
45. J. Farrell, G. Saloner, Converters, compatibility, and the control of interfaces. J. Ind. Econ., 9–35 (1992)
46. J.V. Nickerson, M. zur Muehlen, The ecology of standards processes: insights from internet standard making. Mis Quarterly, 467–488 (2006)
47. V.K. Narayanan, T. Chen, Research on technology standards: accomplishment and challenges. Res. Policy **41**(8), 1375–1406 (2012)
48. International Organization for Standardization (ISO), What is standardization? Accessed 21 Mar 2025
49. Institute of Electrical and Electronics Engineers (IEEE), The role of standards in technology development. Accessed 21 Mar 2025
50. International Telecommunication Union (ITU), Standardization in telecommunications. Accessed 21 Mar 2025
51. World Intellectual Property Organization (WIPO), A FRAND in need: why establishing standardized technologies is so complicated. Accessed 21 Mar 2025
52. Tiedemann (ed.), *The Essential Role of Technology Standards*. Accessed 21 Mar 2025
53. International Telecommunication Union (ITU). ITU-T: Setting the standard. Accessed 21 Mar 2025
54. Institute of Electrical and Electronics Engineers (IEEE). Developing standards with IEEE. Accessed 21 Mar 2025
55. I. Walden, *Telecommunications Law and Regulation* (OUP Oxford, 2012)
56. M.R Gilbert, M. Masucci, Defining the geographic and policy dynamics of the digital divide, in *Handbook of the Changing World Language Map* (2020), pp. 3653–3671
57. R.S. Sharma, E.M. Samuel, E.W.J. Ng, Beyond the digital divide: policy analysis for knowledge societies. J. Knowl. Manag. **13**(5), 373–386 (2009)
58. K. Nicholas, Geo-policy barriers and rural internet access: the regulatory role in constructing the digital divide. Inf. Soc. **19**(4), 287–295 (2003)
59. C. Asrani, Spanning the digital divide in India: Barriers to ICT adoption and usage. J. Public Aff. **22**(4), e2598 (2022)
60. K. Wang, Z. Wang, H. Yaning, D. Chen, Digital forgotten people?: decomposing digital divide in urban China. J. Chin. Govern. **10**(1), 57–79 (2025)
61. Cenerva Universal service—dr. charles jenne. Accessed on 13 Jun 2025
62. A. Paliwala, Digital divide: Globalisation and legal regulation. UTS L. Rev. **6**, 26 (2004)
63. L. Arendt, Barriers to ICT adoption in SMEs: how to bridge the digital divide? J. Syst. Inf. Technol. **10**(2), 93–108 (2008)
64. R. Nicholls, Telecommunications regulation and the global digital divide (2005), Available at SSRN 888842
65. J.E Nuechterlein, H.A. Shelanski, Lessons from telecom regulation for tech competition policy (2025), Available at SSRN 5172586

66. M. Bourreau, Y. Sun, F. Verboven, Market entry, fighting brands, and tacit collusion: evidence from the French mobile telecommunications market. Am. Econ. Rev. **111**(11), 3459–3499 (2021)
67. H.J. Mwakatumbula, G.C. Moshi, H. Mitomo, Consumer protection in the telecommunication sector: a comparative institutional analysis of five African countries. Telecommun. Policy **43**(7), 101808 (2019)
68. G. Finocchiaro, Transparency of digital providers and digital divide, in *Privacy and Data Protection in Software Services* (2022), pp. 3–13
69. S.K.A. Kumar, *5G Network Slicing for Rural Connectivity: Multi-Tenancy in Wireless Networks* (University of Strathclyde, 2023)
70. S.K.A. Kumar, E.J. Oughton, Techno-economic assessment of 5G infrastructure sharing business models in rural areas. Front. Comput. Sci. **5**, 1191853 (2023)
71. C. Swain, T. Pearson, Educators and technology standards: influencing the digital divide. J. Res. Technol. Educ. **34**(3), 326–335 (2002)
72. M.R. Canazza, Global effort on bridging the digital divide and the role of ICT standardization, in *2009 ITU-T Kaleidoscope: Innovations for Digital Inclusions* (IEEE, 2009), pp. 1–7
73. K. Chetty, U. Aneja, V. Mishra, N. Gcora, J. Josie, Bridging the digital divide in the G20: skills for the new age. Economics **12**(1), 20180024 (2018)
74. P. Resta, T. Laferrière, R. McLaughlin, A. Kouraogo, Issues and challenges related to digital equity: an overview, in *Second Handbook of Information Technology in Primary and Secondary Education* (2018), pp. 987–1004
75. M.G. Quibria, S.N. Ahmed, T. Tschang, M.-L. Reyes-Macasaquit, Digital divide: determinants and policies with special reference to Asia. J. Asian Econ. **13**(6), 811–825 (2003)

10 Digital Inclusion: Stakeholders and Future Directions

So far, we have looked at different strategies that help bridge the digital divide and strategies ranging from technological innovation to socioeconomic development to understand the impact of this critical challenge in several countries. However, these investments carry inherent risks that must be carefully assessed to enhance understanding and accurately predict future targets, particularly in areas such as digital skill adoption and affordability. Hence, the digital divide remains one of the most pressing challenges of our time, and billions of people still lack meaningful connectivity to the Internet and the opportunities it enables.

By understanding the complex interplay and overlay between technology, policy, and social factors, this chapter seeks to provide guidelines for researchers, practitioners, and policymakers working toward digital equity. Moreover, it emphasizes the importance of collaborative efforts across sectors to address this global issue comprehensively and sustainably. To provide equitable digital access, it emphasizes the necessity of a forward-looking viewpoint that focuses on underserved areas, developing technology, and inclusive approaches that address the unique challenges faced by different demographic groups and identify important open-ended research challenges.

10.1 Stakeholders' Role Toward Digital Divide

Inequality in digital access continues to persist across geographic, economic, and social dimensions, despite significant progress in extending connectivity [1]. As technological advancements reshape the global telecommunication ecosystem and related sectors, there is a growing need to ensure that these innovations do not widen the divide further but instead act as tools for inclusion. Stakeholders play a crucial role in bridging the digital divide, which is the gap between those who have access to modern ICT and those who don't. Here's how they contribute.

10.1.1 International Organizations

International organizations such as the UN and its specialized agencies, such as UN Development Programme (UNDP) and APT, together with regional and national organizations, play a fundamental role in curbing the digital divide, especially in developing countries. With their extensive resources, professional competencies, and networks, these institutions ensure that digital inclusion is one of the key priorities on the international agenda. They aim to secure appropriate funding, expertise, and technical support to develop policies fostering collaboration among the countries to address the complex issue in question. They also focus on ensuring equitable digital access through funding mechanisms and capacity-building programs.

Another key role of international organizations is to be able to provide sources for the implementation of wide-scale computerized inclusion programs. These entities often partner with state, NGOs, and private actors to implement projects such as the installation of broadband networks, provision of subsidized computers, and training of the population in how to use digital media communication tools. To illustrate, the ITU has implemented several initiatives aimed at enhancing connectivity in underserved areas. By consolidating the contributions and lessons learned from various projects, international organizations enable even the poorest countries to work toward digital moderation. These organizations also focus on addressing the unique challenges faced by marginalized communities, for example, the gender gap, urban-rural disparities, and persons with

S. Koratagere Anantha Kumar, *Wireless Broadband Access to Bridge the Digital Divide*, Textbooks in Telecommunication Engineering, https://doi.org/10.1007/978-3-032-09261-8_10

disabilities. The ITU Broadband Commission has been a key advocate for global connectivity since 2010, and its mission is to promote broadband access as a driver for economic growth, social inclusion, and sustainable development. They are supporting the project, *Giga*, which aims to connect all schools to the Internet by 2030. It helps in building a partnership among the stakeholders.

A Multilateral Development Bank (MDB) is an international financial institution established to promote economic development globally, especially in low- and middle-income nations. They focus on offering long-term sustainable goals and stabilizing the global economies. The reports and the policy recommendations that emanate from MDBs leadership, such as the World Bank and organizations such as the WEF, are important instruments for national governments and other actors because they provide factual evidence and practical perspectives on the challenges that the countries face. They propose guiding frameworks for national governments, policymakers, and various stakeholders by offering data-driven insights, partial policy recommendations, and an investment guide. This would help the countries to reform their policies and attract new investments that align with the national interest.

Apart from these ideas, these organizations also deploy preventive measures for designing and implementing other digital inclusion strategies and account for the risks involved. These organizations also coordinate activities and resources by acting as facilitators to bring together governments, private sector players, NGOs, and academic institutions to avoid duplication and align efforts toward digital connectivity. International organizations use the power to lobby for global policies, standards, and regulations that incorporate anti-discrimination policies on digital inclusion, Internet accessibility, fair telecom competition, and inclusive digital ecosystems.

10.1.2 Government

The government can promote digital services and digital schemes to raise public awareness about the benefits of adopting digital technology, for example, they could also subsidize the cost of end user devices by lowering the import tax and encouraging CSR initiatives to support user adoption initiatives. The government can also help in setting up digital community hubs that can connect to the Internet and provide access to the community in areas where affordability is a key challenge. Governments can establish regulatory and policy guidelines that foster fair competition and expand access to wireless broadband.

To help bridge the digital gap, the government should also encourage collaborations with private businesses and NGOs. Authorities could offer businesses subsidies, tax breaks, and other benefits to entice them to set up and run reasonably priced wireless Internet services in rural and isolated areas. To address the issue of digital literacy among the younger generations, governments should also incorporate digital literacy initiatives into the educational system. In a similar vein, they might provide training and courses to adults and senior citizens to expand their knowledge of digital technology's advantages, risks, and usage. In areas where there is a digital divide, this action would generate demand for digital connectivity and aid in the development of a sustainable business.

10.1.3 Businesses

Businesses play a key role in bridging the digital divide as they play a pivotal role by investing in infrastructure and providing services. The Big Tech companies are the world's largest, leading, and influential technology firms that act as both technology adopters and technology disrupter. They shape global policies, standards, innovation, digital markets, and drive the adoption of usage. Their investments in broadband infrastructure are significantly shaping the global digital market, both in developed and developing markets. TSPs are responsible for deploying broadband infrastructure, laying backhaul, middle-haul, and last-mile technologies by making use of cutting-edge technologies to provide Internet services and mobile connectivity to the end users. Businesses are also key customers for TSPs, who would generate significant revenue for them.

The equipment manufacturers are responsible for producing affordable end user access devices, thereby addressing affordability and digital literacy gaps and promoting digital inclusion among users. The people in underserved regions may find these devices more affordable with flexible payment plans using microfinance or installment payment modes. Additionally, the service providers are offering tiered pricing plans to promote meaningful connectivity. Furthermore, to include digital pedestrians, operators are offering text-based services at minimal to no cost, to lower the barriers and foster digital inclusion [25].

The businesses would also contribute to bridging the digital divide through CSR initiatives and adopting a holistic approach that helps in empowering the marginalized communities, creating a new marketplace to promote economic growth and development, and boosting social progress. The businesses work with stakeholders, such as governments, regulators, consumers, the workforce, investors, shareholders, industry partners, suppliers, competitors, media, academia, research institutions, civil society, and advocacy groups to evolve, expand, and adapt to regulatory, market, and technological shifts. Businesses invest in enhancing end-user digital literacy and promoting equitable access to bridge the digital divide and foster inclusive growth.

10.1.4 Academic and Research Institutions

Academic and research institutions play a critical role in transforming the digital divide by leveraging their expertise, resources, and networks to bridge the digital divide by contributing toward research-driven policy recommendations, training marginalized communities, offering community libraries to their students, and designing and researching innovative solutions in bridging the digital divide. One of the key contributions of academic and research institutions is conducting impactful research on emerging solutions and technologies that include digital adoption, digital literacy, broadening the range of network coverage, and QoS. Academic organizations offer important perspectives that help to inform decision-making at the state and international levels by conducting studies on trends, evaluating the effectiveness of existing policies, and finding innovative alternatives.

For instance, universities may conduct studies to explore community networks or even satellite-based Internet as an alternative solution in connecting the unconnected populations. Research on the socioeconomic impact of connectivity helps policymakers prioritize programs based on their potential benefits to underserved communities. Academic institutions also act as incubators of talent and innovation by establishing research laboratories, innovation hubs, and start-up accelerators for students and researchers to experiment and bring new ideas to the market. They support and fund entrepreneurship in the region and create opportunities that contribute to economic development. These help in bringing innovative, digitally inclusive solutions to the market in a shorter turnaround time.

Academic institutions engage in advocacy work to promote communication among stakeholders in addition to conducting research, developing policymakers, and providing training. Researchers and educators actively participate in shaping the digital future through conferences, published reports, and active participation in policy discussions. Their recommendations create an enabling environment for digital inclusion through affordability, accessibility, and digital literacy.

10.1.5 NGOs and Non-profits

NGOs and non-profits often work on the ground to provide training and support to marginalized communities, empowering them with digital skills and knowledge. They ensure that vulnerable groups, including women, children, old people, and people with disabilities, are not left behind in the digital world and support digital inclusion initiatives. They offer programs to encourage the usage of technology and create an awareness of the benefits offered by wireless broadband such as hands-on training sessions, workshops, mentorship, and reading materials to create a digitally aware community. Program topics can span basic digital tasks such as Internet navigation and email use, as well as advanced areas including online education, e-commerce, and digital applications in education, healthcare, governance, employment, and entrepreneurship.

These organizations also distribute low-cost or refurbished devices in regions where affordability is the key challenge to the digital divide. They partner with businesses, especially tech companies and the government, to ensure that people in underserved areas have access to affordable devices that are capable of connecting to wireless broadband technologies. Some NGOs and non-profit organizations provide free or subsidized Internet connections to underserved communities to lessen the financial burden on end users who must pay high prices for broadband access.

Establishing community-based Internet APs is another essential function of these groups, which are frequently found in community centers, libraries, or schools; providing digital services to tackle affordability and digital literacy challenges. These APs are connected to computers, printers, and other equipment. Furthermore, these community centers become essential lifelines for students, small company owners, and job seekers, and act as locations for offering training sessions, creating a cooperative atmosphere where neighbors may develop and learn from one another. These groups promote digital inclusion in underserved communities by collaborating closely with governments, policymakers, and stakeholders in the corporate sector.

Through conducting research, preparing reports, and organizing advocacy campaigns, they bring attention to the challenges faced by marginalized individuals. These organizations work at the grassroots to foster an inclusive digital world and contribute significantly toward societal well-being, individual empowerment, and the country's economic progress. This digital inclusion strategy would help in empowering a digital future for all and providing fairer access to ICT services.

10.1.6 Educational Institutions

Schools and universities play a pivotal role in promoting digital inclusion by training digital skills and technologies among young and adult students. They provide courses ranging from basic computer skills, such as typing and navigating the Internet, to advanced skills, such as coding, data analysis, cybersecurity, and business skills. Educational institutions assist in incorporating digital literacy into their curricula to provide students with the tools needed to succeed in a digital environment. They can also provide access to digital resources and support initiatives that promote digital inclusion among students and their families [2]. Additionally, this model is beneficial in training a next generation of individuals who are digitally empowered and can tap into the benefits of digital transformation. This would help the end users become informed and responsible digital citizens.

Schools in underserved regions can collaborate with the government, NGOs, and private companies to secure funds for such initiatives. Universities globally provide eduroam connections, enabling students and staff to access secure, global Wi-Fi connectivity to continue their learning beyond traditional classroom settings, and the credentials are valid across any eduroam network worldwide. The database is centrally maintained, ensuring seamless connectivity across participating institutions worldwide. Educational institutions can partner with technology-based companies to provide refurbished devices to their students and enhance their digital skills. This will help to create a market for tech companies to invest in these regions.

Educational institutions also include community libraries that provide access to digital resources such as computers, free or low-cost Wi-Fi zones, books on learning digital skills, lending laptops, tablets, and videos, and ensure access to high-speed Internet within the institution's premises. This is an important community access point in places where individual Internet connection costs are too expensive. These venues foster a collaborative learning environment where community members grow together. Educational institutions also collaborate with other stakeholders to advocate for digital inclusion policies. Their efforts often result in policy amendments, such as implementing affordable Internet services and conducting digital literacy programs. These groups are vital in promoting digital inclusion because they operate at the local level, disseminate necessary resources, and push for structural change.

10.1.7 Community Leaders and Activists

Outreach conducted at the local level using grassroots organizations helps in reducing the digital gap. For local leaders, the challenges and barriers that exist are evident as they have been engaged in the day-to-day social life in their respective regions. As trusted members of society, leaders can accelerate the adoption of ICTs by actively promoting their benefits and advocating for contextually appropriate solutions. They often serve as a conduit between these communities, governments, private companies, and NGOs, who would be able to assist in solving their issues and are a critical part of the solution. When community leaders identify the resource gap, they are empowered to allocate support and resources to those underserved areas. For instance, a community organization might collaborate with a tech partner to provide affordable devices and network access to rural communities. By grounding the initiative in local values and needs, they help ensure its relevance, adoption, and long-term success—meaningful connectivity.

Community leaders also help the people organize and access resources needed for collective action. In addition, active community members such as leaders and activists are significantly useful in tracking and assessing the activities aimed at digital inclusion. As a result, community leaders are in a position to know what works and what does not work, thus ensuring that programs are responsive and enhanced over the years. They also act as watchdogs to ensure the promised funding and support are realized. Their work at the grassroots level not only responds to present problems but also enables the development of integrated and community-based approaches that close the digital gap and ensure that all people have access to technology.

10.1.8 Collaborative Efforts

By working together, these stakeholders can pool their resources, expertise, and influence to bridge the digital divide and create a more inclusive and equitable digital society. Collaborative efforts would help understand the problem from each other's perspective.

Addressing the digital divide requires collaboration among diverse stakeholders. Governments and regulators can enact supportive policies and provide funding to build the necessary infrastructure to ensure that underserved communities have access to affordable and reliable Internet services. Meanwhile, innovation of businesses and enterprises brings cost-effective solutions suitable to the different economic groups, such as cost-effective devices and localization of digital services. Educational institutions contribute by equipping individuals with skills that help them develop their lives in the digital age. Academic and research institutions contribute to new studies, perform policy analysis, and develop innovative technologies that push the boundaries of connectivity.

Adding to the importance are NGOs, non-profits, and community leaders, who focus on grassroots efforts by ensuring that the benefits of digital inclusion can trickle down to marginalized groups through tailor-made programs and advocacy. International organizations add to these efforts by bringing technical expertise, funding, and a platform for global collaboration, which helps in developing the technology on a large scale and driving the cost down. In this manner, when these entities work together, they can create a robust ecosystem in which each plays a complementary role in bridging the gap. The group's endeavor, therefore, addresses not only the immediate challenges of digital exclusion but also opens up the way for sustainable progress in a world where everyone can play a meaningful role in the digital age.

10.2 Open-Ended Research Questions

There is a need to explore the critical challenges that remain unresolved while bridging the digital divide. Emphasizing the complexity of implementing sustainable solutions and building innovative, self-sustaining business models is instrumental in outlining open-ended research questions that need to be addressed to advance our understanding of the digital divide. It would help to develop innovative and scalable interventions. By doing so, this section lays the groundwork for identifying future research directions and action plans. When tackling the digital divide, some of the important concerns raised in this section may be the subject of future research, policy studies, or thought pieces, which would help in narrowing the gap.

10.2.1 Key Technological Questions

- There are different types of network deployment across the globe today, ranging from private to public networks, local to wide area networks, and *No Sharing* to *NHN*. How to integrate all these networks to enhance coverage and support services to the end users on a single plan?
- Emerging technologies show that they have the potential to bridge the digital divide, but the challenge lies in investigating how emerging technologies (e.g., 6G, AI, blockchain) can be leveraged to improve digital access. How can they be commercialized for the technology to become scalable for a WAN, which is affordable? Explore cost-effective solutions of AI, and other emerging technologies in bridging the digital adoption globally.
- In this book, different types of satellites and their role in bridging the digital divide were explored. Investigate in-depth what the long-term impacts of satellite Internet are in remote and underserved areas.
- How can alternative energy sources be effectively integrated into digital infrastructure to reduce the growing energy consumption of the telecom sector, particularly amid increasing base station deployment?
- Investigate the use of renewable energy resources in regions with unreliable electricity. Furthermore, consider the use of renewable energy sources, such as wind, solar, biogas, tidal, and hydroelectric energy; assess their feasibility for powering network infrastructure. Explore how these energy solutions can ensure sustainability, pave the path for newer business models, offer reliable power sources, and be cost-effective while supporting telecommunication operations.
- Service providers would seek to safeguard their exclusive hold on spectrum licenses. However, there are various areas in which the spectrum is underutilized and could be shared with the deployment by local operators or community-driven engagement under the spectrum-sharing mechanism [3]. How would rational spectrum sharing help improve the coverage and QoS in bridging the digital divide?

- Select any location of interest and then explore what network capacity is required to support various technologies while ensuring affordability and efficiency. How can network planning address current needs and anticipate future data demands? What strategies can be implemented to ensure scalability and sustainability over the next decade? How should networks adapt to evolving technological advancements and user behaviors?
- All industries are increasingly prioritizing net-zero targets to mitigate environmental degradation and safeguard our planet. The network design must incorporate environmental impact and sustainability considerations [4]. With this in mind, what specific network requirements are necessary to support the UN SDGs, particularly in terms of ensuring high QoS while minimizing energy consumption?
- Research works have analyzed effective deployment strategies for rural networks, emphasizing resilience against theft, damage, and tampering while ensuring robust security measures. What approaches can balance affordability, scalability, and durability while addressing the unique challenges of rural environments and protecting the infrastructure?
- How can AI-driven resource optimization approaches be leveraged to manage network congestion and promote digital adoption in rural areas, thereby addressing the digital divide and supporting equitable connectivity?
- The network generally falls back to older technologies when QoS is poor, often leading to poor QoE [5]. What capabilities and design requirements are necessary for an intelligent network infrastructure that can dynamically maintain optimal QoS and QoE? What role could AI/ML play in predicting, preventing, and recovering from performance degradation?
- Examine the potential of NTN solutions such as drones and UAVs for accelerating infrastructure development in remote and underserved regions. Which of these technologies can be employed to mitigate logistics issues, cut costs of deployment, and ensure access to reliable connectivity in regions lacking traditional network infrastructure?
- Explore cost-efficient strategies for infrastructure upgrades that would help in lowering the energy consumption in mobile towers, antennas, data centers, RAN, edge networks, and core. Identify innovative solutions for sustainability, which include energy-efficient hardware, smart power management, and renewable energy integration. What are the mechanisms by which upgrades balance operational costs and environmental impact in maintaining network performance and scalability?
- How can advanced technologies such as cloud-based architectures, AI-driven analytics, and blockchain be leveraged to design a cost-effective and scalable spectrum-sharing database and DSA mechanism that ensures efficient, interference-free spectrum allocation for secondary users?
- Discuss further the various ways dynamic spectrum allocation can be used to address changing demands by optimizing bandwidth availability and improving connectivity. How could such mechanisms balance affordability, reliability, and adaptability while meeting the specific challenges in deploying and using rural networks?
- What strategy should be used to narrow the adoption gaps in places where coverage and capacity are within reach, but there is a lack of digital literacy? Should the rollout start with broadcasting mode before moving on to full-scale wireless broadband? In areas where the coverage and the capacity are provided but the basic digital literacy in the target segment is lacking, a phased strategy can be planned. In the initial phase, broadcasting modes could be digital television or radio as a way of raising awareness through digital tools and services [6]. How can this staged approach lead to effective community engagement and sustainable digital inclusion in these regions?
- Edge computing enables affordable and fair sharing of digital resources by enhancing the ease of use when there is limited connectivity. Therefore, the research question will be: "How can edge computing bridge the digital literacy gap and enhance local education, e-governance, e-health, and e-commerce in underserved areas?"
- It would be more cost-effective to train local communities in basic network operations and allow MNOs to centrally manage and allocate network resources due to the high cost of sending technicians to maintain rural wireless broadband networks. The research question could therefore be, "How can a framework incorporating self-organizing network (SON), community-driven management and remote maintenance be designed to optimize the operation of rural wireless broadband networks while reducing costs and empowering local communities?"
- While national operators may oversee and manage some aspects of technological services, a fully unified approach may not be feasible or desirable in many contexts [7, 8]. Instead, collaboration between national operators, regional players, and local communities will likely be the most effective approach to managing and delivering technological services in regions with a digital divide. Analyze the technical and organizational barriers inherent in this model and recommend possible modifications to enhance its effectiveness.
- What are the most effective and widely adopted methods for securely, efficiently, and scalably connecting local and private networks to service provider infrastructures? How do these approaches mitigate challenges such as latency, cost, security, and scalability—and how can they be optimized for underserved or rural environments with distinct infrastructure and usage constraints?

- What is the impact of standards driver research and SDRs on equipment for providing wireless broadband services in rural areas? This research would thus investigate how these technologies could reduce deployment costs, improve service reliability, and enable sustainable broadband access for rural communities.
- Though the concept of community-driven networks is becoming popular, there is little study on the obstacles faced by local communities in maintaining their own networks, such as technical skill gaps, resource limitations, and scalability issues [9]. What are the key challenges and potential recommendations in the ongoing operations and maintenance of community-driven networks, and how can these be addressed to ensure long-term sustainability and reliability?
- What is the impact of slicing and other emerging technologies on resource allocation for end users in rural areas, especially to enhance coverage, QoS, and QoE? There is a need to explore how slicing should be implemented to keep the cost low and accommodate more service providers to maximize resource usage.

10.2.2 Socioeconomic and Policy Questions

- Cultural and regional factors can significantly influence how technology is perceived, utilized, and accepted. And understanding these dynamics helps in developing a tailored deployment solution for each underserved area. Therefore, one of the crucial questions is, how do cultural and regional differences affect technology adoption?
- Achieving equal accessibility in a diverse-demographics social environment is the key to achieving universal connectivity [10]. Policymakers need to know which policies are most effective in allowing equal inclusion and equal access. What policies are most effective for ensuring equitable access across diverse demographics?
- More affordable financial models and pricing strategies could help low-income populations and underserved regions achieve reliable access. How can financial models make connectivity more affordable? There is a need for tailored models of underserved regions of different digital divide levels.
- Deployment will be eased with appropriate frameworks in place. Hence, what legal frameworks are essential to facilitate infrastructure sharing among service providers? The legal framework would be different when using different infrastructure sharing strategies and technologies. Hence, it is critical to standardize the framework for smooth business operations [11].
- Key to sustainable infrastructure is both technical feasibility and socioeconomic inclusion. How do stakeholders and policymakers ensure that the network is techno-economically feasible and digitally inclusive?
- Effective assessment methods will expedite the deployment of reliable networks in underserved areas. What methods can be used to effectively evaluate local or private networks in terms of QoS, QoE, data security, data protection, secure connectivity, and other factors that would accelerate the rollout of NHN?
- Satellite networks, when integrated with terrestrial ones, could help bridge the digital divide in remote areas [12]. Policies supporting such integration can enhance connectivity where traditional infrastructure is not viable. Explore what policies should be formulated to encourage satellite integration with terrestrial networks to bridge the digital divide.
- To develop policies that encourage rural deployments, it is first essential to understand the demand and forecast future growth in the region of interest, thereby estimating spectrum requirements. Next, identifying suitable and available spectrum bands, determining license ownership, and examining associated legal frameworks becomes crucial. These steps inform how policy mechanisms can analyze the supply-demand dynamics of spectrum resources and support the creation of pricing models that incentivize investment in rural areas while maximizing returns across diverse deployment contexts. How can spectrum demand forecasting, band suitability analysis, and licensing frameworks be integrated into policy design to enable investment-friendly and scalable spectrum allocation strategies for rural network deployments?
- Rural telecommunication infrastructure must address more than just the deployment of physical infrastructure [13]. It is essential to explore ways to encourage the adoption and effective use of digital services through Internet connectivity. Without such efforts, there are significant risks to establishing and maintaining a sustainable business model in these areas. What are the risks involved in the deployment of rural connectivity solutions, and how can these risks be mitigated? What policies and strategies would help in mitigating these risks?
- There is a need to understand the impact of technology on bridging the digital divide and formulating regulations and policies to drive the technology and its supporting use case. What role would policymakers, local governance, and other stakeholders play in enhancing the sustainability of rural digital infrastructure?
- Policymakers must establish indicators and frameworks that guide the deployment in rural areas and ensure that the network is viable and future-proof. Without such foresight, the network will require expensive upgrades, undermining the feasibility

of the rural networks. What indicators need to be in place to guarantee that the deployment satisfies the current demand while also being flexible to accommodate future uses?

- Achieving digital inclusion is difficult due to regulatory constraints in terms of market entrance criteria, market competition, pricing approaches, resource allocation, and economic standing. How should policies be formulated for business practices and government to encourage digital inclusion and address urban-rural digital disparities?
- What policy frameworks and financial models, including price differentiation, PPP investments, and reforms, would encourage new industry entrants in bridging the digital divide? How can we ensure the success of these investments?
- The differentiation of price may make services affordable for underprivileged communities through region-specific pricing models. Such differentiation would require careful policy design so that it would be affordable to the users and profitable to the providers [14]. What should be the price differentiation policy that would promote digital inclusivity and market competition?
- It's important to learn which strategies are best suited to provide incentives for this partnership to cross the digital divide. Therefore, there is a need to investigate ways to encourage PPP investments, such as through crowdsourcing, tax exemptions, co-funding, or subsidies, which might aid in closing the digital gap for each country [15].
- New entrants in the broadband market can spur competition, innovation, and affordability [11]. Policies should focus on reducing barriers to entry and fostering innovation in underserved areas, and also understand the impact of entrants on the existing players [16]. What new policies should be introduced, or existing ones reformed, to encourage new entrants in the wireless broadband industry to contribute to bridging the digital divide?
- The lack of affordable access to devices is one of the main barriers to digital inclusion in rural areas. What factors would help end users in rural and remote areas procure low-cost, affordable, and smart gadgets to enhance their digital literacy skills? Are there any policies that are hindering this or could support the procurement of end user devices at affordable prices for underserved communities?
- Reliable methods of QoE assessment are important for developing successful digital inclusion initiatives. Hence, the question here would be how to measure the user QoE in these scenarios. Which models and frameworks are suitable for accessing this factor?
- Continuous monitoring and evaluation of policies are necessary to ensure their effectiveness and adapt them to evolving technological and socioeconomic conditions. What framework would be in place to assess whether the policies are helping in bridging the gap or if there is a need for further reforms?

10.2.3 Data and Metrics

- Current metrics often fail at capturing complex aspects of digital inclusion: digital literacy, societal impact, and equitable access. New metrics have to be formulated to assess inclusive value. These are modeled using the KVI assessment of the technological impact on society. What new metrics should be developed to measure digital inclusion effectively?
- Furthermore, to ascertain the success and scalability of the digital initiatives undertaken, robust means of tracking and analyzing their outcomes need to be adopted for evidence-based decision-making. How can we better track and analyze the impact of digital initiatives as well as emphasize the importance of this tracking mechanism? This would also help to formulate rational policies to bridge the digital divide.
- Infrastructure sharing requires sensitive data exchange among service providers, yet privacy, security, trust, and competitive concerns complicate collaboration across diverse use cases and applications. Key data points thus assume an important role in building collaboration. What data should be shared among the service providers while employing infrastructure sharing strategies? How would policymakers ensure that data sharing adheres to the existing policies and regulations?
- Privacy and security policies would be hard to maintain with users seamlessly roaming between several network technologies used by operators for data offloading purposes or offering worldwide roaming plans. The users would also roam on other service providers' networks where there is an agreement among them. In these settings, how can data privacy and security regulations be developed while enabling smooth roaming for end users across scenarios spanning from one rural region to another or the whole world?
- There are several initiatives and programs across the globe to enhance the digital literacy of communities. Hence, there is a need to model and develop a framework on how to measure the effectiveness of training end users about using digital services. This would help ensure that the funds are being allocated to the community of interest and that the programs achieve the digital inclusion aims.

- Effective teaching strategies are needed to make individuals resilient against cybercrimes and risks. How should this be done? This is also applicable to the tech-savvy end users.
- Data storage decisions also affect privacy, security, and performance. A clear framework is needed to guide the choice between private and service provider networks. Where should end users' data be stored and processed, within a private network or on a service provider's network, and what factors should guide this decision? This choice would impact how to optimize data accessibility, data quality, decision-making, cost, and operational efficiency.
- Standardized reporting plays a critical role in addressing the digital divide by enabling consistent and accurate data collection across diverse regions and demographics [17, 18]. This uniformity provides a clearer picture of gaps in affordability, digital literacy, access, and infrastructure. Once these gaps are identified, policymakers and organizations can allocate resources more efficiently, ultimately promoting equitable access to digital technologies, fostering inclusive growth, and reducing disparities in connectivity and digital skills. How can reporting formats be standardized across rural networks to improve understanding of user behavior and infrastructure gaps, thereby enabling data-driven decision-making for digital inclusion?
- A comprehensive framework is necessary for the holistic understanding of the digital divide and, consequently, to prioritize strategies appropriately. How could a universal framework be developed for measuring the digital divide, including its multifaceted nature?
- There is also a need to periodically upgrade the data compliance framework to understand the impact of emerging technologies in bridging the digital divide. How can the framework for understanding the impact of emerging technologies in bridging the digital divide be periodically upgraded? How would the impact be assessed and analyzed? Would it be a quantitative or qualitative analysis?
- A comprehensive framework will be required to holistically understand the digital divide and formulate appropriate strategies to effectively work on the issue. Which standardized framework can be developed for checking the level of individual digital literacy of end users to accurately and consistently assess the impact of digital inclusion programs?
- There is insufficient access to geospatial data, such as site locations, and frequency bands, and restricted access to data collected on user activity by the service provider [19]. The absence of data results in poor resource allocation and missed chances to close the digital gap. This makes it more difficult for policymakers, stakeholders, and academics to formulate decisions based on evidence [20]. How does limited access to comprehensive geospatial and behavioral data impact the effectiveness of evidence-based decision-making in addressing the digital divide?
- Stakeholders have different practices of data sharing among themselves, and this disparity in data-sharing practices creates significant challenges for coordinated efforts to address the digital divide [21]. Each entity tends to operate according to its own data standards, privacy concerns, and proprietary interests, which causes fragmented and incomplete datasets. This lack of trust and uniformity prevents a comprehensive understanding of digital access and usage patterns. How do differences in the data-sharing approach by stakeholders shape the coordination and effectiveness of initiatives that address the digital divide?

10.2.4 Business Models

- Knowing how to effectively utilize existing resources may offer sustainable solutions to bridge the digital divide. Then the research question lies in how to maximize the usage of the existing resources and obtain the desired QoS to meet the demand of underserved regions.
- In rural areas, innovative strategies for monetizing digital services must be developed to ensure sustainability, profitability for service providers, and value for end users–thus encouraging investment and long-term development. Several aspects need to be considered, such as demand, affordability, KPIs, KVIs, per capita income, cost of living, willingness to pay, energy prices, and rent. What are the innovative strategies that can be developed to monetize digital use cases in rural areas while improving the overall business case for stakeholders?
- In the design and deployment of telecommunication networks, the timely and effective provision of connectivity during disasters or emergencies is often taken into account as a critical responsibility [22, 23]. In case of such scenarios, the stakeholders' roles and their strategies need to be clear to avoid undue delays and inefficiencies. This is an important aspect that influences network rollout by a local operator or through community participation. Who should be responsible for delivering services to enable a prompt response during disaster management and emergencies, for example, a service provider, the local government, or a collaborative effort?

- Pilot projects tend to succeed in small, local areas but face challenges in scaling across broader geographies. Lowering barriers related to cost and logistics is important for the replication of success and broader impact. How do pilot initiatives in a specific region get scaled effectively across broader geographies without a significant increase in cost or logistical burdens?
- Cross-industry collaboration by telecom operators, governments, and energy and transportation sectors can help develop synergies in reducing costs and improving efficiency while bridging the digital divide. What frameworks could be developed to encourage partnerships among stakeholders and end users, individuals and businesses, to bridge the digital divide? Also, evaluate the cost benefits in cross-industry collaborations over the no-collaboration scenario.
- Building models involving local businesses, cooperatives, and community-based organizations is key to creating localized and inclusive solutions. It allows community ownership of solutions developed in this process, ensuring solutions that are relevant and sustainable. It also promotes economic development and social cohesion, since local stakeholders are directly involved in the process of bridging the digital divide and improving overall digital literacy and access. Explore different business models that involve these stakeholders, their roles, and the values they offer to create localized and inclusive digital divide solutions.
- Networks in underserved areas would become feasible only when there is a potential for at least a break-even scenario. Financial incentives can be created to attract investment and innovation to ensure long-term viability. Therefore, there is a need to explore innovative ways to monetize digital services, such as education, healthcare, and agricultural solutions, in rural areas to create sustainable revenue streams for all the stakeholders while the end users reap the benefits of digital connectivity.
- Infrastructure sharing by telecom operators would bring about a reduction in cost as well as enhancement of the deployment efficiency in less-covered regions; however, good business models are a must. How can telecom operators adapt business models to incorporate infrastructure sharing in underserved regions, and what are the key factors in optimizing TEA, pricing, and policy decisions?
- It is a challenging endeavor to strike a balance between government subsidies that prevent dependency on the one hand, while promoting adoption and connectivity on the other hand. Over-reliance on subsidies may prevent service providers from being innovative, self-sufficient, and create distortions in the market. The government must develop appropriate monitoring and governance mechanisms to ensure that subsidies reach the beneficiaries without misallocation or misuse. What is the optimal scope and scale of government subsidies needed to promote infrastructure development and service adoption while avoiding long-term dependency?
- The balancing of affordable access for low-income populations with the sustainability of business models is a very critical challenge. There are several strategies employed to address this balance, such as tiered pricing models, cross-subsidization, innovative business models, and community engagement [24]. There is a need to explore the impact of these strategies on overall sustainability and KVIs for the rural deployment scenario. The key research question here is, how can access to digital services for low-income populations be balanced with the sustainability of telecommunication business models?
- Regions with high, moderate, and low levels of digital divide pose different challenges that require adapted business models for each region in response to the unique socioeconomic and infrastructural requirements. More research needs to be done to understand how these region-specific models can have the maximum impact and sustainability. How do region-specific business models differentiate the degree of digital divide to maximize impact and sustainability in addressing the challenges of socioeconomic and infrastructural disparities?
- Providing affordable and accessible digital services to rural and remote populations, especially individuals with disabilities, is essential for ensuring inclusivity and addressing the unique challenges faced by this underserved demographic. How can digital services be tailored and made more affordable to effectively meet the needs of individuals with disabilities in rural and remote areas?
- Small island states and remote regions face specific geographic and economic challenges. Tailored policies and approaches would be necessary to address their connectivity challenges. Which tailored approaches, business models, and policies should be undertaken to address the peculiar connectivity challenges of low-income, small island states and remote regions?
- Multilingual content and interfaces can significantly improve accessibility. How can multilingual content and interfaces be developed to address linguistic diversity and promote digital inclusion across various regions?
- Sharing infrastructure and resources introduces different challenges that don't exist in other scenarios. What innovative business models can support infrastructure and spectrum sharing to increase access in marginalized regions?
- Dedicated business models and pricing strategies are necessary to cater to the emerging use case of digital pedestrian and cyclist services [25]. What business models and pricing strategies can be developed to support emerging use cases, such as digital pedestrian and cyclist services?

- Digital inclusion should address disparities across gender, age, and ability in creating access to this all-encompassing ecosystem. How can business models be developed to be inclusive of gender, age, and ability and offer equal access to digital services?

10.3 Summary

Achieving digital inclusion by 2030 will require a diversified, multi-stakeholder approach that strategically addresses critical infrastructure, affordability, literacy, and data accessibility gaps. By targeting these barriers through tailored interventions and collaborative policy design, digital inclusion can become more attainable across varied socioeconomic and geographic contexts. Bridging the affordability and digital literacy gaps should be pursued by implementing initiatives to reduce the cost of devices and connectivity, ensuring that underserved groups have access to digital tools and are capable of using them confidently. International organizations, governments, and businesses should collaborate on subsidy programs, tax incentives, and innovative financing models that alleviate financial barriers. Digital literacy calls for scalable, tailored models so that effective, targeted digital-skills training would reach underserved regions. The focus needs to lie in enabling people to put technology to meaningful use, providing education, and assisting them in working and living opportunities.

Resilient infrastructure formation must be central to regions prone to climate-related hazards, as well as for global deployments. For example, climate-resilient and disaster-ready technologies, starting with robust, reliable networks along with decentralized and efficient energy management solutions, help in maintaining access without interruption in such areas. Promoting equality and equity shall be a staple in these strategies. Gender-specific adoption of technology promotes inclusion, attempting to close digital divides within various communities and sociocultural sets. Another critical sector to be addressed through the development of appropriate policies and tailored, localized solutions involves indigenous people and refugees groups whose unique needs and vulnerabilities warrant specific consideration in digital inclusion efforts. In short, these activities align with the Broadband Commission 2030 targets: ensuring universal, affordable, and inclusive access to digital technologies and building a world that leaves no one behind in the digital age.

References

1. S. Lythreatis, S.K. Singh, A.-N. El-Kassar, The digital divide: a review and future research agenda. Technol. Forecast. Soc. Chang. **175**, 121359 (2022)
2. E.L.-C. Law, P. Vostanis, M.J. O'Reilly, Insights from impacts of the digital divide on children in five majority world countries during the COVID-19 pandemic. Behav. Inf. Technol. **42**(15), 2696–2715 (2023)
3. A. Chaoub, M. Giordani, B. Lall, V. Bhatia, A. Kliks, L. Mendes, K. Rabie, H. Saarnisaari, A. Singhal, N. Zhang et al., 6G for bridging the digital divide: wireless connectivity to remote areas. IEEE Wirel. Commun. **29**(1), 160–168 (2021)
4. I.P. Chochliouros, M.-A. Kourtis, A.S. Spiliopoulou, P. Lazaridis, Z. Zaharis, C. Zarakovitis, A. Kourtis, Energy efficiency concerns and trends in future 5G network infrastructures. Energies **14**(17), 5392 (2021)
5. K. Oztoprak, Y.K. Tuncel, I. Butun, Technological transformation of telco operators towards seamless IoT edge-cloud continuum. Sensors **23**(2), 1004 (2023)
6. M.A. Imran, M. Zennaro, O.R. Popoola, L. Chiaraviglio, H. Zhang, P. Manzoni, J. van de Beek, R. Stewart, M.A. Cox, L.L. Mendes, et al., Exploring the boundaries of connected systems: communications for hard-to-reach areas and extreme conditions. Proc. IEEE (2024)
7. W.-K. Chen, V. Nalluri, S. Ma, M.-M. Lin, C.-T. Lin, An exploration of the critical risk factors in sustainable telecom services: an analysis of Indian telecom industries. Sustainability **13**(2), 445 (2021)
8. A. Emami, D.H.B. Welsh, A. Davari, A. Rezazadeh, Examining the relationship between strategic alliances and the performance of small entrepreneurial firms in telecommunications. Int. Entrep. Manag. J. 1–26 (2022)
9. W. Jiang, H. Han, Y. Zhang, X. Li, M. He, W. Gu, Rise of community-driven communication network research. In: *Proceedings of the 2023 9th International Conference on Communication and Information Processing* (2023), pp. 308–312
10. R.P. Fisk, A.S. Gallan, A.M. Joubert, J. Beekhuyzen, L. Cheung, R. Russell-Bennett, Healing the digital divide with digital inclusion: enabling human capabilities. J. Serv. Res. **26**(4), 542–559 (2023)
11. S.K.A. Kumar, *5G network slicing for rural connectivity: multi-tenancy in wireless networks*. PhD thesis, University of Strathclyde (2023)
12. J. Garrity, A. Husar, Digital connectivity and low earth orbit satellite constellations: opportunities for Asia and the pacific. *Asian Development Bank* (2021)
13. D.F. Cabrera-Castellanos, A. Aragón-Zavala, G. Castañón-Ávila, Closing connectivity gap: an overview of mobile coverage solutions for not-spots in rural zones. Sensors **21**(23), 8037 (2021)
14. S.-L. Chen, C.-J. Chen, Y.-H. Lee, The correlation between broadband prices and the digital divide: a survey of Taiwan's north coast. Telemat. Inform. **84**, 102033 (2023)

15. M. Harvey, D.P. Hastings, G. Chowdhury, Understanding the costs and challenges of the digital divide through UK council services. J. Inf. Sci. **49**(5), 1153–1167 (2023)
16. S. Callander, N. Matouschek, The novelty of innovation: competition, disruption, and antitrust policy. Manage. Sci. **68**(1), 37–51 (2022)
17. G. Wikström, N. Bledow, M. Matinmikko-Blue, H. Breuer, C. Costa, G. Darzanos, A. Gavras, T. Hossfeld, I. Mesogiti, K. Petersen et al., Key value indicators: a framework for values-driven next-generation ICT solutions. Telecommun. Policy **48**(6), 102778 (2024)
18. M.M.H. Raihan, S. Subroto, N. Chowdhury, K. Koch, E. Ruttan, T.C. Turin, Dimensions and barriers for digital (in) equity and digital divide: a systematic integrative review. Digit. Transform. Soc. (2024)
19. A. Zilberman, A. Offer, B. Pincu, Y. Glickshtein, R. Kant, O. Brodt, A. Otung, R. Puzis, A. Shabtai, Y. Elovici, A survey on geolocation on the internet. IEEE Commun. Surv. Tutor. (2024)
20. T. Yang, H. Zhang, H. Wang, Addressing territorial digital divide through digital policy: lessons from China's national comprehensive big data pilot zones. Inf. Technol. Dev. 1–27 (2024)
21. R. Harmelink, R. Joosten, E. Topan, A. Adriaanse, J. van Hillegersberg, Data: to share or not to share? a semi-systematic literature review in (rational) data sharing in inter-organizational systems. Discover Data **2**(1), 1–34 (2024)
22. A. Marshall, C.-A. Wilson, A. Dale, New pathways to crisis resilience: solutions for improved digital connectivity and capability in rural Australia. Media Int. Aust. **189**(1), 24–42 (2023)
23. M.J. Murillo, J.A. Paco, D. Wright, Long-distance telecommunication in remote poor areas: from partnerships and implementation to sustainability. IEEE Technol. Soc. Mag. **34**(1), 19–30 (2015)
24. H. Hambly, R. Rajabiun, Rural broadband: gaps, maps and challenges. Telemat. Inform. **60**, 101565 (2021)
25. J. Noll, W.A. Mansour, C. Holst, S. Dixit, F. Sukums, H. Ngowi, D. Radovanović, E. Mwakapeje, G.M.N. Isabwe, A.S. Winkler, et al., Internet lite for sustainable development. J. NBICT (2018)

A Worldwide Initiatives: CTU

Although digital connectivity plays a crucial role, it is not reaching all the people in the world due to economic, geographical, and socio-political factors. Governments and other key stakeholders across the globe are working toward this connectivity and Usage gap, focusing on connecting the unconnected. This section will present some key strategies adopted to address the digital divide, highlighting the main focus areas, technologies adopted, and initiatives aimed toward the digital divide, based on a cross-country analysis, as shown in Table A.1. Each country has a unique set of challenges, and the initiatives showcase this diversity in the chosen approaches.

S. Koratagere Anantha Kumar, *Wireless Broadband Access to Bridge the Digital Divide*,
Textbooks in Telecommunication Engineering, https://doi.org/10.1007/978-3-032-09261-8

Table A.1 Global efforts to bridge the digital divide

Country	Focus areas	Technologies	Initiatives
Advanced economies			
United States	Rural access, high costs, digital literacy, subsidies, internet adoption, gender gap	Infrastructure, rural satellite, 6G, spectrum sharing	BEAD, ACP, Rural Digital Opportunity Fund
Canada	Rural access, critical infrastructure, digital penetration, digital literacy, economic disparities, language barriers	Infrastructure, satellites, 6G, spectrum allocation	Connecting the families, Universal broadband fund, Indigenous communities, Innovation, Science, and Economic Development (ISED), Digital Literacy Exchange Program, subsidies
United Kingdom	Rural, spectrum crunch, 100% coverage, gender gap, digital adoption, resilience, digital sustainability	6G, 5G, satellites, infrastructure, fiber optics, private networks open radio access networks (O-RAN), spectrum, AI, cybersecurity, digital democracy	Project Gigabit, Local full fiber networks, Digital Skills Partnership, Subsidies, Digital public infrastructure, PPP
Germany	Rural access, high-speed connectivity, fiber optics, satellites	Boost high-speed broadband rollout, affordability, and digital literacy, cyber security, data protection	Digital Infrastructure Fund, Digital Strategy 2025, Senior Technology-Experts, Gigabit strategy
France	Provide ultra-fast broadband to all citizens by 2025, Digital literacy, high costs, affordability, affordable end-user devices, cyber security, data protection	6G, private networks, broadband rollout	Digital France 2025, Très Haut Débit Plan, CSR, ARCEP
Japan	Digital literacy, adoption, tailored network	Private 5G, localized 5G solutions, satellites, O-RAN, fiber to remote, AI, user-friendly interfaces, language barriers	Local 5G networks program, Digital Garden City Nation, SDG, Mira-Digi, Official Development Assistance
South Korea	Promote national broadband and inclusive 5G	5G, AI, and smart infrastructure, freedom of content	Digital New Deal, support content access policies, Enhancing Digital Inclusiveness
Australia	Inclusive network, high-speed broadband for all, digital literacy, improved coverage and QoS, supportive policies	Fiber, satellites, 5G, 6G, AI	NBN, Better Connectivity Plan, Regional Connectivity Program (RCP), MBSP, Support for First Nations communities, PPP, Good Things Australia, Digital Inclusion Grants
New Zealand	Rural access, digital literacy, affordability, ultra-fast broadband, Policies	Fiber optics, fixed wireless, satellites, 5G, cybersecurity, resilience infrastructure, future networks, constrained resources, cloud, AI	Digital Strategy for Aotearoa, community-led initiatives, Digital Inclusion Alliance Aotearoa, Aotearoa's Digital Priorities in 2024
Finland	Universal access of 100 Mbps per household, Internet adoption, resilience, security, digital skills	Fiber optic, 5G, 6G, satellite	National Broadband Plan, Digital Decade Program
Norway	Bridge urban-rural digital divide, user-centric services	Fiber optic, 5G, 6G, satellite, AI	Digital Norway Initiative, digital single market
Sweden	Achieve full 1 Gbps broadband connection for all households, digital governance	Fiber optic, smart rural tech, 6G, Shared spectrum, Gigabit networks	Broadband Strategy for Sweden, Digitalidag, Digital Decade strategic roadmap
Denmark	Universal access for all, smart digital solutions in a rural area, rural broadband	Fiber optic, 6G, IoT, cybersecurity, AI	Digital Growth Strategy, ReDI School of Digital Integration, Digital by default, Danish Digital Skills and Jobs Coalition, National Strategy for Cyber and Information Security
Netherlands	Gigabit broadband for both urban and rural areas, digital skills, inclusive digitally, digital services	Fiber optic, 5G, 6G, Digital twin	National Gigabit Network Program, Digital inclusion, Green and Digital Jobs Action plan, Digital economy strategy, CSR, National broadband strategy and policy
Switzerland	Universal access, high-speed broadband, digital services, gender gap	Fiber optic, rural tech, satellites, cybersecurity, climate change, AI, open networks, smart cities, digital twins	Swiss Broadband Initiative, Digital Literacy, Digital Switzerland Strategy, ICT Switzerland, Digital Administration Strategy Switzerland, Digital Federal Administration Strategy

(continued)

Table A.1 (continued)

Country	Focus areas	Technologies	Initiatives
Emerging markets			
India	Universal access, digital divide, affordability, digital literacy, high-speed Internet connectivity	4G, 5G, FWA, satellites, Wi-Fi, mmWave, fiber optics	Digital India, BharatNet, PM-WANI, PMGDISHA, Internet Saathi Program, Universal Connectivity, National Portal of India, Satellite broadband Internet, Optical Fiber Network, National Education Policy
China	Digital knowledge skills, low level of education, digital literacy, e-governance	Fiber, satellites, 5G, 6G, cybersecurity, AI	Digital Village strategy, WomenUp, CSR, Online civilization, Internet Plus
Brazil	Affordability, digital literacy, infrastructure, accessibility, financial inequality, digital governance	Fiber optic, 4G, 5G, future networks, satellites, AI,	Programa Internet Para Todos, Projeto Telecentros.BR, Conecta Biblioteca, National Strategy for Connected Schools, SpaceX
Mexico	Rural connectivity, affordability, education, accessibility, policy, digital literacy, resilient, secure, peaceful, accessible, open cyberspace	Fiber optic, satellites, 4G, 5G, infrastructure, 6G	México Conectado, Red Compartida, Digital Inclusion Program, E-Mexico Initiative, Subsidized Mobile Devices, Estrategia Digital Nacional 2021–2024
Argentina	Affordable broadband, rural access, public Wi-Fi, responsible AI, digital literacy, capacity expansion, resilient infrastructure	Fiber optic, satellites, 4G/5G, AI, cybersecurity	Plan Federal de Internet, ENACOM subsidies, Connecting Equality Program, EDISON Alliance
South Africa	Rural connectivity, education, affordability, e-governance, digital literacy, infrastructure development, policies	Satellite Internet, broadband, fiber optics, GenAI, emerging technologies, future networks	South Africa Connect, National Broadband Policy, Digital Skills Development Program, Public Wi-Fi Expansion, CSR, NGO
Indonesia	Island connectivity, education, public access	Fiber optic, satellite Internet, 5G, Wi-Fi	Palapa Ring Project, Desa Digital Program, Free Wi-Fi for Villages, Smart City Initiatives, Digital Transformation Multi-Stakeholders Partnership Initiative, BAKTI Aksi Programme, Government programs, Digital Village
Philippines	Public Wi-Fi, rural access, disaster resilience, digital resources, skill training, community empowerment	Satellites, 4G/5G, broadband, Wi-Fi	Free Wi-Fi for All, DICT Initiatives, Tech4ED Centers, Digital Transformation Strategy
Malaysia	Affordable Internet, rural gaps, education, policies, investments	Fiber optics, satellites, 4G/5G, subsea cables	JENDELA Plan, MyDigital Blueprint, Rural Broadband Initiative, FIRA, Cyber Security Bill 2024, Digital Ecosystem Acceleration, Labuan Connect, PRIHATIN
Vietnam	Internet affordability, education access, rural inclusion, digital literacy, digital skill development	Public Wi-Fi, fiber optics, 4G/5G, AI	National Digital Transformation Program, E-Government Development Plan, InclusiveTech, Digital Support Program for Youth
Thailand	Smart villages, rural Internet, public Wi-Fi, policies, trade barriers, digital resources, price-sensitive markets	Fiber optic, satellite, 5G, cybersecurity, IoT, advanced computing, AI, 6G, O-RAN	Thailand 4.0, Net Pracharat Project, Smart City, Free Wi-Fi for Schools
Turkey	Rural-urban digital divide, affordability, digital governance, digital literacy, digital skills, user-centric content, terrain challenges	Broadband, fiber optics, mobile networks, e-subscriber identity module (e-SIM), AI	Universal Service Fund, Turkey Digital Plan, Digital Literacy Campaign, FATIH, EBA, National Technology Initiative, Digital Türkiye strategy, CSR
Saudi Arabia	Urban-rural divide, smart infrastructure, digital inclusion, digital skills and training, policies and regulations	5G, satellite Internet, fiber broadband, 6G, future networks, spectrum sharing, AI, UAVs, HAPS	Vision 2030 Digital Initiative, National Digital Transformation Plan, Digital Inclusion of Saudi Rural Areas, Fuel program
United Arab Emirates	Rural access, digital education, government digitalization, digital skills and training, e-governance, community and public engagement	5G, broadband expansion, fiber optics, satellites, 6G, future networks, AI, automation	Smart Dubai, Digital UAE, National Broadband Plan, Mohammed Bin Rashid Smart Learning Programme
Qatar	Digital education, smart infrastructure, universal access, digital literacy, gender gap, empowerment program for all	5G, fiber broadband, AI, future networks, satellites, spectrum allocation	Qatar Smart Nation Program, National Broadband Plan, Digital Transformation Agenda, Digital Inclusion Strategy, e-Accessibility Policy

(continued)

Table A.1 (continued)

Country	Focus areas	Technologies	Initiatives
Lower-income countries			
Kenya	Rural connectivity, affordability, digital literacy, community engagement, affordable devices, low-cost satellite Internet, progressive policies	Mobile broadband, fiber optics networks, satellite Internet, Wi-Fi, mmWave, TVWS	National ICT Masterplan, Konza Technopolis, Mawingu Wi-Fi Networks, Digital Skills at Scale, National Digital Master Plan 2022–2032, AfricaConnect
Uganda	Affordable Internet, rural education, digital skills, affordability, digital education	4G expansion, satellite Internet, Wi-Fi	Rural Communications Development Fund, Digital Uganda Vision, RENU, the Ugandan NREN
Nigeria	Bridging the urban-rural gap, affordability, youth employment, digital literacy, infrastructure access, poverty elimination, inclusive education system	Mobile broadband, fiber optics networks, satellite connectivity, Wi-Fi	National Broadband Plan, Digital Economy Policy, One Million Jobs Initiative, CANi, CSR
Ghana	Rural access, affordability, digital education	Mobile networks, satellite Internet, public Wi-Fi	Ghana Digital Transformation Strategy, ICT for Accelerated Development Policy, eTransform, Ghana Rural Telephony and Digital Inclusion, Ghana Investment Fund for Electronic Communications, USAID/Microsoft Airband Initiative
Ethiopia	Digital literacy, affordable Internet, rural access, digital skills	Fiber optic, mobile broadband, satellite Internet, Wi-Fi, infrastructure access, AI	Digital Ethiopia 2025, Universal Access Fund, Digital Agriculture, SchoolNet, Digital transformation
Tanzania	Rural connectivity, digital literacy, education access, policy and legislation, education initiatives	4G, fiber optics, satellite Internet, cybersecurity, high-speed broadband	National ICT Broadband Backbone, Universal Communications Service Access Fund, Digital Agenda for Tanzania Initiative
Zambia	Affordability, rural Internet access, digital skills	Mobile broadband, satellite Internet, AI, cloud computing, high-speed broadband access	Smart Zambia Project, Universal Access and Service Fund, National Public Free WiFi, UN Capital Development Fund, National Digital Transformation Strategy 2023–2027
Pakistan	Internet affordability, rural access, education connectivity, improved access, digital literacy, cultural resistance, supportive government policies	Fiber optic, 4G, public Wi-Fi	Digital Pakistan Initiative, Universal Service Fund, Smart Villages Program, National Cyber Security Policy
Bangladesh	Bridging urban-rural divide, digital literacy, education access	4G/5G, fiber optics networks, satellite Internet, TVWS	Digital Bangladesh Vision 2021, Info-Sarker Project, e-Quality Centre for Inclusive Innovation, Digital Public Infrastructure, Smart Bangladesh plan, Aspire to Innovate
Nepal	Disaster resilience, digital skills, community engagement, affordability, infrastructure access, digital literacy, language barriers, budget constraints	Satellite Internet, mobile broadband, 4G/5G, Wi-Fi, fiber optics, mmWaves, TVWS	Nepal Digital Framework, Rural Telecommunication Development Fund, Rural Communities Access to Information Society, NGO, CSR, National Broadband Policy
Sri Lanka	Rural access, digital literacy, education connectivity, limited resources, high cost	4G/5G, fiber optics networks, satellite Internet	Digital Economy Strategy 2030, e-Sri Lanka Development Project, Class Saathi, PPP, Nenasa Program
Myanmar	Bridging rural-urban gap, affordable access, education, connectivity, post-conflict infrastructure	Fiber optic, mobile broadband, satellite Internet, solar-powered BSs	Myanmar National Broadband Plan, Digital Myanmar Strategy, Universal Service Strategy, NGOs
Cambodia	Affordable Internet, rural education, digital skills, inclusion	Mobile broadband, satellite Internet, public Wi-Fi	Cambodia ICT Masterplan 2020, USOF, Digital Literacy Initiative, Cambodia Community Innovation Centers
Laos	Rural access, affordability, digital literacy, inclusive education	Fiber optic, satellite Internet, mobile broadband, solar-powered ICT	Digital Transformation Policy, USOF, Khang Panya Lao digital learning platform
Papua New Guinea	Rural connectivity, affordability, disaster resilience, digital economy development	Satellite Internet, mobile broadband, public Wi-Fi, community networks, digital learning platforms	National ICT Policy, Connect PNG Initiative, Universal Access Scheme, PNG Academic and Research Network

(continued)

Table A.1 (continued)

Country	Focus areas	Technologies	Initiatives
Small island developing nations			
Fiji	Rural connectivity, disaster resilience, affordability, digital literacy	Mobile broadband, satellite Internet, fiber optics, AI	DigitalFiji Program, Universal Access Fund, National Digital ID, Semata
Samoa	Rural access, disaster preparedness, digital literacy, resilient	Satellite Internet, mobile broadband, cloud, fiber optics	Samoa National ICT Policy, Samoa Submarine Cable, Digitally Connected and Resilient Samoa Project
Vanuatu	Rural connectivity, digital skills, disaster resilience, community networks	Satellite Internet, mobile broadband, Public Wi-Fi, digital ID	Universal Access Policy, Vanuatu National ICT Development Policy, Smart Villages and Smart Islands, UnBlocked Cash project
Maldives	Digital literacy, access for remote islands, disaster and climate resilience	Fiber optic, mobile broadband, satellite connectivity, digital ID, e-learning	Smart Maldives Initiative, National Broadband Policy, Nationwide Fiber Rollouts
Mauritius	Digital skills, economic diversification, rural access, youth empowerment	Fiber optic, 4G/5G, cloud computing, IoT	Digital Mauritius 2030, National Broadband Plan, free mobile data packages for citizens aged 18–25
Seychelles	Affordability, rural connectivity, digital transformation, digital skills	Fiber optic, mobile broadband, satellites, cloud	Seychelles National ICT Policy, Submarine Cable Project, GigaNet
Tonga	Rural access, disaster resilience, affordability, gender inclusion	Satellite Internet, mobile broadband	Tonga Digital Government Strategy, Tonga Submarine Cable System, Digicel's beehive, women empowerment
Kiribati	Remote island connectivity, affordability, disaster resilience	Satellite Internet, mobile broadband, ICT in education	Kiribati National ICT Policy, World Bank Digital Development Programs, Smart Island
Solomon Islands	Rural access, disaster resilience, affordability	Satellite Internet, mobile broadband, submarine cables	Solomon Islands National ICT Policy, Coral Sea Cable System, National Data Center for secure operations
Barbados	Digital skills, e-government, affordability, inclusive	Fiber optic, mobile broadband, Wi-Fi	Barbados National Digital Strategy, e-Government Initiatives, Smart Barbados
Jamaica	Rural access, digital literacy, e-commerce, affordable	Fiber optic, mobile broadband, Public Wi-Fi	Vision 2030 Jamaica, Universal Access Fund, Public Wi-Fi program
Trinidad and Tobago	Digital literacy, e-governance, rural access, inclusion	Fiber optic, 4G/5G, mobile broadband	National ICT Plan 2022, Fast Forward Initiative
Saint Lucia	Rural access, disaster resilience, digital skills	Satellite Internet, mobile broadband, Smart infrastructure	Government ICT Policy, Universal Service Fund, National broadband Policy and Strategy
Dominica	Rural connectivity, disaster resilience, affordability, job creation	Satellite Internet, mobile broadband	National ICT Policy, Climate Resilient Digital Infrastructure Projects, National Digital Transformation Strategy 2022–2026
Antigua and Barbuda	Digital transformation, disaster resilience, rural access, PPP	Mobile broadband, satellite Internet, fiber optics	Antigua ICT Development Strategy, CARICOM ICT Initiatives, Ethical Data Usage, Caribbean Digital Transformation Project
European nations			
Italy	Digital literacy, rural access, affordability, digital skills	Fiber optic, 5G, satellite, AI, FWA	National Recovery and Resilience Plan (NRRP), Digital Italy Initiative, Italy 5G Plan, EOLO-Thales Partnership
Spain	Rural access, affordability, digital skills	Fiber optic, 5G, satellite	Spain Digital Agenda 2026, UNICO Program, Digital Spain 2025
Portugal	Universal access, digital literacy, rural coverage, integrated networks	Fiber optic, 5G, satellite, AI, IoT, AR/VR	Portugal Digital Transition Plan, Agenda Portugal Digital
Greece	Digital skills, rural access, affordability, high-speed network expansion	Fiber optic, 5G, satellite, fixed very high speed networks	Greece Digital Strategy 2027, Ultra-Fast Broadband Initiative, Digital Transformation Bible 2020–2025

(continued)

Table A.1 (continued)

Country	Focus areas	Technologies	Initiatives
Austria	Rural broadband expansion, affordability, digital literacy	Fiber optic, 5G	Austria Broadband Strategy 2030, Digital Austria Initiative
Poland	Digital literacy, rural access, e-government, digital skills	Fiber optic, 5G, satellite, AI, ultra-fast broadband	Operational Program Digital Poland, Broadband Expansion Programs, European Funds for Digital Development 2021–2027, Google AI hub
Czech Republic	Rural broadband, affordability, digital skills	Fiber optic, 5G, satellite	Czechia Digital Strategy, National Broadband Plan, Digital Czech Republic Programme
Hungary	Digital skills, rural broadband, affordability	Fiber optic, 5G, satellite, AI	Digital Success Program 2030, Hungary National Broadband Plan
Bulgaria	Rural connectivity, affordability, digital skills	Fiber optic, 5G, satellite, AI	National Broadband Plan, Digital Bulgaria 2030
Romania	Rural broadband, affordability, digital literacy, digital rights	Fiber optic, 5G, satellite, AI	Digital Romania Agenda 2020, Broadband Development Strategy
Croatia	Rural access, affordability, e-government	Fiber optic, 5G, satellite	National Broadband Plan, Digi-Inclusion Network, Digital Services Act Implementation
Slovakia	Rural connectivity, digital literacy, affordability, digital skills	Fiber optic, 5G, satellite, digital platforms	Digital Slovakia Strategy, Broadband Expansion Program, EIB Loan for Green and Digital Projects, Digital Coalition
Ireland	Rural broadband, digital literacy, e-government, digital skills, inclusive, assistive technology	Fiber optic, 5G, satellite, assistive technology	National Broadband Plan Ireland, Connecting Communities Initiative, Local Government Digital & ICT Strategy 2030, WHO Collaboration on Assistive Technology
Estonia	Digital literacy, e-governance, rural broadband, public service digitization	Fiber optic, 5G, satellite, AI	e-Estonia Initiative, National Broadband Plan, AI Leap, Digital Public Services
Lithuania	Digital skills, rural broadband, e-government, inclusive networks	Fiber optic, 5G, satellite, assistive technology	Digital Lithuania Program, Rural Broadband Expansion Plan, EdTech Initiative
Latin America and the Caribbean			
Chile	Rural access, digital skills, affordability, digital literacy, rights, infrastructure development	Fiber optic, satellite, 5G	National Digital Inclusion Plan, Fibra Óptica Austral, Digital Teacher Competencies, Chile Digital 2035, Zero Digital Gap Plan, Kodea
Peru	Rural broadband, digital literacy, infrastructure	Fiber optic, satellite, 4G	National Broadband Plan, FITEL (Telecommunications Investment Fund), Digital Basic Basket Program
Colombia	Rural connectivity, affordability, digital skills, inclusion, digital services	Fiber optic, 4G, 5G, satellite, digital platforms, AI	Colombia Vive Digital, Internet for All, AI Policy Framework
Ecuador	Digital literacy, rural access, affordability, infrastructure	Fiber optic, satellite, 4G, AI, digital learning tools	Ecuador Digital Strategy, Connected Ecuador Plan, Future Teacher's Kit Project
Uruguay	Universal broadband, digital literacy, e-government	Fiber optic, satellite, 5G	Plan Ceibal, Antel's Fiber Optic Program, Mercosur Digital Citizen Initiative
Paraguay	Rural connectivity, affordability, digital skills, digital identity, human rights	Fiber optic, satellite, 4G, digital finances infrastructure	National Telecommunications Plan, Paraguay Digital Agenda, Informadas y Resilientes Project
Bolivia	Rural access, affordability, infrastructure	Fiber optic, satellite, 4G, digital platforms	Bolivia Digital Agenda, Tupac Katari Satellite Program, Bolivia Digital Inclusion Program, InternetBolivia.org Foundation
Guatemala	Rural broadband, affordability, digital literacy, infrastructure, inclusion	Fiber optic, satellite, 4G, Wi-Fi	National Digital Agenda, Connectivity for Development Plan, Open Government Challenge, Digital Transformation Initiative
Honduras	Rural connectivity, digital literacy, affordability, STEM skills	Fiber optic, satellite, 4G, mobile technology	Digital Transformation Plan, Rural Connectivity Programs, Tinkering with Tech

(continued)

Table A.1 (continued)

Country	Focus areas	Technologies	Initiatives
El Salvador	Digital literacy, rural broadband, digital infrastructure and payments	Fiber optic, satellite, 4G, digital platforms	El Salvador Digital Agenda, National Broadband Plan, Digital Infrastructure Expansion, Starlink
Costa Rica	Universal broadband, digital skills, affordability, employability skills	Fiber optic, 5G, satellite, digital platforms	National Telecommunications Plan, Connected Communities Program, FONATEL, Hogares Conectados
Panama	Rural broadband, digital literacy, affordability, accessibility, user-centric approach, gender divide	Fiber optics, satellite, 4G, AI, digital platforms	Internet for All Plan, Digital Agenda Panama 2030, Digital Agenda 2020 and Panama 4.0
Cuba	Affordability, infrastructure, digital literacy, inequalities, gender divide, digital literacy	Fiber optics, satellite, 4G, digital platforms, open networks, net neutrality	Nauta Wi-Fi Expansion, Connectivity Development Program, Promoting Open Access, Cuba Digital
Haiti	Digital literacy, rural access, affordability, gender gap, geographic disparity	Satellite, 4G, wireless networks, Wi-Fi	Haiti Broadband Connectivity Program, Digital Inclusion Plan, Community Internet Networks
Dominican Republic	Digital literacy, rural connectivity, affordability, digital skills, gender gap	Fiber optic, satellite, 4G, green networks	Digital Agenda 2030, Rural Connectivity Initiative, GODR, Renewable Energies
Other active nations			
Israel	Digital innovation, cybersecurity, digital literacy, integrated networks	Fiber Optic, Satellite, 5G, AI, smart infrastructure, UAVs, HAPS	Digital Israel, National Broadband Plan, ISOC-IL
Russia	Rural connectivity, digital education, affordability, digital literacy, AI readiness	Fiber Optics, Satellite, 5G, AI, NTFPs	Digital Economy Program, National Digital Literacy Program, Digital Innovation and AI in Experimental Legal Regimes
Belarus	Expanding broadband access, e-governance	Fiber Optic, 4G	National Digital Development Strategy, Affordable Internet Program, Digital and AI Readiness Assessment
Ukraine	Rural broadband, digital skills development	AI, fiber Optics, Satellite, 4G, 5G, UAVs, HAPS	Digital Ukraine 2030, Digital Education Projects, Mriia, Brave1, CodeUA, Digital State UA, YMCA and HP Partnership
Kazakhstan	Smart cities, digital literacy, rural access, inclusion	Fiber Optic, Satellite, 4G, 5G, digital infrastructure	Digital Kazakhstan Program, Affordable Connectivity Initiative, Digital Acceleration for an Inclusive Economy (DARE), Biometric Digital IDs
Uzbekistan	E-government, digital economy development, gender divide, digital literacy, rural development	Fiber Optic, Satellite, 4G, Wi-Fi	Digital Uzbekistan 2030, Universal Internet Access Plan, Empowering Education Project
Mongolia	Rural connectivity, digital literacy, inclusivity, affordability	Fiber Optic, Satellite, 4G, Wi-Fi, future networks, UAVs	E-Mongolia Project, Universal Service Obligations, Girls Code Programme, Digital Inclusion through CSOs Empowerment
Bhutan	Digital skills development, rural access, affordability	Fiber Optic, Satellite, 4G, Wi-Fi, UAVs, NTFPs	Digital Drukyul Flagship Program, Rural Connectivity Projects
Sierra Leone	Digital literacy, affordable connectivity, inclusivity	Fiber Optic, Satellite, 4G, Wi-Fi, UAVs,	National Digital Transformation Plan, Community-Led Internet Initiatives, GIGA, Leh Wi Lan
Rwanda	Digital literacy, gender inclusion, rural connectivity	Fiber Optic, Satellite, 4G, AI, Wi-Fi, UAVs, NTFPs	Smart Rwanda Master Plan, Digital Ambassadors Program, UNESCO and Rwanda Development Board Partnership

Integrated Practice Set B

B.1 Using historical data for 1G through 5G adoption rates, develop a graph showing technology adoption curves. Predict when 6G adoption might surpass 50% based on previous patterns.

B.2 Using data for internet penetration, literacy rates, and affordability indices for different regions, calculate a "Digital Divide Index" and rank countries based on their scores.

B.3 Given data on deployment and operation costs for satellite (LEO/MEO/GEO) versus fiber optics for a 1,000 km area, perform a cost-benefit analysis over a 10-year period.

B.4 Formulate an optimization model to minimize the cost of deploying a broadband network in a rural area while meeting a minimum internet speed requirement. **Decision Variables**: Number and locations of wireless towers. **Objective Function**: Minimize installation and maintenance costs. **Constraints**: Signal coverage area, user demand, and bandwidth capacity.

B.5 Develop a mathematical model to estimate data traffic in a city using Internet usage per user (GB/day), number of active users, and peak usage times. Additionally, derive a formula to predict traffic for a future network upgrade (e.g., from 4G to 5G).

B.6 Using the parameters of LEO, MEO, and GEO satellites (orbital altitudes and coverage radius), calculate the required number of satellites to cover a given geographic region. Use the formula for the surface area of a sphere and signal beam coverage: $A_{\text{coverage}} = 2\pi r h$ where (r) is Earth's radius and (h) is the satellite's altitude.

B.7 Create a cost function to compare backhaul technologies (fiber optics, free-space optical, and microwave) based on deployment cost per km, maintenance cost per year, and data throughput. Model a scenario to determine the most cost-effective solutions for a city of 10,000 users.

B.8 Write an algorithm using a greedy approach to select the most suitable locations for 6G base stations based on population density and expected data usage.

S. Koratagere Anantha Kumar, *Wireless Broadband Access to Bridge the Digital Divide*,
Textbooks in Telecommunication Engineering, https://doi.org/10.1007/978-3-032-09261-8

Solution Manual

Chapter 1

1.1 The answer should define what broadband access is, and how the digital divide is causing a disparity among the people with and without access to the internet.
1.2 Look up the ITU Dataport website and find the details on the digital divide based on gender. This would help in understanding why the gender divide exists and how it impacts them.
1.3 African and Asian countries have significant challenges in terms of connectivity. Examine why only these regions have a higher number of unconnected users.
1.4 Focus on the difference between the two technologies to better understand their role in addressing the digital divide.
1.5 Broadband access can be a powerful enabler for individuals who have never been online or are frequently disconnected. It opens doors to education, employment, healthcare, and financial services, empowering them to participate in the digital economy and engage more fully in civic and social life. Explore the answers along these perspectives.
1.6 **(a)** The current rural population with broadband access is 25% of 12 million, which is 3,000,000 people. The target is to provide broadband access to 80% of the rural population, which is 9,600,000. Therefore, the additional people needing broadband access are 6,600,000. **(b)** The cost per person for providing broadband access is $30. Hence, the total funding required is $198,000,000 to achieve the target.
1.7 **(a)** Current unconnected population is 2.8 billion. An additional population to connect by next decade is 2 billion. **(b)** Cost calculations for people in rural is $500 billion, and urban is $200 billion, and the overall cost is $700 billion.
1.8 If rural costs increase by 20%, then the new cost per rural user would be $600 which would result in a total cost for rural investment of $600 billion and would result in an increase of $100 billion.
1.9 **Hint**: To address the digital divide, consider the challenges faced in underserved regions. Reflect on how limited access to **electricity** impacts the deployment of telecommunication infrastructure and explore renewable energy solutions. Think about the **logistical and cost-related barriers** in infrastructure development and identify innovative, cost-effective methods. For **security issues** like theft and tampering, consider strategies involving technology-based monitoring and community engagement to safeguard assets.
1.10 Discuss the benefits offered by each technology, what parameters they are sensitive to, range, QoS, data rates, and scenarios where they would be suitable as a digital divide solution.

Chapter 2

2.1 Briefly explain the purpose of the case study. Mention that the case studies are designed to reinforce learning, enhance critical thinking, and provide practical insights. Also, keep in mind that case study solutions are flexible, and alternative approaches may also be valid. Every case study has a template form that needs to be followed: **Case Study Title**: Clearly state the case study title. **Problem Statement**: Restate the original problem and briefly describe the problem and critical issues. **Proposed Solution**: Identify the challenge from different perspectives—operators, technical, economic, and social issues. Then outline the proposed strategy solution, technical details, infrastructure choice, and implementation steps. Next, explain why the selected strategy is optimal in terms of efficiency, cost, implementation, and scalability. Following this, outline a deployment strategy to practically rollout the solution. Finally, describe the future scope of the proposed solution and strategy. **Key Takeaways**: Summarize the main learnings and describe briefly the other alternative strategies that were considered.

S. Koratagere Anantha Kumar, *Wireless Broadband Access to Bridge the Digital Divide*,
Textbooks in Telecommunication Engineering, https://doi.org/10.1007/978-3-032-09261-8

2.2 Follow the same steps as the previous case study solution steps. This problem exists globally and is a hurdle for government organizations. Similar to the previous question, there is no single solution, but justify the selected approach.
2.3 Substitute the given values into the logistic growth model formula and calculate $U(t)$ for a specific time t (e.g., one year, two years, etc.). **Interpretation**: The result gives the estimated number of new internet users over time. Perform the analysis for 10 years and understand the growth in the number of users. This helps in understanding the user growth and is also applied in the diffusion models to understand the uptake of a newer technology.
To explore further, model different scenarios by setting different K, U_0, and r values and visualize them using data modeling tools.
2.4 By substituting the corresponding values, we will get a propagation delay of 0.0167 ms, a transmission delay of 1.2 ms, a processing delay of 2 ms per node, and a queuing delay of 1.5 ms per node. Therefore, total latency is 4.7167 ms.
2.5 **(a)** Free-space path loss is estimated as 100 dB; **(b)** Received power is 55 dBm; **(c)** Link margin is 10 dB; **(d)** 115 dB
2.6 Compare deployment costs, latency, and scalability using hypothetical or real-world data. – Example: Fiber optics may be costlier to deploy in rural areas due to terrain challenges. **Key Metrics**: Quantify differences such as cost per km, average latency, and capacity to scale for growing user demands. **Revenue**: This would result in a different revenue for the same 3,000 people in urban and rural areas, which helps you better understand the ROI difference in both deployment regions.
2.7 Compare the energy requirements and carbon emissions of LEO, MEO, and GEO satellites versus fiber networks. – Satellites have ongoing power needs for propulsion and communication. – Also consider the usage of renewable energy resources for the ongoing operations. **Key Metrics**: Calculate energy consumption (in kWh) per GB of data transmitted for each technology.
2.8 Define potential risks (e.g., flooding, damaging terrestrial infrastructure).– Propose alternatives such as satellites, HAPS, or UAVs to maintain connectivity during floods. **Key Strategies**: – Use redundancy (e.g., backup power, multiple network paths). – Opt for resilient technologies—satellite-based communication.
2.9 We aim to model the transition of users from 4G to 5G in India using the Bass diffusion model and study the transition of ARPU. **Model assumptions**: p = 0.02, q = 0.38, initial 5G users in 2023 to be 15 million, and the targeted number of 5G users to be 900 million in the time range from 2023 to 2030. Let the ΔZ be INR 70, and the current value of Z be set to INR 150. **Results**: Draw inferences and conclusions from the results obtained.
2.10 Define an optimization problem with the objective function. Minimize Deployment Cost: $C = \sum_{i=1}^{n} c_i x_i$ where c_i is the cost of deploying HAPS i, and x_i is a binary decision variable. Add constraints for coverage, bandwidth, and budget. **Result**: Use algorithms like linear programming to find the optimal deployment plan.

Chapter 3

3.1 **(a)** Total data rate = Low-band + Mid-band + High-band = 4430 Mbps. **(b)** Percentage contribution = $\frac{\text{Band Rate}}{\text{Total Data Rate}} \times 100$. Low-band: 0.68%, Mid-band: 9.03%, High-band: 90.29%.
3.2 **(a)** Initial deployment cost = Tower Installation + Backhaul Cost = \$450, 000. **(b)** Total cost = Initial cost + Operational costs over 5 years = \$700, 000.
3.3 **(a)** 6666666.67 bits/joule. **(b)** Recalculate efficiency with 2 kW power = 5000000 bits/joule.
3.4 Evaluate spectrum effects on range and speed: Low-band for coverage, High-band for speed. Compare FWA benefits (cost, scalability) against fiber deployment challenges in rural areas.
3.5

```
Input: Total area A = 10 km^2
Input: Coverage radius per tower R (based on 700 MHz, e.g., 1.5 km)
Input: Cost per tower C_t (includes installation, backhaul, power, etc.)
Input: Terrain factor T_f (e.g., 1.2 for farms and hills)
Output: Total deployment cost C_total

Step 1: Calculate effective coverage area per tower
    A_tower = 3.14 * R^2 / T_f

Step 2: Estimate number of towers needed
    N = ceil(A / A_tower)
```

```
Step 3: Compute total deployment cost
    C_total = N * C_t

Return C_total
```

3.6 **(a)** Total cost = Fixed + Recurring over 10 years. Option A: $\$400,000 + \$200,000 = \$600,000$, Option B: $\$455,000$. **(b)** Cost efficiency = $\$7583.33/\text{km}^2$.
3.7 **(a)** Net Profit = Revenue − Expenses over 5 years = $\$1,000,000$. **(b)** ROI = 100%.
3.8 **(a)** Total cost = $\$1,000,000$. **(b)** Average cost = $\frac{\text{Total Cost}}{\text{Users}} = \$100/user$.
3.9 **(a)** Energy efficiency = 2000000 bits/joule. **(b)** With 2.5 kW power: 2400000 bits/joule.
3.10 **(a)** Total emissions = 500 metric tons CO_2. **(b)** 20% reduced power emissions = 400 metric tons CO_2. Reduction: 100 metric tons CO_2.
3.11 **(a)** NPV = Discounted cash flows − Initial investment = $\$99,916$. **(b)** Project is viable as NPV > 0.
3.12 Discuss how different spectrum bands of 5G offer various data rates, and then estimate the cost of provisioning services using the low, mid, and high bands for the same village. Then explore innovative solutions to overcome these challenges.
3.13 Perform scenario planning and highlight the key uncertainties that impact the 5G deployment in rural areas, and also the potential growth opportunities.
3.14 Draft a business model canvas for 5G rural networks.
3.15 Modify the existing value network configuration of 5G rural network using NHN presented in the chapter to show the active sharing strategy.
3.16 Discuss how spectrum sharing and infrastructure sharing would help lower the cost, and present some studies that have explored these strategies for rural connectivity. Then analyze what the key challenges were and potential solutions to mitigate those.
3.17 Select a country of interest, and find the data on the cost of equipment, civil works, and spectrum to deploy a 5G network in the rural area of interest. Then perform the TEA and sensitivity analysis.
3.18 Total energy depends on base station count, power per site, operational hours, and auxiliary system loads like cooling.
3.19 Potential renewable energy sources at 5G cell sites include solar, wind, biomass, and hydrogen fuel, enabling sustainable rural deployment.

Chapter 4

4.1 **(a)** Maximum achievable data rate: Data Rate $= 500\,\text{MHz} \times 3.5\,\text{bps/Hz} = 1750\,\text{Mbps}$. **(b)** New data rate with reduced spectral efficiency: New Data Rate $= 500\,\text{MHz} \times 2.8\,\text{bps/Hz} = 1400\,\text{Mbps}$. Percentage reduction: Reduction $= \frac{1750-1400}{1750} \times 100 \approx 20\%$.
4.2 **(a)** Total deployment cost: Cost $= 120 \times (\$5\,\text{million} + \$10\,\text{million}) = 120 \times \$15\,\text{million} = \$1800\,\text{million}$. **(b)** Total cost over 5 years: Total Cost = Deployment Cost + 5 × Operational Cost. Total Cost $= 1800 + 5 \times 50 = \$2050\,\text{million}$.
4.3 **(a)** Received signal power: $P_{\text{received}} = \text{EIRP} - \text{Path Loss} + G_{\text{receiver}} = 50 - 200 + 30 = -120\,\text{dBW}$. **(b)** If path loss increases by 10 dB: $P_{\text{received}} = 50 - 210 + 30 = -130\,\text{dBW}$.
4.4 **(a)** Advantages and disadvantages: Ku-band offers better coverage and availability in poor weather, while Ka-band provides higher throughput but is more susceptible to rain attenuation. **(b)** Examples: – Ku-band: SES Networks for broadband. – Ka-band: SpaceX Starlink for high-speed internet. Explain these in detail, and compare their real-time performance metrics.
4.5 **(a)** Energy efficiency $= \frac{10\,\text{Gbps}}{5\,\text{kW}} = 2\,\text{Mbps/Joule}$. **(b)** With power consumption of 6 kW Efficiency $= \frac{10}{6} = 1.67\,\text{Mbps/Joule}$.
4.6 Develop a business model canvas and highlight all the parameters. Additionally, focus on: **(a)** Revenue streams: Subscription fees, government subsidies, and partnerships with ISPs. **(b)** Partnerships with terrestrial network providers help extend coverage and reduce costs in underserved regions.
4.7 **(a)** Total emissions: $E_{\text{total}} = \text{Number of Satellites} \times \text{Emissions per Launch} = 100 \times 200,000 = 20,000,000\,\text{kg } CO_2$. **(b)** With a 15% reduction: $E_{\text{new}} = E_{\text{total}} \times 0.85 = 20,000,000 \times 0.85 = 17,000,000\,\text{kg } CO_2$.
4.8 Develop a SWOT analysis table for integrating satellite networks with terrestrial networks Strengths: Wide coverage, high reliability. Weaknesses: High latency, high cost, and integration technology. Opportunities: Rural and remote area coverage. Threats: Competition from terrestrial networks and regulatory challenges.
4.9 **(a)** Probability of at least 3 ground stations in 50 km^2: $\lambda = 5$ stations per 100 km^2 $= 0.05$ stations per km^2, $\mu = 0.05 \times 50 = 2.5$.

Using Poisson distribution:
$P(X \geq 3) = 1 - P(X < 3) = 1 - (P(X = 0) + P(X = 1) + P(X = 2))$.
$P(X \geq 3) = 1 - \left(e^{-2.5}\frac{2.5^0}{0!} + e^{-2.5}\frac{2.5^1}{1!} + e^{-2.5}\frac{2.5^2}{2!}\right)$.
$P(X \geq 3) \approx 1 - (0.082 + 0.205 + 0.256) \approx 0.457$.
(b) Expected number of ground stations in 200 km^2: $E = 0.05 \times 200 = 10$ stations.
4.10 **(a)** Deployment cost per Mbps: Satellites: High initial costs, scalable for wide coverage. Terrestrial networks: Lower deployment costs but limited in remote regions. **(b)** Long-term sustainability: Satellites have higher operational costs but lower emissions per user in rural areas. Terrestrial networks are cost-effective in dense areas.
4.11 **(a)** Free-space path loss $=\approx 196.4$ dB. **(b)** With frequency at 20 GHz, path loss increases by≈ 4.4 dB.
4.12 **(a)** Received signal power $= 50 - 200 + 30 = -120$ dBW.
(b) With an additional 5 dB receiver gain $= -120 + 5 = -115$ dBW.
4.13 **(a)** Noise power: $P_{\text{noise}} = k \cdot T \cdot B = 1.38 \times 10^{-23} \times 290 \times 20 \times 10^6 \approx 8.004 \times 10^{-15}$ W. Convert to dBW: $P_{\text{noise(dBW)}} = 10\log_{10}(8.004 \times 10^{-15}) \approx -141$ dBW. **(b)** SNR $= -100 - (-141) = 41$ dB.
4.14 **(a)** Total rain attenuation: $A_{\text{rain}} = 0.25$ dB/km $\times$ 10 km $= 2.5$ dB. **(b)** With a 20% increase in rain attenuation: $A_{\text{rain(new)}} = 2.5 \times 1.2 = 3.0$ dB.
4.15 **(a)** Actual SNR: SNR(actual) $= P_{\text{received}} - P_{\text{noise}} = -90 - (-100) = 10$ dB **(b)** Link margin: Link Margin $=$ SNR(actual) $-$ SNR(required) $= 10 - 10 = 0$ dB. **(c)** Implications: A negative link margin indicates the communication link cannot meet performance requirements, resulting in possible data loss or service degradation.

Chapter 5

5.1 **(a)** Radius of the circular coverage area of a conical base: $R = 500 \cdot 0.577 = 288.5$ meters. **(b)** Total area covered: Area $= \pi R^2 = \pi(288.5)^2 \approx 261,450\,\text{m}^2$.
5.2 **(a)** Initial deployment cost = Number of UAVs × Manufacturing Cost $= 10 \times \$20,000 = \$200,000$. **(b)** Total operating cost over 5 years = Number of UAVs × Maintenance Cost per Year × Years Operating Cost $= 10 \times \$5,000 \times 5 = \$250,000$. Total cost = $450,000.
5.3 Maximum communication range: FSPL = 110 dB $d = 10^{\frac{\text{FSPL}-[\log_{10}(f)+\log_{10}(\frac{4\pi}{c})]}{20}}$. Simplify and solve for d.
5.4 **(a)** Minimum solar panel area: Rearrange for A: $A = \frac{10{,}000}{0.22 \cdot 1{,}000} = 45.45\,\text{m}^2$.
5.5 **(a)** Received signal power:$P_{\text{received}} = 60 - 145 + 25 = -60$ dBW.
5.6 **(a)** Total annual carbon emissions: E_{total} = Number of UAVs × Emissions per UAV $= 20 \times 50 = 1,000$ kg CO_2. **(b)** Total emissions over 10 years: $E_{\text{total (10 years)}} = 1,000 \times 10 = 10,000$ kg CO_2.
5.7 **(a)** Spectral efficiency $\eta_{\text{spectrum}} = \frac{50\,\text{Mbps}}{10\,\text{MHz}} = 5$ bps/Hz. **(b)** Impact of increasing bandwidth by 20 MHz: $\eta_{\text{spectrum}} = \frac{\text{Data Rate}}{\text{Bandwidth}} = \frac{50}{20} = 2.5$ bps/Hz. Spectral efficiency decreases, the system may support more users or handle interference better.
5.8 **(a)** Advantages: Rapid deployment, independent of damaged infrastructure. High mobility and adaptability in dynamic disaster scenarios. **(b)** Challenges: Limited operational duration due to energy constraints. Vulnerability to severe weather conditions.
5.9 **(a)** Daily energy consumption: $E_{\text{daily}} = P \times \text{Time} = 500\,\text{W} \times 6\,\text{hours} = 3,000$ Wh/day. **(b)** Total energy for 10 UAVs over 30 days: $E_{\text{total}} = 10 \times E_{\text{daily}} \times 30 = 10 \times 3,000 \times 30 = 900,000$ Wh $= 900$ kWh.
5.10 **(a)** Cost per Mbps: HAPS systems have higher upfront costs but are scalable, especially in areas with low population density. **(b)** Benefits: Long-term sustainability, reduced digital divide in rural areas, and integration with terrestrial networks to improve coverage.

Chapter 6

6.1 **(a)** Total coverage area of the mesh network: Each node has a coverage radius of $R = 50$ m, so the area covered by one node is $A_{\text{node}} = \pi R^2 = 7853.98\,\text{m}^2$. For 5 nodes, assuming no overlap: $A_{\text{total}} = 5 \times 7853.98 = 39,269.9\,\text{m}^2$. **(b)** With 2 additional nodes (total 7 nodes): $A_{\text{new}} = 7 \times 7853.98 = 54,977.86\,\text{m}^2$. Percentage increase: Increase $= \frac{54{,}977.86 - 39{,}269.9}{39{,}269.9} \times 100 \approx 40.0\%$.
6.2 **(a)** Initial deployment cost: Cost $= 10 \times \$2,000 = \$20,000$. **(b)** Total 5-year cost: Operating Cost $= 10 \times \$500 \times 5 = \$25,000$

Total Cost = Deployment Cost + Operating Cost $= 20,000 + 25,000 = \$45,000$.

6.3 (**a**) Spectral efficiency: $\eta = \frac{\text{Data Rate}}{\text{Bandwidth}} = \frac{200\,\text{Mbps}}{20\,\text{MHz}} = 10\,\text{bps/Hz}$. (**b**) Additional bandwidth required: To achieve a data rate of 400 Mbps: Bandwidth $= \frac{\text{Data Rate}}{\eta} = \frac{400}{10} = 40\,\text{MHz}$. Additional bandwidth: Additional Bandwidth $= 40 - 20 = 20\,\text{MHz}$.
6.4 (**a**) Maximum number of users: Users $= \frac{\text{Backhaul Data Rate}}{\text{Data Rate per User}} = 100$ users. (**b**) With backhaul reduced to 800 Mbps: Users $= \frac{800\,\text{Mbps}}{10\,\text{Mbps}} = 80$ users. Reduction = 20 users.
6.5 (**a**) Daily energy consumption: $E_{\text{daily}} = P \times \text{Time} = 15\,\text{W} \times 24\,\text{hours} = 360\,\text{Wh/day} = 0.36\,\text{kWh/day}$. (**b**) Total energy for 20 access points over a year: $E_{\text{total}} = 20 \times 0.36 \times 365 = 2,628\,\text{kWh}$.
6.6 (**a**) Total annual carbon emissions: $E_{\text{annual}} = P \times$ Time per Day $\times$ Days per Year $\times$ Number of Access Points. $E_{\text{annual}} = 15\,\text{W} \times 24\,\text{hours} \times 365\,\text{days} \times 15 = 1,971,000\,\text{Wh} = 1,971\,\text{kWh}$. Carbon emissions: Emissions $= E_{\text{annual}} \times$ Carbon Intensity $= 1,971 \times 0.5 = 985.5\,\text{kg}\ CO_2$. (**b**) With efficient access points ($P = 10\,\text{W}$) - Carbon emissions: Emissions(new) $= 1,314 \times 0.5 = 657\,\text{kg}\ CO_2$. Reduction $= 328.5\,\text{kg}\ CO_2$.
6.7 (**a**) Probability of 3 users in a 50 m^2 area: $\lambda = 5$ users per 100 m^2 $= 0.05$ users per m^2, $\mu = 0.05 \times 50 = 2.5$. Using Poisson distribution: $P(X = 3) = e^{-2.5}\frac{2.5^3}{6} \approx 0.213$. (**b**) Expected number of users in 200 m^2: $E = \lambda \times 200 = 0.05 \times 200 = 10$ users.
6.8 (**a**) Advantages: TVWS has a better range due to lower frequencies. Cost-effective for rural connectivity with minimal infrastructure. (**b**) Challenges: Regulatory restrictions on TVWS usage. Interference management with existing broadcasts.
6.9 (**a**) Payback period: Revenue per Year $= 500 \times 2 \times 12 = \$12,000$. Payback period: Payback Period $= \frac{\text{Initial Investment}}{\text{Revenue per Year}} = \frac{10,000}{12,000} \approx 0.83$ years. (**b**) Total revenue over 5 years: Revenue (5 years) $= 12,000 \times 5 = \$60,000$.
6.10 (**a**) Technical constraints: limited backhaul, coverage issues, interference. (**b**) Economic and social barriers: high costs, maintenance, and scalability, lack of digital literacy.

Chapter 7

7.1 (**a**) Reflect on 6G's URLLC and its role in telemedicine and remote surgery. Consider applications in real-time patient monitoring and healthcare delivery in underserved regions. (**b**) Explore how 6G can connect mIoT systems in agriculture, smart cities, and environmental monitoring. Examine its impact on autonomous vehicles, drones, and robotics in improving rural connectivity and logistics.
7.2 (**a**) Discuss how mmWave frequencies support high-speed internet for remote areas. Reflect on its ability to handle large data rates and reduce connectivity gaps in rural regions. (**b**) Consider signal attenuation over long distances and its susceptibility to weather conditions. Reflect on potential infrastructure and cost challenges for sparsely populated areas.
7.3 (**a**) Reflect on how Li-Fi can improve connectivity in rural schools and clinics. Discuss its advantages in environments with limited access to traditional wireless networks. (**b**) Consider the dependency on lighting infrastructure for Li-Fi's functionality. Reflect on its challenges for outdoor environments or long-range applications.
7.4 (**a**) Consider IoT applications in crop monitoring, precision irrigation, and soil analysis. Reflect on how connected sensors and devices optimize resource use and boost productivity. (**b**) Explore the economic challenges faced by small-scale farmers in adopting IoT. Discuss issues, such as energy constraints and technical expertise in rural deployments.
7.5 (**a**) Explore how RIS enhances signal strength and coverage in challenging terrains. Reflect on how it improves the efficiency and reliability of networks in underserved areas. (**b**) Consider the cost-effectiveness of deploying passive RIS in low-resource settings. Examine technical challenges like maintenance and adaptability to rural scenarios.
7.6 (**a**) Reflect on how quantum cryptography ensures secure financial transactions in rural areas. Explore its role in safeguarding sensitive data for rural banking and commerce. (**b**) Discuss the cost and complexity of setting up quantum networks in underserved regions. Reflect on the lack of technical expertise and infrastructure in remote areas.
7.7 (**a**) Reflect on how AI can optimize bandwidth allocation and predictive maintenance, as well as resource optimization for remote connectivity networks. (**b**) Look into case studies where AI-driven solutions have enhanced network reliability and efficiency.
7.8 (**a**) Reflect on its potential for transmitting large datasets in rural research centers and for high-speed collaboration between rural and urban hubs. (**b**) Discuss range limitations and infrastructure costs for THz technologies. Reflect on its suitability for short-range, high-speed applications only.
7.9 (**a**) Explore hybrid approaches combining satellite, terrestrial, and renewable energy systems. Reflect on the importance of AI-based predictive systems for network reliability. (**b**) Highlight the significance of power backups and redundant communication links during disruptions. Also, highlight the roles of air layers in providing temporary services.

7.10 **(a)** Reflect on how energy-efficient 6G solutions reduce operational costs in rural setups. Consider the role of AI-driven optimization for resource-efficient networks. **(b)** Examine how solar or wind-powered infrastructure supports network sustainability in off-grid areas.
7.11 Use the following formulas to estimate the range and QoS:
$d = 10^{(P_t+G_t+G_r-P_{r,min}-32.44-20log_{10}(f))/20}$.
$\text{SINR} = \frac{P_r}{I+N_0}$.

Chapter 8

8.1 **(a)** Reflect on how community-led projects empower locals through affordable and manageable connectivity solutions, using low-cost infrastructure, local content creation, and skill-building programs. **(b)** Explore how education is transformed through e-learning tools and increased access to information. Examine healthcare improvements enabled by telemedicine and digital health solutions. Investigate how digital marketplaces and job opportunities foster economic growth.
8.2 **(a)** Look at specific initiatives targeting digital literacy and education for women. Identify subsidized programs that make connectivity and devices accessible to women. **(b)** Think about how digital access boosts participation in education, the workforce, and entrepreneurship. Reflect on how these initiatives foster equity and social mobility for women and girls.
8.3 **(a)** Focus on initiatives like BharatNet or Digital India that aim to bridge rural-urban connectivity gaps. Mention how these programs promote digital literacy and affordable broadband access. **(b)** Discuss collaborations between the government and private sector to improve infrastructure. Highlight examples of community Wi-Fi projects, rural affordable devices, as well as rural mobile connectivity initiatives.
8.4 **(a)** Consider how satellite technology overcomes barriers like geography and distance. Reflect on how it provides essential connectivity in disaster-prone or remote island regions. **(b)** Examine education initiatives enabled through satellite connectivity. Look into telemedicine advancements and enhanced disaster response systems.
8.5 **(a)** Focus on programs that establish free public Wi-Fi or subsidize internet access for low-income communities. Highlight digital literacy campaigns or skill training initiatives driven by corporations. **(b)** Examine how these CSR initiatives have improved access to information, education, and jobs. Reflect on their long-term impact on socioeconomic inclusion.
8.6 **(a)** Look at how community networks are established and managed by local populations. Reflect on the role of open-source and affordable technologies in achieving self-reliance. **(b)** Investigate how connectivity fosters education, healthcare access, and local employment opportunities. Explore how it strengthens social cohesion and community development.
8.7 **(a)** Look into broadband mandates, affordable access subsidies, and advanced technologies like 5G or fiber optics. **(b)** Explore how scalable strategies like PPP and digital literacy programs can inspire global adaptation. Reflect on the importance of addressing both infrastructure and education needs.
8.8 **(a)** Explore how public Wi-Fi enhances affordability and access for underserved communities. Examine its impact on education, small businesses, and e-government services. **(b)** Reflect on issues like sustainability, funding, and network security in public Wi-Fi deployments.
8.9 **(a)** Examine programs like Isarel's National Digital Israel Initiative and rural broadband projects. **(b)** Investigate collaborations with technology providers and foreign investors to modernize digital infrastructure.
8.10 **(a)** Look into how digital exclusion affects rural economies, education, and healthcare outcomes. **(b)** Reflect on how research findings inform national broadband plans and targeted interventions for underserved communities. **(c)** Select any four large-scale research projects such as A4AI, Giga, AfricaConnect, BharatNet, UNDP's Digital Inclusion Initiatives, OECD's G20 recommendations, and more.

Chapter 9

9.1 **(a)** Refer to pilot trials and studies discussed in the previous chapter, like community networks and gender-focused programs. Highlight how these findings provide data for identifying connectivity gaps and solutions. Discuss how evidence-based policymaking can tailor policies to specific regional needs. **(b)** Pilot trials test the practicality of policies and gather feedback for refinement. Use examples of trials in Africa or Oceania that led to improved policies. Highlight the importance of adjusting policies to local socio-economic and geographic conditions.
9.2 **(a)** Key regulations for community networks: Discuss enabling policies like spectrum allocation for small-scale networks. Highlight initiatives that encourage community participation and ownership. **(b)** Regulatory barriers and solutions: Address

issues like high licensing fees and lack of legal frameworks for community networks. Suggest solutions like simplifying licensing processes and promoting open access policies.
9.3 **(a)** Standards for satellite connectivity: Explore interoperability standards ensuring seamless communication between satellite networks. Discuss how performance metrics like latency and reliability are maintained through standardization. **(b)** Impact on remote regions: Provide examples of improved educational access via satellite-enabled e-learning platforms. Mention telemedicine initiatives that benefit from standardized satellite links.
9.4 **(a)** Targeted policies in Africa: Mention specific programs focusing on digital literacy for women and girls. Highlight efforts to provide affordable devices and connectivity. **(b)** Measuring long-term impacts: Discuss challenges like lack of consistent data and metrics to evaluate progress. Suggest approaches such as longitudinal studies and community feedback mechanisms.
9.5 **(a)** Adapting successful initiatives: Use examples from Europe (CSR programs) or Asia (public-private partnerships) to guide strategies for Latin America. Discuss how regional customization is critical for effective implementation. **(b)** Role of standards and partnerships: Highlight how global standards ensure scalability and interoperability. Discuss how partnerships with international organizations can drive sustainable adoption.
9.6 **(a)** Public-private collaborations: Mention partnerships that improve infrastructure, like India's BharatNet or Saudi Arabia's Vision 2030. Highlight contributions of corporate players in funding and technical expertise. **(b)** Effectiveness of regulations: Discuss how these regulations ensure affordability and universal access. Provide examples of successful implementations in rural and underserved areas.
9.7 **(a)** Barriers to commercialization: Identify financial constraints, lack of awareness, and technological readiness as key challenges. Explore difficulties in scaling successful pilot trials globally. **(b)** Overcoming barriers: Suggest subsidies, government incentives, and collaborations with private entities. Highlight the importance of cost-effective technology solutions for resource-limited regions.
9.8 **(a)** Lessons from Nordic countries: Discuss policies promoting universal broadband access and digital literacy programs. Highlight the focus on equitable access and infrastructure development. **(b)** Integrating research findings: Explore examples of how global studies informed policy frameworks, such as tailored education or healthcare solutions. Mention the role of continuous data collection in shaping effective policies.
9.9 **(a)** Role of international organizations: Explore contributions from organizations like ITU in setting global connectivity standards. Highlight how standardization fosters interoperability and scalability of digital solutions. **(b)** Impact on underserved populations: Provide examples of improved connectivity due to standardized infrastructure in rural and remote areas. Discuss how standardization reduces costs and ensures reliable access.
9.10 **(a)** Examples of regulations: Discuss regulations like those in the United States for public Wi-Fi access and broadband expansion. Highlight policies in Europe targeting universal connectivity and digital inclusion. **(b)** Improving regulatory processes: Suggest evidence-based approaches for policy design, incorporating feedback from pilot trials. Highlight the need for flexible and adaptive regulations that address local challenges.

Appendix B

B.1 Plot historical adoption rates (use exponential growth models for adoption curves). And then forecast the 6G adoption model using the 5G adoption rate data.
B.2 For computing the digital divide index, use the formula: Digital Divide Index $= \frac{1}{\alpha} + Z + a$ where α is the Internet penetration rate, Z is the cost per broadband package, and a is the literacy rate adjustment factor. Once the digital divide index is computed for the countries of interest, sort them based on the computed indices.
B.3 First, explore the papers and datasets to obtain a ballpark number for the cost of satellite technologies and the cost of deploying a fiber optics network. Then apply the concepts of techno-economic assessment and study their feasibility over the 10 years. Note: There is no requirement to estimate the revenues.
B.4 **Objective Function**: Minimize cost: Min $C = \sum_{i=1}^{n} (C_c + C_o)$ Where C_c is the CAPEX, C_o is the OPEX, and n is the number of towers. **Constraints**: Signal coverage radius (r) ensures all users are covered. Solve using linear programming.
B.5 To model the data traffic in a mobile communication system: **Formula**: Total Traffic (GB/day) $= u \cdot N$ Where u = usage per user (e.g., 3 GB/day) and N = number of active users (e.g., 500,000 users). **Future prediction for 5G**: Increase u due to higher speeds.
B.6 To study the signal coverage analysis for satellite communication, use the following method for all the different types of satellites: For example, LEO satellites at h altitude and radius, coverage area per satellite: $A_{\text{coverage}} = 2\pi r h$. The overall area

to be covered by satellite, divided by the area coverage of the type of satellite, would then help in estimating the required number of satellites.

B.7 To compare the cost of several backhaul technologies, there is a need to understand the installation cost per km^2—β, $C_{maintenance}$ the yearly maintenance cost, η is the data throughput efficiency, and C_{BW} is the cost per unit bandwidth. The cost model here would be: $C_{backhaul} = \beta + C_{maintenance} + \frac{\eta}{C_{BW}}$ Compare values for fiber, free-space optical, and microwave to determine the most cost-effective.

B.8 To understand the optimized locations of 6G BSs, the system model needs to be given certain inputs, such as the population density map and data usage statistics. Then, the greedy algorithm is applied to select locations, maximizing user coverage within a budget constraint. To apply the budget constraints, there is a need to estimate the TCO for the required number of 6G sites to provide maximum coverage and offer a certain minimum level QoS during the congestion. **Output**: Optimized BSs locations are obtained.

Index

S. Koratagere Anantha Kumar, *Wireless Broadband Access to Bridge the Digital Divide*,
Textbooks in Telecommunication Engineering, https://doi.org/10.1007/978-3-032-09261-8

MIX
Papier aus verantwortungsvollen Quellen
Paper from responsible sources
FSC® C105338

If you have any concerns about our products,
you can contact us on
ProductSafety@springernature.com

In case Publisher is established outside the EU,
the EU authorized representative is:
Springer Nature Customer Service Center GmbH
Europaplatz 3, 69115 Heidelberg, Germany

Printed by Libri Plureos GmbH
in Hamburg, Germany